4th Scientific and Technical Conference on Modern Technologies and Energy Systems (WTiUE 2016)

E3S Web of Conferences Volume 13 (2017)

Cracow, Poland
12-14 October 2016

Editors:

Jan Taler
Tomasz Sobota
Bohdan Weglowski

ISBN: 978-1-5108-3653-2

Printed from e-media with permission by:

Curran Associates, Inc.
57 Morehouse Lane
Red Hook, NY 12571

Some format issues inherent in the e-media version may also appear in this print version.

Printed by Curran Associates, Inc. (2017)

For additional information, please contact EDP Sciences – Web of Conferences at the address below.

EDP Sciences – Web of Conferences
17, Avenue du Hoggar
Parc d'Activité de Courtabœuf
BP 112
F-91944 Les Ulis Cedex A
France

Phone: +33 (0) 1 69 18 75 75
Fax: +33 (0) 1 69 28 84 91

contact-edps@webofconferences.org

Additional copies of this publication are available from:

Curran Associates, Inc.
57 Morehouse Lane
Red Hook, NY 12571 USA
Phone: 845-758-0400
Fax: 845-758-2633
Email: curran@proceedings.com
Web: www.proceedings.com

TABLE OF CONTENTS

Statement of Peer review

In submitting conference proceedings to E3S Web of Conferences, I certify to the Publisher that I adhere to the **Policy on Publishing Integrity** of the journal in order to safeguard good scientific practice in publishing.

1. All articles have been subjected to peer review administered by the proceedings editors.
2. Reviews have been conducted by expert referees, who have been requested to provide unbiased and constructive comments aimed, whenever possible, at improving the work.
3. Proceedings editors have taken all reasonable steps to ensure the quality of the materials they publish and their decision to accept or reject a paper for publication has been based only on the merits of the work and the relevance to the journal.

Title, date and place of the conference

IV Konferencja Naukowo-Techniczna Współczesne Technologie i Urządzenia Energetyczne WTiUE 2016
4th Scientific and Technical Conference on Modern Technologies and Energy Systems, WTiUE 2016
October 12-14, 2016
Cracow, Poland

Proceedings editor(s):

Prof. Jan Taler, PhD, DSc
Prof. Bohdan Węglowski, PhD, DSc
Dr Tomasz Sobota, PhD

Date and editor's signature

January 31, 2017

Comparative study of operation of condensing and traditional boilers equipped with the ORC module for electricity generation

Dariusz Mikielewicz [1,*], *Jarosław* Mikielewicz[2]

[1]Gdańsk University of Technology, Faculty of Mechanical Engineering, ul. Narutowicza 11/12, 80-233 Gdańsk, Poland
[2]Institute of Fluid-Flow Machinery PAS, ul. Fiszera 15, 80-231 Gdańsk, Poland

Abstract. Condensing technology applied to boilers is to make full use of thermal energy contained in the fuel. That means that additionaly the heat from condensation of exhaust gases can be used for the purposes of heating the domestic hot water and to cover the demand for central heating. The study analyzed the operation of the "traditional" boiler equipped with the ORC module as the similar arrangement but with the condensing boiler. In the case of a conventional boiler there is noted a greater fuel consumption and the greater power generated than in the case of the unit with the condensing boiler. Postulated is the indicator in the form of a ratio of turbine power to the mass flow rate of fuel, which in turn gives a higher value for the condensing boiler, thus demonstrating that the operation of condensing boiler ORC module will be more economical. Perspective domestic micro CHP with ORC should be installed in boilers with recovery of heat from condensation from the exhaust gases.

1 Introduction

Condensing technology in boilers is to make a full use of thermal energy contained in the fuel. A full utilisation of thermal energy means that the energy contained in the water moisture in exhaust gases can be recovererd by means of condensation. That is obtained by cooling of the exhaust gases so that the moisture undergoes condensation removing heat to the heated water.

In condensing boilers mostly the natural gas is used characteristic for the fact that it contains some hydrogen and hence the significant amount of moisture is produced in the combustion process. Similar situation holds in the case of the thermal oil.

The normalised thermal efficiency of the condensing boiler is usually reaching 105÷109 %. That stems from the assumptions to the standards which did not foresee the possibility of heat recovery from the moisture in exhaust gases. The moisture contained in the exhaust gases of traditional boilers is directed to the chimney without moisture condensation. Hence the reference value (100%) was assumed as the energy contained in the dry exhaust gases E_{OUT} and it was that value to which the efficiency was referring:

$$\eta = \frac{E_{OUT}}{E_{IN}(=100\%)} \tag{1}$$

In that definition the heat contained in the moisture was totally neglected and the attainable efficiencies smaller than about 90%. Due to the fact that the standard related to determination of the boiler efficiency has not changed over the time and the fact that for some time there are on the market the condensing boilers enabling condensation of the moisture from exhaust gases, hence recovering that heat, the output energy E_{OUT} is supplemented with additional component E_{H2O} obtained from moisture condensation from exhaust gases:

$$\eta = \frac{E_{OUT} + E_{H_2O}}{E_{IN}(=100\%)} \tag{2}$$

In such case the nominator can attain higher values than the denominator due to the fact that efficiencies of the boilers are higher than 90% and additional component resulting from the condensation of moisture contributes with extra 10% of energy. In effect the efficiencies exceeding 100% are obtained, namely of the order 105-107%. Due to application of such condensing boilers for heating purposes obtained are attractive economical benefits. An increased boiler efficiency coverts to the reduction of annual demand for gaseous fuel and hence the reduction in heating costs.

The moisture content in exhaust gases confirms that contained there is energy which, if not managed, is lost to the chimney. The exhaust gases leaving the traditional boiler have a temperature reaching even 200°C. One of the components of the boiler is the heat receiver known as the exhaust gases/water heat exchanger. Its task is to receive heat from the exhaust gases and subsequently to transfer it to water. In the condensing boiler the heat exchanger is very different than in traditional boilers. First of all it is designed in such a way that it can cool exhaust gases below the dew point temperature, rendering in such a way condensation of moisture, which it related to the transfer of heat to cooling water. That

* Corresponding author: Dariusz.Mikielewicz@pg.gda.pl

DOI: 10.1051/e3sconf/20171301001

process is relatively simple but requires application of adequate materials, which on other hand are rather expensive. It may happen that moisture has an acidic indicator with pH at the level of 4-5. Due to that fact the material of the tubes used for that purpose in condensing boilers should withstand and resist the action of acid condensate. By application of acid resistant steels and special cast alloys it is possible to recover a significant amount of heat from the exhaust gases. Appropriate selection of the material does not lead to material damages of other equipment. The amount of produced condensate depends primarily on the size of boiler as weel as the operational conditions. The condensation process occurs when the exhaust gases will be cooled below specified low temperature, which must be preceeded by reduction of temperature of water. Under normal conditions the condensation occurs at temperature of exhaust gases of about 50°C. Every condensing boiler is additionally equipped with the system for removing of the excess of condensate.

2 The boiler model cooperating with the micro CHP ORC module

The concept of the cogenerative boiler of the so called IIIrd generation is under development at the Department of Energy and Industrial Apparatus for some time now, Fig. 1. Such domestic microCHP, cooperating with the ORC module is scheduled to produce heat. Electricity is generated as a byproduct, when there is the demand for heat. In Authors oppinion the considered cogenerative boiler will replace the traditional and condensing boilers in the future for providing heat to single-family houses, multi-family houses or even entire etates, [1]. As far as the size is concerned the boiler with the ORC module will only be bigger a little, not more than one third of the original size. However, apart from the heating function it will additionally generate electricity. The cogenerative boiler will operate in significantly lower temperatures than the combustion engine or the gas turbine. Hence it will require less of expensive materials and the technology of manufacturing is simpler. With its help it is possible to generate electricity with the price comparable to electricity produced in professional power plants of large power. Better utilisation of fuel energy in cogenerative boilers leads to reduction of emissions accompanying the fuel combustion process. The small cogenerative boiler can by fully automated and does not require support. The basic elements of the cogenerative boiler are : boiler (evaporator), vapour turbine, condenser, electricity generator and the circulation pump.

The concept of the cogenerative boiler requires solving of new problems such as the selection of optimal thermodynamic cycle [2,3], selection of the adequate working fluid [4,5], development of relevant calculation algorithms of heat exchangers (evaporator, condenser and in some cases regenerator) and the thermodynamic cycle for the cogenerative system [6,7]. Interesting is also the performance of the gas boiler cooperating with the ORC or whether in the latter case it would be possible to condense moisture from the exhaust gases.

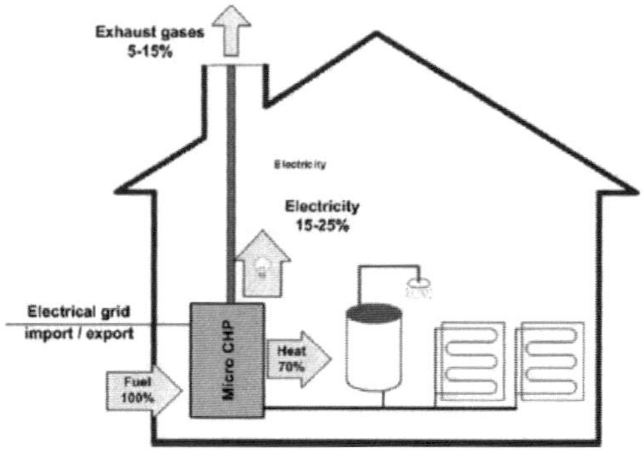

Fig. 1. Schematic of cogenerative boiler.

Integrated calculation model of the boiler and the ORC module has been developed to a simple cogenerative system with internal regeneration, has been schematically presented in Fig. 2.

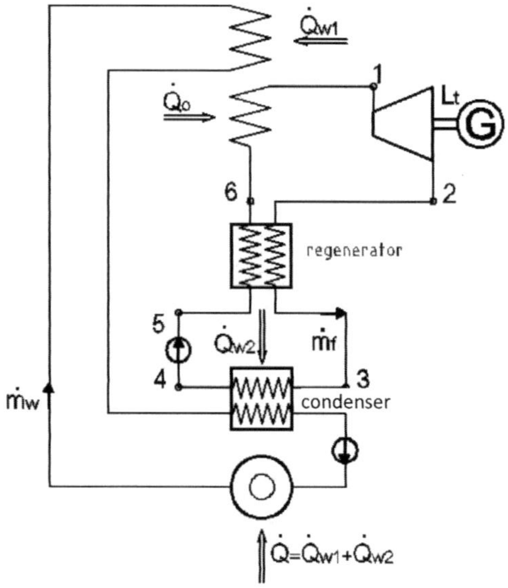

Fig. 2. Scheme of cogenerative boiler with intermediate heating.

In the present paper it has been assumed that the heat losses in the case of traditional boiler are about 10%, whereas in the case of condensing boiler these can be assumed as negligibly small, hence for the study these were assumed as 0%. Considered in the paper turbine power are not including circulation pump of ORC. In case of the fluid considered here it is negligible. Also not considered in the analysis are circulation pumps of the central heating and domestic hot water circuits, as such pumps are present in the case of the boiler with and without the ORC. They are of comparable capacities.

2.1 Traditional boiler with ORC module without moisture condensation

In the paper it has been assumed, that the demand for central heating and domestic hot water is equal to $\dot{Q}=\dot{Q}_{w1}+\dot{Q}_{w2}$=20kW. The boiler is supplied with chemical energy in the form of the rate of heat resulting from combustion of gaseous fuel with the flow rate $\dot{m}_{p,1}$ of specified low calorific value W_p:

$$\dot{Q}_{0,1}=\dot{m}_{p,1}W_p \tag{3}$$

The heat is used for production of work in the expansion device as well as for heating of water in the condenser which is next directed to the heat receivers. In the case of the boiler without condensation of moisture from the exhaust gases the heat produced in the boiler is removed from the condenser in the amount $\dot{Q}_{w2,1}$ (scheduled for covering the demand for central heating and domestic hot water), as well as in the form of work produced in the turbine, $\dot{L}_{t,1}$. No condensation of moisture from the exhaust gases is considered at the moment, i.e. $\dot{Q}_{w1,1}$=0. There is also some loss of energy through the chimney, which is related to higher than the dew point temperature. The energy balance yields :

$$\dot{Q}_{0,1}=\dot{m}_{p,1}W_p=\dot{Q}_{w1,1}+\dot{Q}_{w2,1}+\dot{L}_{t,1}+\dot{Q}_{stack1} \tag{4}$$

The work which is obtained in the turbine is determined from the knowledge of enthalpy before and after the turbine and the mass flow rate of ORC working fluid:

$$\dot{L}_{t,1}=\dot{m}_{f,1}(h_1-h_2) \tag{5}$$

In the case of traditional boiler the amount of working fluid circulating in the ORC installation is determined from the condenser balance, where the demand for heat in known and equal to the total heat demand for central heating and hot water ($\dot{Q}=\dot{Q}_{w2,1}$=20kW) :

$$\dot{Q}=\dot{Q}_{w2,1}=\dot{m}_{f,1}(h_2-h_4) \tag{6}$$

Taking into consideration the simplifying assumption that the heat transferred through the chimney is about 10% of supplied heat to the boiler we can determine the mass flow rate of the fuel required to cover heat demand:

$$\dot{m}_{p,1}=\frac{\dot{Q}+\dot{L}_{t,1}}{0.9W_p} \tag{7}$$

The thermodynamic efficiency of the cycle yields:

$$\eta_1=\frac{\dot{Q}+\dot{L}_{t,1}}{\dot{m}_{p,1}W_p} \tag{8}$$

At the assumptions set in the present work the total efficiency is 0.9 (other losses are neglected).

2.2 Traditional boiler with ORC module with moisture condensation

In the case when we are dealing with condensation of moisture from the exhaust gases we supply to the condenser a smaller rate of heat, as part of utility hot water will receive the heat from condensation of moisture in the exhaust gases. Smaller amount of heat supplied to the evaporator of ORC will render a reduced production of electricity in ORC in comparison to the traditional boiler as a smaller mass flow rate of working fluid will circulate. Presence of the regenerator in the system causes that the central heating and hot water systems are supplied only due to the heat of phase change in the condenser, h_{lv}. That influences the mass flow rate of working fluid in ORC installation. The reduced heat demand in the condenser is due to the fact that part of heat demand is covered by heat of condensation of moisture from the exhaust gases, $\dot{Q}_{w1,2}$.
In the considered case the return water from central heating instalation can be heated and respectively less heat will be required to be supplied to condenser:

$$\dot{Q}_{0,2}=\dot{m}_{p,2}W_p=\dot{Q}_{w1,2}+\dot{L}_{t,2}+\dot{Q}_{w2,2}+\dot{Q}_{stack2}=\dot{Q}+\dot{L}_{t,2}+\dot{Q}_{stack2} \tag{9}$$

In that case the heat in the form of the mass flow rate of exhaust gases is removed to atmosphere having a small enthalpy, as it is assumed that the exhaust gases will feature temperature smaller than 50°C and the condensation of moisture will be present. Hence it can be assumed that heat removed to the chimney will be equal zero, $\dot{Q}_{stack2}$=0.
In such case the mass flow rate of fuel required to supply the rate of heat to the cycle will be lower than in the case without condensation. It will yield:

$$\dot{m}_{p2}=\frac{\dot{Q}_{w1,2}+\dot{Q}_{w2,2}+\dot{L}_{t,2}}{W_p}=\frac{\dot{Q}+\dot{L}_{t,2}}{W_p} \tag{10}$$

In such case the turbine power will yield:

$$\dot{L}_{t,2}=\dot{m}_{f,2}(h_1-h_2) \tag{11}$$

Amount of working fluid circulating in the installation will also be determined from the energy balance of condenser. In such case it is assumed however that only 90% of heat will be supplied to the condenser:

$$\dot{Q}_{w2,2}=0.9\dot{Q}=\dot{m}_{f,2}h_{lv} \tag{12}$$

Additionally in the present analysis it has been assumed that ORC is realised in the same temperatures as in the case without condensation and therefore the enthalpy drop in the expansion device will be the same.
The total efficiency in the case of heat recovery is determined from the relation:

$$\eta_2=\frac{\dot{Q}_{w1,2}+\dot{Q}_{w2,2}+\dot{L}_{t,2}}{\dot{m}_{p,2}W_p}=\frac{\dot{Q}+\dot{L}_{t,2}}{\dot{m}_{p,2}W_p} \tag{13}$$

DOI: 10.1051/e3sconf/20171301001

In line with the assumptions other heat losses have been neglected in the analysis. In case of condensing boiler cooperating with the ORC the heat transferred to the central heating installation consists of two components. The first one is the heat removed from the condenser ($\dot{Q}_{w2,2}$=0.9$\dot{Q}$) and the rest of heat, which results from teh condensation of moisture from exhaust gases. In the analysis it has been assumed that $\dot{Q}_{wl,2}$=0.1$\dot{Q}$.

3 Results of calculations

In order to show the performance of the two cases an appropriate algorithm has been developed and calculations were carried out in the MathCAD15 environment. It has been assumed that the temperature of vapour before turbine is t_l=160°C, whereas temperature of condensation is 50°C. Rate of heat in the amount $\dot{Q}$=20kW is given away to the central heating and domestic hot water installation. In the case of traditional boiler that heat is entirely generated in the ORC and is removed in the condenser to the heat receivers. In the case of condensing boiler in the condenser the rate of heat $\dot{Q}_{w2,2}$=0.9$\dot{Q}$ is transferred, as the remainder is supplied to the central heating installation with the heat due to condensation of moisture from exhaust gases. That heat has been estimated to be equal to ten percent of the total heat demand, $\dot{Q}_{wl}$=0.1$\dot{Q}$. It should be remembered that in case of dry fluids in ORC installation there is the regenerator in the system, in which vapour after expansion is cooled to the saturation temperature. The liquid subcooling is not considered.

Table 1. Results of calculation for the traditional boiler (only the heat of phase change is considered h_{lv}).

Fluid	$\dot{m}_{p,1}$	$\dot{m}_{f,1}$	$\dot{L}_{t,1}$	$\dot{Q}_{0,1}$	$\dot{L}_{p,1}$	$\eta_{ORC\,1}$	$\dot{L}_{t,1}/\dot{m}_{p,1}$
	kg/s×10-3	kg/s	kW	kW	kW		[J/kg]×10-6
R123	0.853	0.125	6.421	29.36	0.204	0.219	7.531
D4	1.167	0.113	16.16	40.18	0.008	0.402	13.85
MDM	1.220	0.103	17.79	41.99	0.015	0.424	14.59
HFE7100	1.165	0.175	16.46	32.90		0.175	14.13
Etanol	0.807	0.022	5.018	27.80	0.035	0.181	6.215

Table 2. Results of calculation for the condensing boiler (only the heat of phase change is considered h_{lv}).

Fluid	$\dot{m}_{p,2}$	$\dot{m}_{f,2}$	$\dot{L}_{t,2}$	$\dot{Q}_{0,2}$	$\dot{L}_{p,1}$	$\eta_{ORC\,2}$	$\dot{L}_{t,2}/\dot{m}_{p,2}$
	kg/s×10-4	kg/s	kW	kW	kW		J/kg×10-6
R123	0.749	0.112	5.779	25.78	0.183	0.229	7.718
D4	1.003	0.101	14.54	34.54	0.007	0.421	14.50
MDM	1.046	0.092	16.01	36.01	0.014	0.424	15.31
HFE7100	1.011	0.157	14.81	28.65		0.410	14.65
Etanol	0.712	0.020	4.516	24.52	0.032	0.184	6.343

It has been assumed that the system is supplied with a high methane content natural gas GZ-50 with a low calorific value of 34430 kJ/m^3. Considered have been 5 working fluids, ie. R123, ethanol, D4, MDM and HFE7100. The physical data for first four fluids have been taken from fluid data base Refprop9.0 [8]. The data for HFE7100 have been taken from the data base in Engineering Equation Solver EES [9]. Selected fluids are the most probable to be implemented in the ORC installation, which can be commercialised. The results of the calculations have been presented in Table 1 for the case of traditional boiler and in Table 2 for the condensing boiler.

4 Conclusions

A special indicator, formulated as the ratio of turbine power to the mass flow rate of fuel has been postulated in the analysis. The value of the indicator has been evaluated both for the case of the traditional boiler with the ORC module and the condensing boiler equipped with the ORC module. In the case of traditional boiler the greater consumption of fuel is noticed as well as a higher level of electricity generation than in the case of the condensing boiler. The developed indicator returns higher values for the case of the condensing boiler showing that the operation of condensing boiler with the ORC module should be more economical.

In the case of condensing boiler still open is the case how the moisture condensation can be made. For the assumptions made in the paper it seems that moisture condensation can only be possible in case of the low-temperature heating installation.

The cogenerative boiler of the third generation featuring the electricity generating module should be installed in units featuring the heat recovery in the form of condensing the moisture from exhaust gases.

References

1. D. Mikielewicz, J. Mikielewicz, Cogenerative micro power plants – a new direction for development of power engineering? Archives of Thermodynamics, 29(4), 2008, 109-132.
2. J. Mikielewicz, D. Mikielewicz, A thermodynamic criterion for selection of working fluid for subcritical and supercritical domestic micro CHP, Applied Thermal Eng., Vol. 30, 2010, 2357-2362.
3. D. Mikielewicz, J. Wajs, J. Mikielewicz, Gas boiler as a heat source for the domestic micro-CHP, J. of Power Technologies, vol. 94(4), 2014, 317 - 322.
4. J. Mikielewicz, D. Mikielewicz, Comparative study of selected fluids for use in supercritical ORC. Arch. of Thermodynamics, Vol. 30(2), 2009, 3-14.
5. J. Mikielewicz, D. Mikielewicz, Optimal boiling temperature for ORC installation, Archives of Thermodynamics, Vol. 33(3), 2012, 27-37.
6. D. Mikielewicz, J. Mikielewicz, E. Ihnatowicz, T. Muszyński, J. Wajs, W. Rybiński, Selected aspects of designing and testing of heat exchangers for ORC in the domestic microCHP, IMP PAN, 2013 (in PL).
7. J. Mikielewicz, D. Mikielewicz, Thermal-hydraulic issues of flow boiling and condensation in organic Rankine cycle heat exchangers, Archives of Thermodynamics, vol. 33(1), 2012, 41-66.
8. Refprop 9.0. National Institute of Standards, 2010.
9. Engineering Equation Solver (EES) website F-Chart Software http://www.fchart.com.

E3S Web of Conferences **13**, 01002 (2017) DOI: 10.1051/e3sconf/20171301002
WTiUE 2016

Exploitation problems and diagnostic of heat recovery steam boiler OU-192

Grzegorz Olewicz[1], *Adrian* Sobczyszyn[2,*], *Kamil* Staszałek[3], and *Przemysław* Syty[4]

[1]EDF Polska S.A., Cracow, Poland
[2,3,4]Research and Technological Services "Pro Novum" Ltd.

Abstract. Combined Cycle Gas Turbine (CCGT) use exhaust gases from gas turbine (GT), which in basic systems are ejected to environment as discharge loss, for steam generation in Heat Recovery Steam Generator (HRSG). Higher unit efficiency achieved inter alia by using in boiler low and high pressure circuit allow to greater use of turbine exhaust gases enthalpy. Modular design solutions with tight structure, preferred on construction site, may be a problematic for maintenance services during operation and overhauls. This paper presents general operational problems of HRSG as well as detailed solutions for boiler OU-192. We discuss scope and results of diagnostic tests required to assess technical condition of the pressure elements, after 100 000 hours operation, with the support of the software LM System PRO+.

1 Combined Cycle Gas Turbine and Heat Recovery Steam Generators in power engineering

Technical and thermodynamic connection between gas and steam cycles allows a significant increase in the efficiency of energy conversion, particularly for cogeneration. In open gas turbine cycle power supplied to circuit is significant, but due the high temperature of exhaust gases, large part of it is discharged (as loss) to environment. In steam circuit, maximum process temperature is relatively low but from thermodynamic point of view, important is low temperature of exhaust gases [1]. Connecting these two circuits and using GT exhaust gases enthalpy for steam generation in HRSG the losses are minimalized. This is very important in both economic and environmental point of view. The main advantages of CCGT are:

- low investment cost compared to conventional coal-fired units,
- shorter time of investment due the modular structure of main elements,
- high flexibility,
- high efficiency affecting the operation costs,
- lower pollutant emission compared to conventional coal-fired units.

High gas prices on Polish fuel market are main disadvantages of using CCGT and significantly affect economic balance.

Construction of HRSG is similar to the construction of conventional coal-fired boiler – it consists of economizer, evaporator and superheaters. The differences arise from principles of HRSG operation and its thermodynamic properties. Generally, these boilers can divided by:

- directions of exhaust gases flow,
- number of separate circuits with different working pressure.

In horizontal oriented boilers exhaust gases flow through horizontal channel with vertical located heating surfaces tubes, while in vertical oriented boilers exhaust gases flow through vertical channel with horizontal located heating surfaces tubes.

The use of more than one separate circuits with different working pressure is thermodynamically justified and it is intended to reduce losses. Apart from one pressure circuit we can find two- or even three-pressure circuits. In this case, each circuit has a separate economizer, evaporator (also with separate drums) and superheaters.

Each of heating surfaces together with headers has a modular construction. Such structure can be problematic in certain situations, which occurs during operation. Modular construction which is favourable during investment stage and warranty full module exchange becomes problematic when one of elements fails. Constructor did not assume enough space required for local repairs within one module (eg. Evaporator). When element fails it requires extensive intervention in structure.

2 HRSG exploitation problems

Long-term operation of HRSG in power industry and industry all over the world, provides great operational

* Corresponding author: a.sobczyszyn@pronovum.pl

experience. In this chapter, due to limited space, we will present two interesting cause of the boiler failure.

One of the most common and often commented in literature problem is flow accelerated corrosion (FAC) on inner surfaces of heat exchangers tubes. FAC mechanism is based on destructive effect of vigorous flow of water or steam which damages and destroys protective oxide layer. In the place where the oxide layer is destroyed, exposed metal corrodes by creating a new layer of oxide. This process is then repeated until the thinning of the wall which leads to failure. In early stage, before failure, characteristic and extensive damages of tube inner surface can be observed. FAC mechanism depends on:

- water/steam quality,
- water/steam temperature,
- disturbances in the flow,
- metal chemical composition.

FAC in HRSG, as mentioned before, is a leading cause of damage. Frequent start-ups and work in cyclic mode has a negative influence over water quality which results in increase the risk of FAC. Also, corrosion products are transported with water/steam to further parts of circuit and they become deposits which can be another source of corrosion problems [2].

Another complex problem in HRSG is lack of or insufficient compensation of thermal expansion of tubes. It leads not only to tube damages, but also can result in serious failures of stubs and headers. Tube exposed on high temperature increases its length, which in spite of a theoretical heat extension margin, can result in buckling. This affects additional stress on header stubs. During operation, tube repeatedly changes its length, which generates an alternating stress and leads to fatigue failures [3,4]. Also, excessive constant stress, caused by bend ovalization together with corrosive environment leads to intensification of destructive corrosion and stress processes. In the result, micro-pitting occurs on privileged locations and elements such as bend with excessive ovalization (>8%) [5].
These reasons of failures are only small part of mechanisms which can occur during HRSG operation. It is worth mentioning, that cyclic operation in case of HRSG, which were designed for stationary operation, can lead to intensification of negative processes leading to failure [4].

3 Description of OU-192 type boiler

Heat recovery steam boiler is a part of combined cycle gas turbine (CCGT). It uses the exhaust gas which flows in horizontal channel over each components inside. The boiler consists of two separate modules – low-pressure (LP) and high-pressure (HP) circuits with natural flow. Major parts of boiler:

- HP superheater II,
- HP superheater I,
- wall-tubes with HP drum,
- HP economizer II,
- LP superheater,
- wall-tubes with LP drum,
- HP economizer I,
- LP economizer.

Only in high-pressure circuit - superheater and economizer are two-graded. Both of modules are ended by main outlet headers.

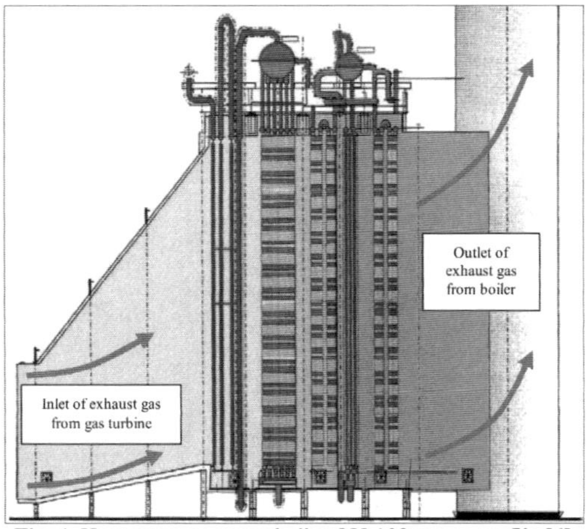

Fig. 1. Heat recovery steam boiler OU-192 type – profile [6].

4 Working conditions analysis supported by life management software on the OU-192 type boiler

Software platform LM System PRO+® was installed in 2010 yr. It allows to register history of exploitation, working parameters and failure analysis needed to determine technical conditions of boiler elements. Based on the program data are charts of pressure and temperature for low- and high-pressure drums and superheater II (fig. 2÷4). Construction and exploitation values are placed in table 1, comparison between register data and assumed values are placed in table 2.

Table 1. Construction and exploitation data of OU-192 type boiler [7].

Construction data	
Performance, t/h	192
Steam pressure (HP), MPa	7,51
Steam temperature (HP), °C	505
Steam pressure (LP), MPa	0,71
Construction data	
Steam temperature (LP), °C	213,5
Performance, t/h	192
Steam pressure (HP), MPa	7,51
Exploitation data	
Total work time, hours	ca. 100 000
Number of start-up	246
Number of failures	8
Number of pressure tests	30

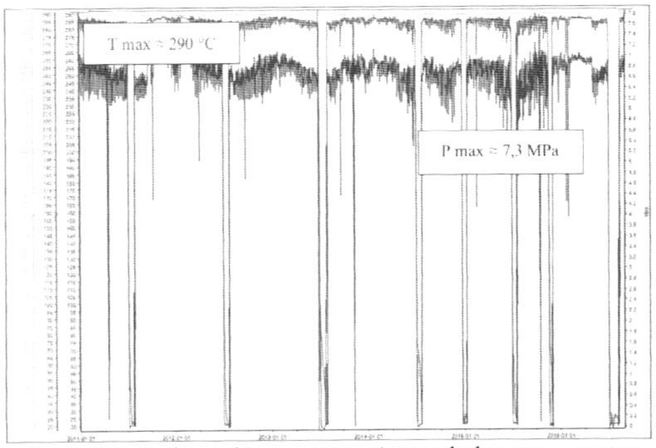

Fig. 2. Chart of metal temperature (upper holes area – center, left side, right side) with steam pressure – HP drum.

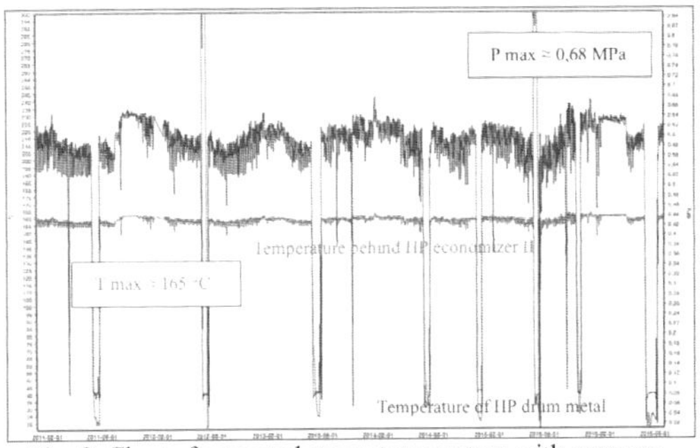

Fig. 3. Chart of saturated steam temperature with pressure – LP drum.

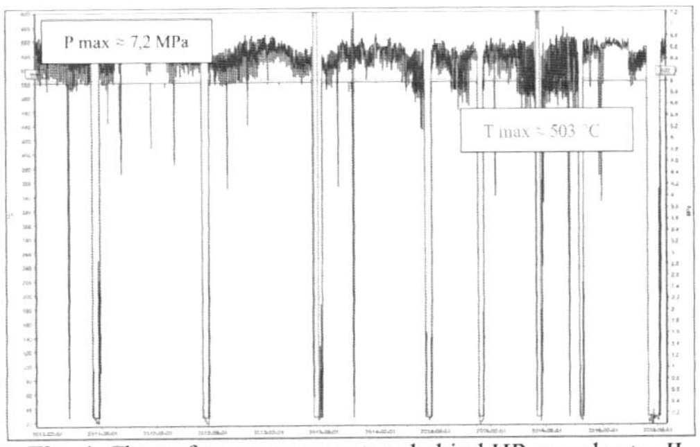

Fig. 4. Chart of steam temperature behind HP superheater II with pressure.

Table 2. Calculated and registered data.

HP drum		
Parameter	*Assumed values*	*Register data*
Pressure, MPa	8,4	7,3
Temperature, °C	300	290
LP drum		
Parameter	*Assumed values*	*Register data*
Pressure, MPa	1,1	0,68
Temperature, °C	200	165
LP drum		
Parameter	*Assumed values*	*Register data*
Pressure, MPa	8,4	7,2
Temperature, °C	540	503

Steam temperature and pressure values, registered by LMSP+®, are not exceeded for following boiler elements:

- LP drum LP,
- HP superheater II,
- HP drum.

It has also been prepared additional charts which show work of CCGT in two-shifted adjustment and change of feed water and boiler water pH values (low-pressure circuit). Results of analysis report show that pH values are close. It's very important because of magnetite layer protection durability. It allows to stop dosing boiler water alkalizer (sodium hydroxide). Charts are shown in figures 5 and 6 below.

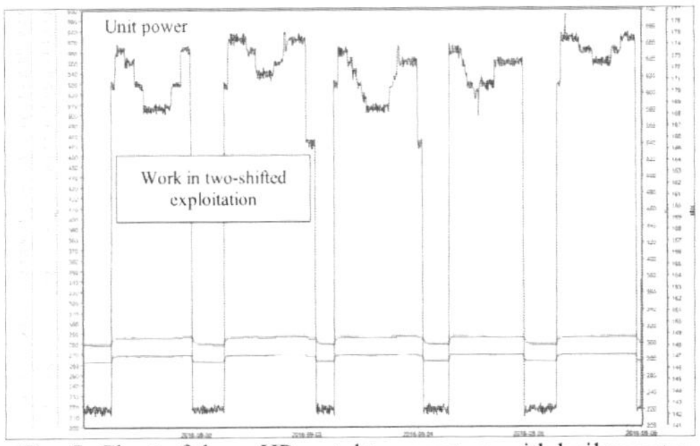

Fig. 5. Chart of drum HP metal temperature with boiler water temperature and unit power .

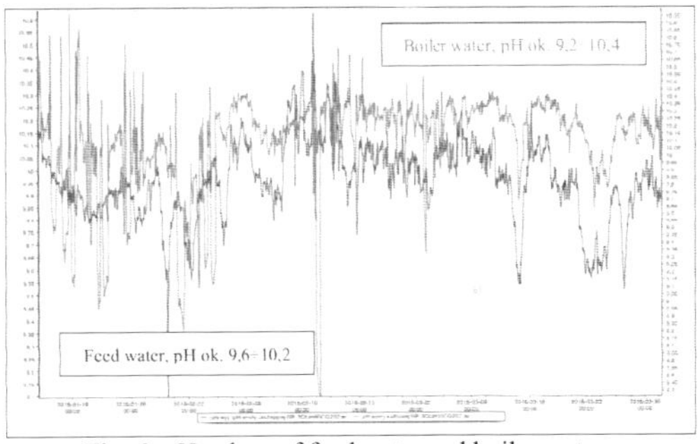

Fig. 6. pH values of feed water and boiler water – low-pressure circuit.

5 Retrospection of OU-192 type boiler

After summary boiler work time, failure mainly occurred on top evaporator headers [9] and HP superheater II headers [10, 11, 12] areas. Damages were on evaporator bend tubes (fig. 7) and welded joints connecting tubes with HP superheater II headers (fig. 8,9,10). This helped to create a diagnosis scope of above elements. For this purpose all available evaporator bend tubes were tested by ovalization measurement and all fillet welds were tested by magnetic particle test. Repairs were done by replacement and welding. Example damages placed below.

DOI: 10.1051/e3sconf/20171301002

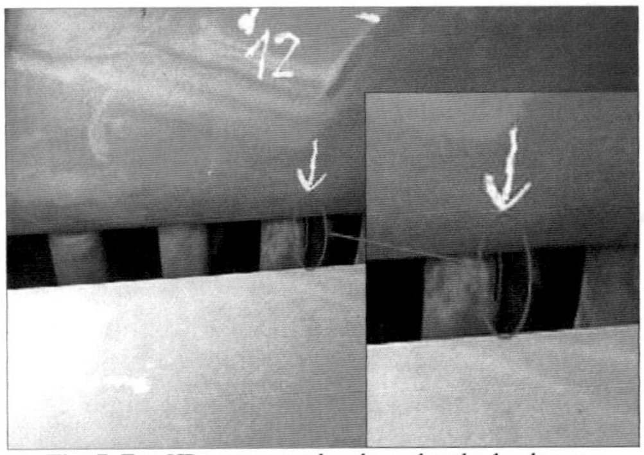

Fig. 7. Top HP evaporator headers – bend tube damage.

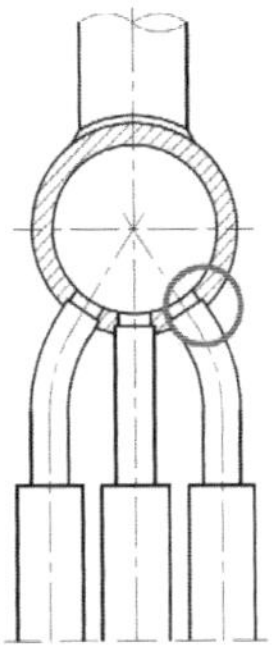

Fig. 8. Place of HP evaporator damage (bend tube section).

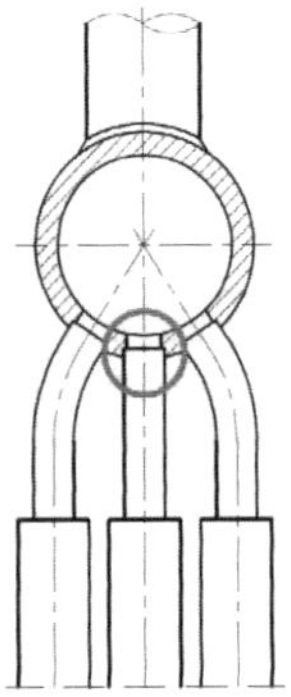

Fig. 9. Place of HP evaporator damage (straight tube section).

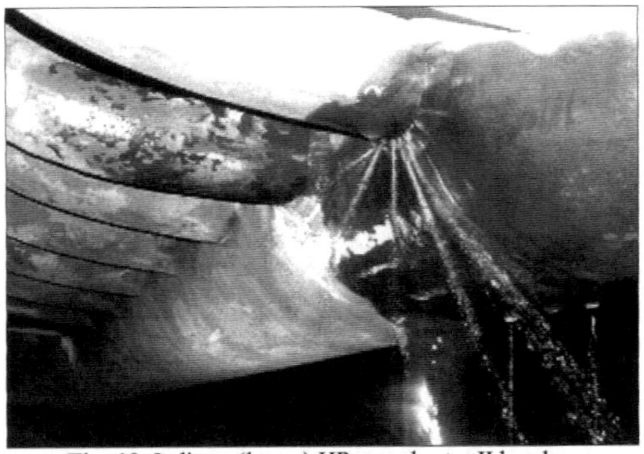

Fig. 10. Indirect (lower) HP superheater II header – fillet weld damage.

6 Overhaul of the OU-192 type boiler

General overhaul of the boiler took place in 2016. During the 45-day repair downtime were done vitality studies on drums and thick-walled parts of the boilers due to the operating time of the boiler of approx. 100 000 hours. In addition, the scope of the renovation included the action of eliminating the most frequently occurring faults of boiler pressure segments and auxiliary installations.

Overhaul scope:

- *70 bends (of all pipe coils) of the superheater SH2 were exchanged at the intersections with the lower indirect headers.*

Renovation works have been conducted in accordance to executive project approved by UDT (Office of Technical Inspection). After replacing bends performed testing of 100% welded joints. Bends ø51.6×3.6 mm were replaced by bends ø51.6×6.3 mm with material 13CrMo4-4.

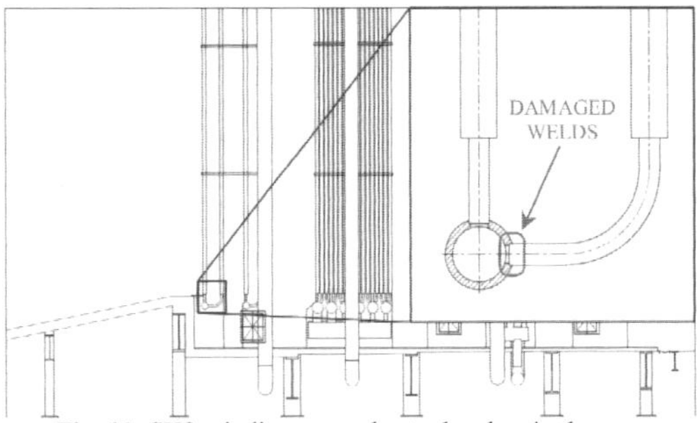

Fig. 11. SH2 – indirect superheater headers in the area of damage.

- *Completing the installation of inspection hatches on the top and bottom of the boiler (design and installation).*

Installation of the external inspection hatches has been completed (fig. 12) due to the difficult access to the individual modules of the boiler during previous shutdowns and when troubleshooting an evaporator in 2014.
As a result, in the years 2015-2016, 3 hatches was installed:

- made 2 new hatches: the level +18 m side wall, the level +24 m ceiling under HR drum,
- expanded inspection hatch at 0 m level, which will allow to build scaffolding.

DOI: 10.1051/e3sconf/20171301002

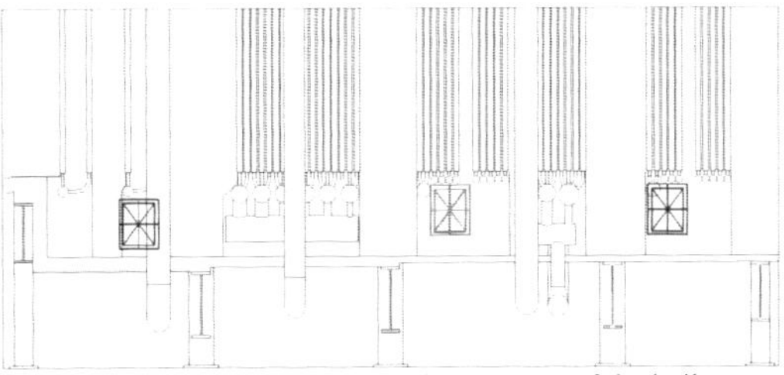

Fig. 12. Enlarged hatch in the bottom part of the boiler.

- *Installation of surveying benchmarks on steam pipelines for the possibility of measuring the position of pipelines.*

Work proceeded according to prepared by "Pro Novum" benchmarks deployment project for:

- low-pressure pipeline from the outlet header (including the same header) to the hull of HP/LP steam turbine,
- pipeline from the outlet header to the tee and from the tee to the reduction-boot station.

In total 31 benchmarks installed.

- *Replacement of fabric expansion joints connecting the boiler to the gas turbine and boiler to the chimney.*

During the renovation listed two fabric expansion joints with dimensions of 7606×20108 mm (turbine – boiler) and 3552×7942 mm (outlet duct – chimney). Also the internal plating in the intake duct between the gas turbine and the boiler was repaired. After completion of the work the tightness test during the commissioning of the boiler was performed.

- *Completion of the modernization of boiler blowdown system.*

Modernization of the boiler blowdown system and the discharge of desalination was made due to determining the presence and transport of solids called wandering oxides in water-steam system of the boiler.

- *Installation of surveying benchmarks on steam pipelines for the possibility of measuring the position of pipelines*

Work proceeded according to prepared by "Pro Novum" benchmarks deployment project for:
- low-pressure pipeline from the outlet header (including the same header) to the hull of HP/LP steam turbine,
- pipeline from the outlet header to the tee and from the tee to the reduction-boot station.

In total 31 benchmarks installed.

- *Replacement of fabric expansion joints connecting the boiler to the gas turbine and boiler to the chimney.*

During the renovation listed two fabric expansion joints with dimensions of 7606×20108 mm (turbine – boiler) and 3552×7942 mm (outlet duct – chimney). Also the internal plating in the intake duct between the gas turbine and the boiler was repaired. After completion of the work the tightness test during the commissioning of the boiler was performed.

- *Completion of the modernization of boiler blowdown system.*

Modernization of the boiler blowdown system and the discharge of desalination was made due to determining the presence and transport of solids called wandering oxides in water-steam system of the boiler.

7 Diagnostics of the OU-192 type boiler

Analysis of the parameters recorded by the LM System PRO+® for adherence to established parameters and possible overruns the limit values, as well as reports from previous diagnostic tests and collected information about failures allowed to select the range of diagnostic tests involving the critical elements and places of their potential damage. The area of diagnostic testing carried out during the planned shutdown included LP and HP circuits of the boiler, as well as LP and HP steam pipelines.
Elements selected for testing [13]:

- LP drum,
- LP main outlet header,
- HP evaporator,
- HP steam superheater I,
- HP steam superheater II,
- HP main outlet header.

Scope of non-destructive testing of boiler elements [13]:

- visual and endoscope inspection,
- magnetic particle inspection,
- metallographic testing (replica),
- hardness measurement,
- ultrasonic measurement of wall thickness,
- ultrasonic measurement of the oxide layer thickness on the inner surface,
- roundness measurement.

Magnetic particle testing showed cracks on the cylindrical surface of one of the holes in water area of LP drum, on the fillet welds of the separator elements in LP and HP drums, on the fillet weld of the HP evaporator header and on the main steam header. The results of other tests have confirmed accordance with the limit values recommended by standards. Life fraction consumed level obtained on the basis of the microstructure is as follows [9]:

- 20% for the LP and HP drums,
- 20% for the LP main steam header,
- 30% for the HP steam superheater I,
- 30% fo the HP main steam header,
- 35% for the HP steam superheater II.

8 SUMMARY

Gas-steam units, especially those that use fuel with an attractive price and are well-suited to the local heat market can be an attractive part of the national energy system (KSE). It is possible due to the high efficiency of generation of electricity and heat, less nuisance to the environment and high flexibility, which means that it can also act as stabilizer of the KSE. Flexible work has, however, always a price in the form of reduction in durability, especially the elements of a heat recovery boiler. To fully exploit advantages of gas-steam systems we should ensure their reliability and availability through the overhaul and diagnostics policy adapted to the conditions of specific units. Not only working conditions can be monitored continuously, but also the current technical conditions which may contribute the reduction of failure. That means lower maintenance costs and less loss of production.

Planned and systematic diagnostics and overhaul should be conducted in accordance with the knowledge of the device. Based on the analysis of the failure, you can type nodes with increased fault level under their individual program of research and modernization, in order to eliminate design errors leading to increased reliability and availability.

References

1. Kehlhofer R. : Combined-cycle gas & steam turbine power plants. PennWell Publishing Company 1997.
2. Anderson J.: Flow-accelerated corrosion in HRSGs. AEP Plant engineering programs, 2009.
3. Jackson P., Moelling D., Anderson F., Malloy J.: Operating experience of large reheat HRSGs in merchant service. Tetra Engineering Group/Tetra Engineering Europe. Source: www.tetra-eng.com
4. Pearson M., Anderson R. : Measurement of damaging thermal transients in F-Class horizontal HRSGs. ETD International Seminar in Cyclic Operation of Heat Recovery Steam Generators (HRSGs), June 24 2005, in London, UK.
5. EPRI - Field Guide: Boiler Tube Failure 1017471 - Final Report, December 2009.
6. www.wydawnictwo-gorscy.pl/pdf/bgp_folder.pdf – website.
7. www.ec.zgora.pl/technologia/urzadzenia-wytworcze – website.
8. Pro Novum Ltd. – Report LM System PRO+® no 21.1591/2015 – unpublished elaboration.
9. Pro Novum Ltd. – Technical Opinion no 022.1081/2015 – unpublished elaboration.
10. Pro Novum Ltd. – Report no 014.1973/2007 – unpublished elaboration.
11. Pro Novum Ltd. – Report no 014.2000/2007 – unpublished elaboration.
12. Pro Novum Ltd. – Report no 072.2952/2013 – unpublished elaboration.
13. Pro Novum Ltd. – Report no 080.3395/2016 – unpublished elaboration.
14. IMIA Working group paper 91 (15): Combined Cycle Power Plants. IMIA Annual Conference 2015, Merida (Yucatan), Mexico.
15. Bain D., Christophersen D. : Some common mechanisms leading to failures in heat recovery steam generators Source:www.veoliawatertech.com
16. Wojtkiewicz R.: Difficulties in removing failures and in examination of heat recovery boilers for example of OU 192 heat recovery boiler failure. „Pro Novum" Conference – Katowice, October, 2015.

Thermal comfort conditions and Polish legal requirements regarding temperatures in residential rooms

Muniak Damian[1,*]

[1] Cracow University of Technology, ul. Jana Pawła II 37, 31-864 Kraków, Poland

Abstract: The article discusses the parameters of thermal comfort in residential environments. The analysis is referred to the Polish legal requirements, pointing out the differences between them and some of their effects, both from the ergonomic, and economic points of view.

1 Introduction

Conditions of thermal comfort in residential environments should be ensured in both the heating season, when the building must be heated and in the summer, when it often requires cooling. These needs are the consequence of climatic conditions. Moreover, the obligation of application installation or heating devices results directly from the provisions of Polish Construction Law [1]. In one of its implementing acts, Regulation of the Minister of Infrastructure of 12 April 2002 on the technical conditions to be met by buildings and their location, Dz.U. (Journal of Laws) 02.75.690 as amended (in Polish: Rozporządzenie Ministra Infrastruktury z dnia 12 kwietnia 2002 roku w sprawie warunków technicznych, jakim powinny odpowiadać budynki i ich usytuowanie, Dz.U.02.75.690 z późniejszymi zmianami) [2], in paragraph § 132. 1, it is said that: "the building, which due to its destination requires heating, should be equipped with the heating system or other appliances heating systems, other than furnaces, fireplaces or kitchen shafts."

2 Thermal comfort

The basic aim of any building heating installation is to ensure appropriate temperature conditions for the people in it. The temperature conditions are a component of what is referred to as thermal comfort, which, in turn, affects the room microclimate, i.e. the set of a given room all physical and chemical parameters that have an impact on the human organism. Some of these parameters, such as the air pressure and ionization, the noise level, lighting, the concentration of carbon dioxide or air pollutants, are not directly related to the heating installation operation. Therefore, analysing the performance of such a system, the considerations come down to an analysis of a narrower issue – the thermal comfort conditions.

The operation of radiators in a given room has both a direct and indirect impact on thermal comfort parameters. For example, the temperature values of air, partitions and elements of the room equipment are affected directly. The radiator operation indirect impact is visible for example in the case of the relative and absolute values of humidity, the air flow velocity and circulation in a given zone.

In order to compare indoor environments characterized by a different combination of parameters that affect thermal comfort, an appropriate comparative criterion has to be used. Many such criteria and methods of their evaluation have been developed over the years. Depending on the region, some of them are more popular than others. It is believed that the first researchers who focused on the issue of determining the thermal comfort parameters were Englishmen Sheppard and Hill [3]. In 1913 they proposed mathematical formulae between the values of temperature and relative humidity, which provide a comfortable environment for people. One of the first widespread parameters was the effective temperature (ET) introduced in the USA in the 1920's by Yaglou et al. [4]. More research was done in the field in the next few years and new more reliable parameters were put forward, including the new (modified) effective temperature ET* [5] and the standard effective temperature (SET) [5, 6, 7] The ET index defines the combination of the air temperature and relative humidity at the level of 100%, which causes the same thermal sensation as a combination of a real environment temperature and humidity assuming an identical air flow velocity [8, 7]. ET* defines the temperature of a temperature-homogeneous black environment with a 50% relative humidity of motionless air ($w<0.1$m/s) for which a given person experiences the same thermal and thermoregulation load as that occurring under real conditions, at identical values of parameters met, clo [8, 5]. SET, the parameter currently used by the ASHRAE (American Society of Heating, Refrigerating and Air-conditioning Engineers) [9], defines the temperature of a temperature-homogeneous black environment with a 50% relative humidity of motionless air ($w<0.1$m/s) for which a given person, compared to standard parameters

* Corresponding author: damian.muniak@gmail.com

of a given physical activity (e.g. met=1.0÷1.1 and clo=0.6), experiences the same thermal and thermoregulation load (skin temperature, sweating) as that occurring under real conditions (real values of met and clo) [9, 8, 10, 11, 6, 7, 12]. The methodology of the SET parameter calculation is laid out in the ASHRAE Standard 55-2013 [9]. In the first half of the twentieth century, there were also published extensive monographs on thermal comfort. The author of one of the first was a Polish internist, Nowakowski, who published his work in 1935 [13].

In order to assess the temperature felt by a human in a given room, the temperature of air, partitions and equipment of the room (the radiant temperature) all have to be taken into consideration. The resultant value is referred to as the operative temperature or operating temperature. Different methods of the quantity determination have been developed and proposed over the years. The manner in which it is currently calculated is regulated by the ISO 7726 international standard [13]. According to it, the parameter should be calculated using the following formula:

$$t_{\mathrm{oper}} = \frac{\alpha_{\mathrm{k}} \cdot t_{\mathrm{k}} + \bar{\alpha}_{\mathrm{pr}} \cdot \bar{t}_{\mathrm{pr}}}{\alpha_{\mathrm{k}} + \bar{\alpha}_{\mathrm{pr}}}. \tag{1}$$

The standard also lays out the methodology of determination of the formula individual elements.

The operative temperature is the temperature of a temperature-homogenous black environment (in terms of the heat transfer) where a human, due to radiation and convection, exchanges the same amount of heat as in a comparable temperature-heterogeneous real environment [14, 15]. If the difference between the temperature individual values is smaller than 4oC, it may be calculated, with a certain approximation, as the mean arithmetic value of the air and radiant temperatures, ignoring individual surface film conductance coefficients [16]. The temperature sensation is thus affected by both the air temperature and the temperature of surrounding partitions and equipment.

However, using operative temperature as the only index of thermal comfort evaluation may in some situations prove insufficient because the thermal comfort sensation is also affected by individual parameters, unrelated to a given environment.

Thermal comfort is defined as a state in which a given environment, such as a room for example, is neutral in its effect on the thermal sensation of a human. This means that neither locally, on some body parts, nor globally does the person feel too high or too low a temperature. This particular state is achieved for the human body thermal balance and may be described as follows [17, 18, 19]:

$$\dot{M} - \dot{W} = \left(\dot{q}_{\mathrm{konw}} + \dot{q}_{\mathrm{pr}} + \dot{q}_{\mathrm{pot}}\right) + \left(\dot{q}_{\mathrm{odd,konw}} + \dot{q}_{\mathrm{odd,par}}\right) + \left(\dot{q}_{\mathrm{ak,sk}} + \dot{q}_{\mathrm{ak,org}}\right) \tag{2}$$

If:

$$\dot{M} - \dot{W} > \left(\dot{q}_{\mathrm{konw}} + \dot{q}_{\mathrm{pr}} + \dot{q}_{\mathrm{pot}}\right) + \left(\dot{q}_{\mathrm{odd,konw}} + \dot{q}_{\mathrm{odd,par}}\right) + \left(\dot{q}_{\mathrm{ak,sk}} + \dot{q}_{\mathrm{ak,org}}\right)$$

then there is an increment in the energy accumulated in the human body and the temperature rises. The human feels warmth.

If:

$$\dot{M} - \dot{W} < \left(\dot{q}_{\mathrm{konw}} + \dot{q}_{\mathrm{pr}} + \dot{q}_{\mathrm{pot}}\right) + \left(\dot{q}_{\mathrm{odd,konw}} + \dot{q}_{\mathrm{odd,par}}\right) + \left(\dot{q}_{\mathrm{ak,sk}} + \dot{q}_{\mathrm{ak,org}}\right)$$

then there is a decrease in the energy accumulated in the human body and the temperature drops. The human feels cold.

As the equation above indicates, due to metabolism the organism produces energy, thus building up the body temperature, and gives up energy through convection and radiation from the skin surface, sweat evaporation, breathing and performing work. The human warm-blooded organism strives to maintain a certain temperature value which, however, may change in a certain range depending on a number of factors (excluding diseases). The body thermoregulation system is responsible for keeping the range of these changes within certain boundaries. For example, if the temperature needs to be reduced, the system triggers the sweating process. Evaporating from the skin surface, sweat absorbs the skin evaporation heat and lowers the temperature. If heat losses need to be reduced, the thermoregulation system may restrict peripheral circulation, thus diminishing the drop in the blood temperature. A physical effort always involves a rise in the energy produced by the organism, and the thermoregulation system activates mechanisms aiming at a reduction in temperature.

The heat output $\dot{M}$ produced by the organism due to metabolism depends on the type of activity and is proportional to the volume of inhaled air, i.e. to the intensity of breathing. This parameter varies in a wide range. For example, for a sitting person who is not performing any physical work it is assumed at the level of 58W per $1\mathrm{m}^2$ of the skin surface area. The parameter related to energy generation due to metabolism is expressed in metabolic units *met*, and 1 *met*=58 $\mathrm{W/m^2}$.

Parameter $\dot{W}$ represents the heat output converted to mechanical work and not accumulated in the body. For this case the value is close to zero. As physical activity becomes more intensive, the heat output resulting from metabolism rises and so does the amount of accumulated energy because not all of it is converted to work. In the case of hard physical labour or for some sports, the metabolic rate may exceed 1000 $\mathrm{W/m^2}$. The values of the parameter for different activities may be found for example in [17, 18, 15]. The most common values are listed in table 1.

Tab. 1. Metabolic rates [15]

Activity	Metabolic rate	
	W/m²	*met*
Reclining	46	0,8
Seatem, relaxed	58	1,0
Sedentary activity (office, dwelling, school, laboratory)	70	1,2
Standing, light activity (shopping, laboratory, light industry)	93	1,6
Standing, medium activity (shop assistant, domestic work, machine work)	116	2,0
Walking on level ground:		
2 km/h	110	1,9
3 km/h	140	2,4
4 km/h	165	2,8
5 km/h (typical walk)	200	3,4

The amount of heat given up to the environment is also affected by clothing. The higher the clothing heat resistance, the more heat may be accumulated in the body, and the body temperature may rise. In order to determine the impact of clothing on the thermal comfort sensation, the *clo* (clothing) parameter is introduced [20], which defines the model heat resistance of clothing ($1clo=0.155(m^2K)/W$). Calculating the heat resistance value and the *clo* parameter, the heat resistance values of individual items of clothing have to be summed up. A list of values of these parameters may be found for example in [17, 18, 15]. The most common values are listed in table 2.

Tab. 2. Thermal insulation for typical combinations of garments [15]

Work clothing	R_{cl}	
	clo	$(m^2K)/W$
Underpants, boiler suit, socks, shoes	0,7	0,110
Underpants, shirt, boiler suit, socks, shoes	0,8	0,125
Underpants, shirt, trousers, smock, socks, shoes	0,9	0,140
Underwear with short sleeves and legs, shirt, trousers, jacket, socks, shoes	1,0	0,155
Underwear with long legs and sleeves, thermo-jacket, socks, shoes	1,2	0,185
Underwear with short sleeves and legs, shirt, trousers, jacket, heavy quilted outer jacket and overalls, socks, shoes, cap, gloves	1,4	0,220
Underwear with short sleeves and legs, shirt, trousers, jacket, heavy quilted outer jacket and overalls, socks, shoes	2,0	0,310
Underwear with long sleeves and legs, thermo-jacket and trousers, Parka with heavy quitting, overalls with heave quilting, socks, shoes, cap, gloves	2,55	0,395

Daily wear clothing	R_{cl}	
	clo	$(m^2K)/W$
Panties, T-shirt, shorts, light socks, sandals	0,3	0,050
Underpants, shirt with short sleeves, light trousers, light socks, shoes	0,5	0,080
Panties, petticoat, stockings, dress, shoes	0,7	0,105
Underwear, shirt, trousers, socks, shoes	0,7	0,110
Panties, shirt, trousers, jacket, socks, shoes	1,0	0,155
Panties, stockings, blouse, long skirt, jacket, shoes	1,10	0,170
Underwear with long sleeves and legs, shirt, trousers, V-neck sweater, jacket, socks, shoes	1,30	0,200
Underwear with short sleeves and legs, shirt, trousers, vest, jacket, coat, socks, shoes	1,50	0,230

Apart from the thermal balance condition the thermal comfort sensation means that a given person is not exposed to draughts or excessive air dryness or humidity. Such a perspective combined with the other parameters proves that thermal comfort is a subjective matter which depends on a given person's needs and preferences as well as on his/her clothes and the work he/she does. Therefore, it is impossible to formulate a universal set of parameters mentioned above that could be a guarantee of thermal comfort for everybody. However, an analysis may be conducted to determine such parameter values that will be satisfying for the greatest possible number of tested individuals.

Danish scientist Ole Fanger was the first to do research in this field. His extensive theoretical analysis, the proposed computational algorithms and the experimental verification are presented e.g. in [17, 18].

Fanger distinguished the following parameters affecting the global sensation of thermal comfort:

• Environmental parameters, unrelated to a given person directly:
- air temperature t_i
- average (radiant) temperature of partitions t_{pr}
- relative humidity of indoor air φ_i
- relative air velocity v_{ar}

• Parameters related to a given person directly:
- clothing heat resistance *clo*
- a person's metabolic rate *met*
- age, sex, ethnic and geographical factors, nutrition habits, meal frequency, etc.

He gathered all these factors in two parameters being the basis of thermal comfort evaluation:
- the *PMV* (predicted mean vote) index. This parameter provides information about the mean rating of a given thermal environment given by the assessing group. It relates to a thermally moderate environment.

- the *PPD* (predicted percentage dissatisfied) index. This parameter represents the percentage of people dissatisfied with thermal conditions they are exposed to in a given environment.

Fanger's works became the basis for European and international standards concerning thermal comfort and they were used to determine its criterial parameters, for example – the EN ISO 7730:2006 Standard. This standard, which is currently the basic normative act concerning the methods of evaluation of thermal comfort of rooms with a moderate microclimate, defines the methods of calculating and determining the *PMV* and *PPD* parameters both for steady-state conditions and for conditions varying in time. Moreover, it specifies categories that indoor environments are classified into based on variations in a series of thermal comfort parameters, additionally including quantities related to local thermal comfort conditions. These additional parameters are as follows:
- Draught rate *DR*. This parameter, expressed in percentages, informs about the draught exposure risk. It therefore represents the risk of local convective cooling.
- Vertical air temperature difference. This parameter provides information on thermal discomfort resulting from a difference in temperature between the head and ankles, for the case where an upward drop in temperature occurs.
- Warm and cool floors. This parameter informs about dissatisfaction with the floor temperature if light home shoes are worn.
- Radiant asymmetry. This parameter provides information about thermal discomfort caused by the partition temperature asymmetry, i.e. by differences in the temperatures of individual partitions (walls and ceilings).

For these parameters the *PD* (percentage dissatisfied) index is defined, which informs about the percentage of people dissatisfied with given conditions. The room categories specified in the standard and the values of thermal comfort parameters required for them are listed in Table 3.

Tab. 3. Room categories related to values of thermal comfort parameters [15]

Room category	**Thermal state of the body as a whole**		
	***PPD*, %**	***PMV***	**Operative temperature permissible variations °C**
A	<6	-0.2÷0.2	±2.0
B	<10	-0.5÷0.5	±2.0
C	<15	-0.7÷0.7	±3.0
Local discomfort caused by			
***DR*, %**	***PD*, %**		
	Vertical air temperature difference	**Warm or cool floor**	**Radiant asymmetry**
<10	<3	<10	<5
<20	<5	<10	<5
<30	<10	<15	<10

In addition to that, the permissible range of the floor temperature (19÷29°C for Category A and Category B), the maximum vertical gradient of temperature and other parameters are defined for the categories.

The *PMV* index may be used if the following requirements are satisfied [15]:
- air temperature in the room under analysis: 10–30°C,
- average radiant temperature of partitions in the room: 10–40°C,
- air velocity in the room: 0–1 m/s,
- water vapour partial pressure in the room: 0–2700 Pa.
- metabolic rate of people in a given room: 0.8-4.0 *met* (46.6–232.8W/m^2),
- a given person's clothing heat resistance: 0–2 *clo*.

This parameter is given on a seven-point scale of the thermal sensation:

- (+3) hot
- (+2) warm
- (+1) slightly warm
- (0) neutral
- (-1) slightly cool
- (-2) cool
- (-3) cold

An environment where -2<*PMV*<2 is considered as moderate. In respect of the foregoing, the following equation is right [15]:

$$PMV = \left[\left(0,303 \cdot \exp\left(-0,036 \cdot \dot{M}\right) + 0,028\right)\right] \cdot \Big\{\left(\dot{M} - \dot{W}\right) - 0,00305 \cdot \left[\left(5733 - 6,99 \cdot \left(\dot{M} - \dot{W}\right) - p_{a}\right)\right] - 0,42 \cdot \left[\left(\dot{M} - \dot{W}\right) - 58,15\right] - 0,000017 \cdot \dot{M} \cdot \left(5867 - p_{a}\right) - 0,0014 \cdot \dot{M} \cdot \left(34 - t_{k}\right) - 3,96 \cdot 10^{-8} \cdot f_{cl} \cdot \left[\left(t_{cl} + 273\right)^{4} - \left(\bar{t}_{pr} + 273\right)^{4}\right] - f_{cl} \cdot \alpha_{k} \cdot \left(t_{cl} - t_{k}\right)\Big\} \tag{3}$$

where:

$$t_{cl} = 35,7 - 0,028 \cdot \left(\dot{M} - \dot{W}\right) - R_{cl} \cdot \Big\{3,96 \cdot 10^{-8} \cdot f_{cl} \cdot \left[\left(t_{cl} + 273\right)^{4} - \left(\bar{t}_{pr} + 273\right)^{4}\right] + f_{cl} \cdot \alpha_{k} \cdot \left(t_{cl} - t_{k}\right)\Big\} \tag{4}$$

$$\alpha_{k} = \begin{Bmatrix} 2,38 \cdot \left|t_{cl} - t_{k}\right|^{0,25} & \text{for } 2,38 \cdot \left|t_{cl} - t_{k}\right|^{0,25} > 12,1 \cdot \sqrt{v_{ar}} \\ 12,1 \cdot \sqrt{v_{ar}} & \text{for } 2,38 \cdot \left|t_{cl} - t_{k}\right|^{0,25} < 12,1 \cdot \sqrt{v_{ar}} \end{Bmatrix} \tag{5}$$

$$f_{cl} = \begin{Bmatrix} 1,0 + 1,290 \cdot R_{cl} \text{ for } R_{cl} \leq 0,078\,(\mathrm{m^2K/W}) \\ 1,05 + 0,645 \cdot R_{cl} \text{ for } R_{cl} > 0,078\,(\mathrm{m^2K/W}) \end{Bmatrix} \tag{6}$$

According to the following formula, a certain value of the *PPD* parameter may be assigned to a given value of the *PMV* index [15]:

 DOI: 10.1051/e3sconf/20171301003

$$PPD = 100 - 95 \cdot \exp\left(-0{,}03353 \cdot PMV^{4} - 0{,}2179 \cdot PMV^{2}\right) \qquad (7)$$

It is obvious that the farther the *PMV* value is from zero, the higher the number of people dissatisfied with thermal comfort of a given environment will be, i.e. the value of the *PPD* parameter will be higher. The relation between the two quantities is presented graphically in Fig. 1. It can be seen that at the average rating at the level of *PMV*=0, the predicted percentage of dissatisfied people is the lowest and totals *PPD*=5%. This also means that irrespective of the conditions of a given room environment, if the number of evaluating people is sufficiently high, there will always be individuals with a thermal discomfort sensation. In order to achieve thermal comfort, it is recommended that -0.5<*PMV*<0.5. Then *PPD*=10%.

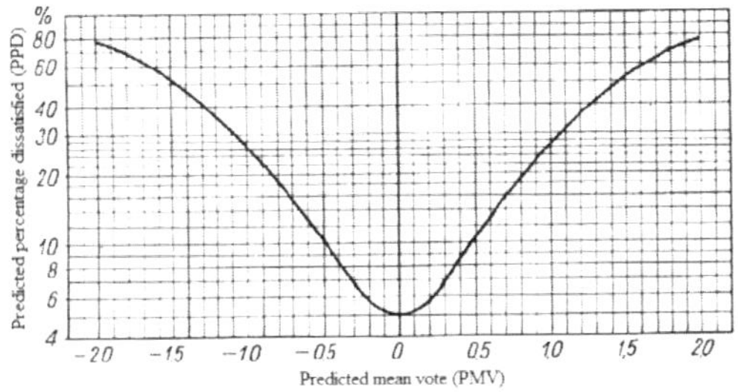

Fig. 1. Relation between PMV and PPD parameters [17, 18, 15]

Fanger carried out a comprehensive study aiming to determine the impact of thermal comfort individual components, and combinations thereof, on the *PMV* and *PPD* parameters. For habitable and office rooms some parameter values may be assumed at a certain level. Owing to that and based on the expected *PMV* value as close to zero as possible, the required values of the air temperature as well as of radiant and operative temperatures may be found by means of the formulae mentioned above. It is also possible to use charts made based on these relations. The charts plotted for selected values of the required parameters are presented in Fig. 2.

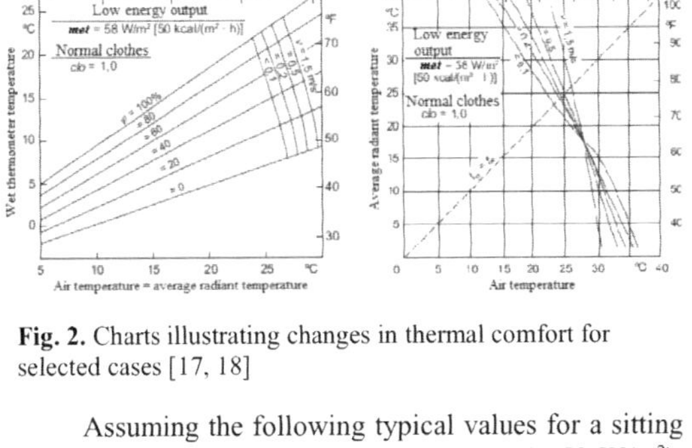

Fig. 2. Charts illustrating changes in thermal comfort for selected cases [17, 18]

Assuming the following typical values for a sitting person resting at home of *clo*=1.0, *met*=1.0 (58 W/m^2), v_{ar}<0.1 m/s, φ_i=50%, the optimum operative temperature (i.e. one at which *PMV*=0) is obtained at the level of about 23°C. For the same conditions in the case of typical office work (*met*=1.1÷1.2) the value amounts to approximately 22°C.

As indicated by Formula (1), the optimum operative temperature may be achieved by different combinations of the air temperature and radiation. For instance, raising the radiant temperature of partitions, the air temperature may be reduced and vice versa. However, it should be remembered that too big an asymmetry of the air or radiant temperatures will make it impossible to achieve the optimum *PMV* value even if the resultant operative temperature equals the required value. According to Table 1 and assuming the room Category B with -0.5<*PMV*<0.5 as the reference point, the permissible variations in the operative temperature are included in the range of ±2.0°C compared to the required temperature level, which is conditioned by the *clo*, *met*, *w* and φ_i parameters. For the previous level of these parameters, i.e. *clo*=1.0, *met*=1.2, v_{ar}<0.1 m/s and φ_i=50%, the optimum operative temperature is 22°C. Considering the assumed thermal category of the room, the permissible operative temperature value is thus included in the range of 20÷24°C. Knowing the permissible range of the operative temperature variability, a set of permissible variations in the air and radiant temperatures may be established according to Formula (1). As stated above, if it is assumed that the difference between individual temperatures is smaller than 4°C, then an assumption may also be made of the equality of individual components of the surface film conductance coefficients in the formula. The range of permissible temperatures will then be as presented in Fig. 3.

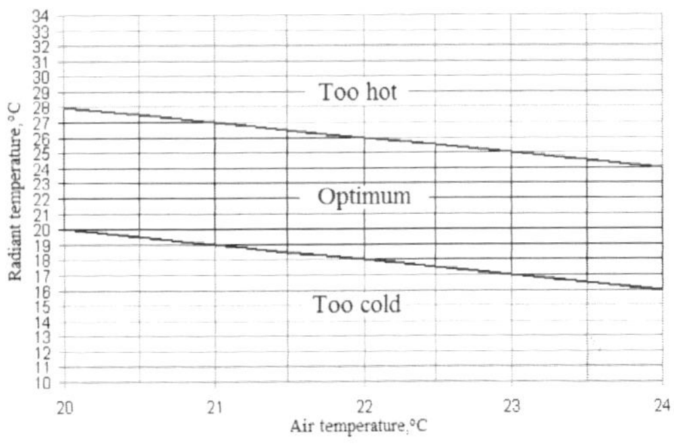

Fig. 3. Optimum (permissible) temperature range

In practice, calculations of thermal comfort conditions using the *PMV* and *SET* indices give similar results.

Knowledge of the unit thermal power (energy output) emitted by the human body, referred to the 1m^2 of the skin is needed to determine the thermal comfort parameters. It may also be an important component of the overall heat balance of the room, conditioning the size of the radiator. To determine it, it is needed not only a knowledge of the number of people staying in the room and emitted unit thermal power, but also the surface of the skin of the human body. There are many alternative formulae to calculate this parameter. One of the most frequently used is the DuBois formula [21]:

$$A_s = 0{,}20248 \cdot m_{cz}^{0{,}425} \cdot h_{cz}^{0{,}725} \quad (6)$$

It applies to adults. For a male of an average body structure the formula results with a value of A_s=1.9 m^2, and for female A_s=1.6 m^2. It follows that, for example, for a male in a sitting position who is not performing any physical work, the value of the emitted thermal power is $A_s \cdot$ 1 met = 1.9 m^2·58 W/m^2 = 110W.

Based on unit thermal power emitted by the human body in the metabolism process and because of the physical work performed, it is also calculated the daily caloric need. It is assumed that a daily average value of this parameter for a typical male (A_s = 1.9 m^2) with a standard physical activity is approx. 2500 kcal, which gives approx. 10,500 kJ = 10500 kWs. This results with a unit thermal power at a level of 10500000/24 (hours per day)/60 (minutes in an hour)/60 (seconds in a minute)/1.9 m^2 (skin surface) = 64 W/m^2. This is a typical daily average unit thermal power emitted.

3 Polish legal requirements regarding the temperatures in a residential environments in relation to the conditions of thermal comfort

Currently the main Polish legislative act defining the temperature values that must be preserved in a residential environments, is quoted earlier document [2]. The selected requirements contained in it, taken with minor changes from the archive Polish Standard PN-82/B-02402P [22], are presented in Table 4.

Tab. 4. The required temperatures in a residential environments [2]

Design temp.*	Purpose or use of the premises	Exemplary rooms
+20°C	- intended for a permanent human occupancy without outerwear, not performing out a continuous physical labor	living rooms, hallways, kitchens equipped with individual gas or electric furnaces, office rooms, meeting rooms
+24°C	- intended to undressing, - intended to accommodate people without clothing	bathrooms, changing rooms, washrooms, shower halls, swimming pools, medical offices intended for patients undressing, infants and meeting rooms for children in nurseries, surgery rooms
*) It is allowed to take other design temperatures for the heated rooms than specified in the table, if this is due to the technological requirements.		

This standard also indicates the need to consider the influence of radiation temperature felt by humans. It is given therein that:

"Indoor temperature t_{ix} in residential and office buildings should be in °C:

- when using convection radiators (panel radiators, column radiators, etc)
- if there is one cooling partition $t_{ix} = t_i$
- if there are two cooling partitions $t_{ix} = t_i + 1$
- if there are three cooling partitions $t_{ix} = t_i + 2$
- if there are four cooling partitions $t_{ix} = t_i + 3$
- when using surface radiators (floor radiators, wall radiators, ceiling radiators) $t_{ix} = t_i - 1$"

Unfortunately, in the current legislation [2] these recommendations have been omitted.

The optimum temperature ranges are also given in the archival Polich Standard PN-B-03421:1978 [23]. The standard defines parameters for three cases:

• Low metabolic rate (e.g. sewing, bookkeeping, type-writing):
- indoor room temperature of air in winter: 20-22°C; in summer: 23-26°C,
- relative humidity (irrespective of physical activity) in winter: 40-60%; in summer: 40-55%,
- air velocity in winter: up to 0.2 m/s; in summer: up to 0.3 m/s.

• Medium metabolic rate (e.g. nail driving, plastering):
- indoor room temperature of air in winter: 18-20°C; in summer: 20-23°C,
- relative humidity in summer: 40-60%,
- air velocity in winter: up to 0.2 m/s; in summer: up to 0.4 m/s.

• High metabolic rate (e.g. carrying heavy objects):
- indoor room temperature of air in winter: 15-18°C; in summer: 18-21°C,
- relative humidity in summer: 40-60%,
- air velocity in winter: up to 0.3 m/s; in summer: up to 0.6 m/s.

In practice most often taken into account are temperatures values 20° C and 24°C, correct for rooms, living rooms, bedrooms, kitchens and bathrooms. Such temperatures are adopted in practice by the designers, architects and installers, both in the calculation of the design thermal heat load of the building and its various rooms, and in the case of designing of the heating system, for determining the operating temperature of system components, the size, power, and type of boiler, type and size heaters, valves settings, pipe diameters, etc. Over the years, Polish legal requirements regarding the indoor temperatures were changing. These data were given in previous versions of the Polish Standard PN-82/B-02402P [22] (with the status of mandatory for use at that time). A significant change, resulting in a requirement to ensure higher indoor temperatures than was the case in previous years, was introduced in the Polish Standard in 1974. This was due to the publication of the Fanger work [17, 18] and conclusions, which he formulated. The values of this parameter for earlier versions of this standard are summarized in Table 5.

Tab. 5. Indoor temperatures in the winter time – Polish normative requirements

Type of the room	Number of the standard/year				Temp. °C
	1*	2*	3*	4*	
Living rooms, hallways	18	18	18	20	20
Kitchens	15	15	15	20	20
Bathrooms	22	22	22	24	24
1*) PN-B-02402:1954; **2***) PN-B-02402:1959; **3***) PN-B-02402:1964; **4***) PN-B-02402:1974					

It is thus seen that in all the Polish recommendations and legal standards, recommended temperatures do not correspond with optimal temperatures of thermal comfort and are lower. In addition, these records do not distinguish between an operative temperature, the air temperature and the radiation temperature, and usually refer to the temperature of air. Most often the operative temperature is lower than air temperature due to the occurrence of cooling partitions (windows, exterior walls - product of the heat transfer surface and the temperature prevails over the product of the heating elements). Differences, if you take into account the subjective feelings of the people staying in the given conditions, can be significant. If, for example, calculate the values of *PMV* and *PPD* using formulas (3) and (7) for a room intended for permanent people occupancy, who do not perform any continuous physical labor, in normal clothing in winter (clo=1,0, met=1,0, v_{ar}<0,1 m/s i φ_i=50%), for which, according to Polish law required temperature should be at the level of t_i=20°C (assuming that the air temperature is equal to the radiation temperature), it turns out that ir results with the values PMV=-0,85 and PPD=20.25%. In addition, as shown in Table 3, that temperature value means that room should be classified to the lowest category (category C) from the point of view of providing thermal comfort. In fact, in practice, often users tend to obtain higher temperatures, the value of t_i = 20°C is in fact perceived as far too low for a large percentage of the respondents.

4 Conclusions

Assuming in the design process and operation of the heating system temperatures compatible with Polish legal requirements, not complying with the optimal conditions of thermal comfort, can have several effects:

• Ergonomic:
- reduced thermal comfort category of the room
- increased percentage of people dissatisfied with the conditions of the microclimate and the thermal comfort of the room
- the need for additional clothing

• Economic:
- reduction of investment costs of installation. A lower temperature results in a lower design thermal load of the building and, therefore, lower power demand for the heat source (eg. a boiler) and radiators. This lowers the purchase price of these devices.
- reduce the operating costs of installation. The requirement to provide smaller heat power to the object can reduce fuel costs, as well as electricity costs for driving the circulation pump.

In both cases of the economic aspects, the differences are conditioned both with the a particular climate zone to which the building belongs, as well as the price of equipment. The lower external design and average year-round temperature (during the heating season) adopted for the calculation, the greater the differences. Design temperature affects primarily on the investment costs, because it affects the size and cost of the devices. The average year-round temperature in the heating season has an impact primarily on the operating costs, because it affects the amount of fuel consumed and energy use. These costs are approximately proportional to the design temperatures difference - external and internal. For example, for climatic zone III (intermediate zone in Poland) and the location of Krakow, for which the design temperature is t_e=-20°C, and the average temperature during the heating season is t_{em}=3,12°C (calculated in accordance with meteorological data contained in the archive Polish PN-B-02025:2001P [24]), an increase of design temperature from t_i=20°C to t_i=23°C results in an increase in operating costs at the level of {[(23-3,12)-(20-3,12)]/(20-3,12)}·100%=17,8%. The increase of the design thermal load is at the level of {[(23-(-20))-(20-(-20))]/(20-(-20)}·100%=7.5%. The actual difference in the cost will naturally depend on the the current weather conditions and the actual average temperature during the heating season for a given year. it can be seen, also in Poland, that the climatic conditions in recent years have changed significantly. The average year-round temperature and the minimum, design temperature, increased. This fact, however, is not included in Polish legal requirements related to the climate zones and design temperatures. With the increase of the design and the year-round average temperature during the heating season also increase differences in investment and operating costs when assuming a higher indoor temperature. But this is not an important argument against the adoption of higher indoor temperatures, because at the same time decreasing then both operating and investment costs for a given indoor temperature - higher external temperatures, in a natural way, decrease heating costs.

Symbols list

A_s – the surface area of the skin, m^2,
α_k – convection term of the surface film conductance (convective heat transfer coefficient), W/(m^2·K),
α_{pr} – radiation term of surface film conductance (radiative heat transfer coefficient), W/(m^2·K),
f_{cl} – clothing surface area factor, -,
φ_i – relative humidity of indoor air, -, %,
h_{cz} – the human's height, m,
$\dot{M}$ – unit heat output generated due to metabolism, W/m^2,
m_{cz} – body mass, kg,
p_a – water vapour partial pressure, Pa,

R_{cl} – clothing insulation, (m^2·K)/W,
t_{cl} – clothing surface temperature, °C,
t_e – design temperature of the outdoor (surrounding) air, °C,
t_{em} – year-round average temperature of the outdoor (surrounding) air, °C,
t_i – conventional temperature of the indoor (surrounding) air, °C,
t_k – convection temperature (air), °C,
t_{oper} – operative temperature, °C,
t_{pr} – radiant temperature (partitions and elements of fittings), °C,
$\dot{q}_{ak,sk}$ – unit heat output accumulated in skin, W/m^2,
$\dot{q}_{ak,org}$ – unit heat output accumulated inside the organism, W/m^2,
$\dot{q}_{konw}$ – unit heat output given up to the surroundings through convection, W/m^2,
$\dot{q}_{pr}$ – unit heat output given up to the surroundings through radiation, W/m^2,
$\dot{q}_{pot}$ – unit heat output given up to the surroundings to evaporate sweat from the skin (latent heat), W/m^2,
$\dot{q}_{odd.konw}$ – unit heat output given up to the surroundings due to breathing, by means of convection, W/m^2,
$\dot{q}_{odd.par}$ – unit heat output given up to the environment due to breathing, by means of evaporation, W/m^2,
v_{ar} – relative air velocity, m/s,
$\dot{W}$ – unit heat output converted to mechanical work, W/m^2,

References

1. Construction Law of 7 July 1994, Dz.U. (Journal of Laws) 94.89.414, as amended (in Polish: Prawo budowlane z dnia 7 lipca 1994 r., Dz.U. Nr 89, poz. 414, z późn. zm.),
2. Regulation of the Minister of Infrastructure of 12 April 2002 on the technical conditions to be met by buildings and their location, Dz.U. (Journal of Laws) 02.75.690 as amended (in Polish: Rozporządzenie Ministra Infrastruktury z dnia 12 kwietnia 2002 roku w sprawie warunków technicznych, jakim powinny odpowiadać budynki i ich usytuowanie, Dz.U.02.75.690 z późniejszymi zmianami),
3. T. Bedford, Basic principles of ventilation and heating, Lewis, London, (1948),
4. F.C. Houghten, C.P. Yaglou, Determining lines of equal comfort, ASHVE Trans., pp. 163-176 and 361-384, **28** (1923),
5. A.P. Gagge, J.A.J. Stolwijk, Y. Nishi, An effective temperature scale, based on a simple model of human physiological regulatory response, ASHRAE trans.: pp. 247-262, **77** (1971),
6. P. Gagge, Y. Nishi, Effective temperature scale useful for hypo and hyperbaric environments, Aviation, Space, and Environmental Medicine, pp. 97-107, February, **48** nr.2, (1977),
7. R.R. Gonzalez, Y. Nishi, A.P. Gagge, Experimental Evaluation of Standard Effective Temperature. A New Biometeorological Index of Man's Thermal Discomfort, International Journal of biometeorology, pp.1-15, **18** (1974),
8. K. Cena, J.A. Clark, Studies in Environmental Science. Bioengineering, Thermal physiology and comfort, Elsevier, (1981),
9. American Standard ASHRAE standard 55-2013: Thermal Environmental Conditions for Human Occupancy,
10. A.R. Chandra, Refrigeration and Air Conditioning", PHI Learning Private Limited, New Delhi, (2010),
11. A.P. Gagge, J. Nishi, R.R. Gonzales, Standard Effective Temperature, Symposium Thermal Comfort and Moderate Heat Stress, CIB Commission W45, Building Research Establishment, (1972),
12. W.P. Jones, Air Conditioning Engineering (fifth edition), Elsevier Butterworth Heinemann, (2001),
13. B. Nowakowski, Principles of ventilation and heating of working places (In Pplish: Zasady wietrzenia i ogrzewania zakładów pracy), Instytut Spraw Społecznych, Warsaw (1935),
14. International Standard ISO 7726:2002: Ergonomics of the thermal environment. Instruments for measuring physical quantities,
15. International Standard ISO 7730:2005: Ergonomics of the thermal environment -- Analytical determination and interpretation of thermal comfort using calculation of the PMV and PPD indices and local thermal comfort criteria,
16. Z. Kabza, Parameter and thermal power engineering equipment measurements (in Polish: Pomiary parametrów i urządzeń termoenergetycznych), a report prepared within the project "The Czestochowa University of Technology Development Plan" co-financed from the EU funds under the European Social Fund (Project No POKL. 04.01.01-00-059/08),
17. P.O. Fanger, Thermal comfort (In Polish: Komfort cieplny), Arkady, Warsaw, (1974),
18. P.O. Fanger, Thermal Comfort. Analysis and Applications in Environmental Enginnering, Danish Technical Press, Copenhagen, (1970),
19. World Wide Web of Centralny Instytut Ochrony Pracy CIOP: www.ciop.pl.
20. A.P. Gagge, A.C. Burton, H.C. Bazett, A practical system of unit for the description of heat exchange of man with his environment, Science, pp. 428-430, **94** (1941).
21. D. Du Bois, E.F. Du Bois, A formula to estimate the approximate surface area if height and weight be known, Archives of Internal Medicine, pp. 863-871, **17** (1916),
22. Polish Standard PN-82/B-02402P: Heating – Indoor temperatures in buildings (in Polish: Ogrzewnictwo - Temperatury ogrzewanych pomieszczeń w budynkach),
23. Polish Standard PN-B-03421:1978P: Ventilation and air-conditioning. Calculation parameters of indoor air in rooms intended for permanent

occupation by humans (in Polish: Wentylacja i klimatyzacja. Parametry obliczeniowe powietrza wewnętrznego w pomieszczeniach przeznaczonych do stałego przebywania ludzi),
24. Polish Standard PN-B-02025:2001P: The calculation of seasonal heat demand for heating of residential buildings (in Polish: Obliczanie sezonowego zapotrzebowania na ciepło do ogrzewania budynków mieszkalnych i zamieszkania zbiorowego).

E3S Web of Conferences **13**, 02001 (2017) DOI: 10.1051/e3sconf/20171302001
WTiUE 2016

Experimental stand for investigation of fluid flow in heat exchangers with cross-flow arrangement

Stanisław Łopata[1,*], *Paweł* Ocłoń[1], and *Marzena* Nowak-Ocłoń[1]

[1]Cracow University of Technology, al. Jana Pawła II 37, 31864 Kraków, Poland

Abstract. The operation analysis of high-performance heat exchanger with tubes elliptical indicated that the heat exchangers can be subject to damage. The reason for this is probably improper distribution of working fluid in tubular space of heat exchanger. Therefore, a part of the tubes may be improperly cooled and subject to compressible stresses. The paper presents an experimental stand allowing to confirm the given assumption. The experimental investigation enables to examine the mass flow rate in heat exchanger tubes. Also, it is possible to assess the impact of the construction of inlet, intermediate and outlet chambers on the flow distribution within the heat exchanger tubes.

1 Introduction

The operation of the high-performance heat exchanger with finned elliptical tubes in cross-flow working fluid arrangement (e.g. flue gas-water), indicates a possibility of failure. This situation occurs mainly in the case when a large difference between the working fluid temperatures exists. The failure can be seen as a certain deformation of tubes in the form of buckling, or as a tube rupture. Examples of the construction of heat exchangers which are mentioned here, often refer to high-performance heat exchangers are shown in Fig. 1 [1, 2]. Figure 2 shows a structure (the basic elements) and the principle of operation.

Fig. 1. Examples of construction of high-performance heat exchangers heater/cooler

Due to the external fins, the efficiency of fin-and-tube heat exchangers in high. The fins are densely assembled on heat exchanger tubes. Thus the heat transfer area is large. The advantages of the heat exchanger, together with the high efficiency, are relatively small size (compactness of construction) at a relatively high power (e.g. in comparison with conventional shell-and-tube structures with the same capacity, but with smooth tubes).

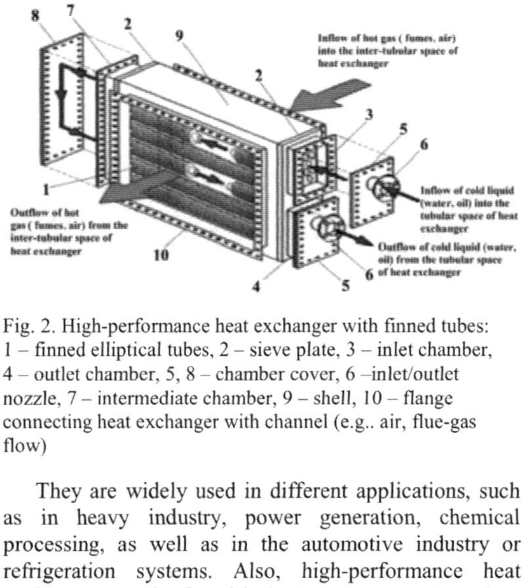

Fig. 2. High-performance heat exchanger with finned tubes: 1 – finned elliptical tubes, 2 – sieve plate, 3 – inlet chamber, 4 – outlet chamber, 5, 8 – chamber cover, 6 –inlet/outlet nozzle, 7 – intermediate chamber, 9 – shell, 10 – flange connecting heat exchanger with channel (e.g.. air, flue-gas flow)

They are widely used in different applications, such as in heavy industry, power generation, chemical processing, as well as in the automotive industry or refrigeration systems. Also, high-performance heat exchangers can be found in heat recovery systems (e.g. to cool down the exhaust gas) in air-conditioning units, among others.

2 Heat exchanger with finned elliptical tubes – exploitation/construction problems

The exploitation of equipment is accompanied by failures and damage [3-7], and in the field of heat exchangers, there are no exception. This statement is

* Corresponding author: lopata@mech.pk.edu.pl

evident, but it is often problematic to find a reason of the failure, which usually requires a thorough analysis of many factors including the flow and heat transfer, as well as the distributions of stress and deformation of the structure [8-12]. Moreover, these issues are gaining importance in cases of repeated operational problems, especially damage [7, 12]. Examples of heat exchanger failures (with the cross-flow arrangement of flue gas and water), are shown in Fig. 3. By the local inspection, the nature of the damage was found. The buckling of several tubes (in both heat exchanger passes) and fractured tube nearby the sieve plate, shows that the compressible stresses were the reason of failure.

Fig. 3. Buckling and fracture of heat exchanger tubes

Figure 3 presents damages on a heat exchanger; that is structurally similar to the one shown in Fig. 2, but vertically disposed and had a profiled chamber: inlet, intermediate, and outlet. The chambers are formed from the large pipes Fig. 4). The heat exchanger operated in one of the heat recovery units in Poland. The problems already appeared after three months from the date of heat exchanger launch. It should be noted that a new device (which was inserted in place of the previous) also suffered a failure, after eight months of operation and both the location of damage, as well as their nature was similar [7, 11].

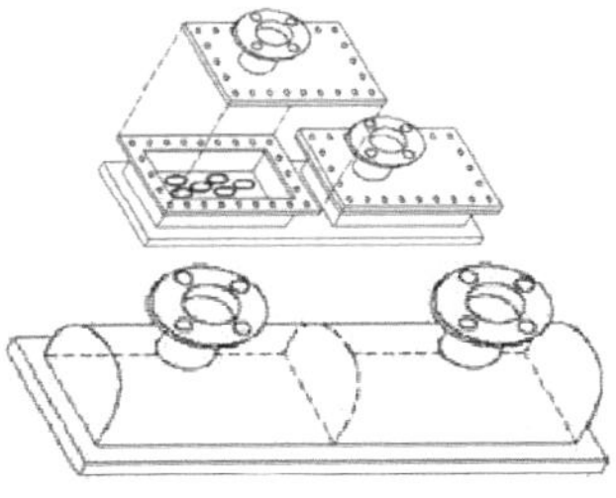

Fig. 4. The shape of heat exchanger headers used in cross-flow heat exchangers

The reason of tube damage can be improper operational conditions, such as raw water (with limescale) formed as a result of the internal surfaces impurities whose presence contributes to a significant increase in temperature and stresses, even exceeding the limit value [12-15]. Also, the small volume of the heat exchanger headers hampers the proper flow distribution to the tubes located in both passes, what has been confirmed by numerical modeling [9-12]. Fig. 5 shows the scheme of the analyzed heat exchanger, and in Fig. 6 the value of the mass flow rate in tubes of both passes (for the first row of tubes, i.e. from the flue gas flow) are depicted. The CFD simulations [12, 16] were performed in ANSYS CFX software for various numbers of finite elements and indicated an abnormal flow distribution in heat exchanger tubes.

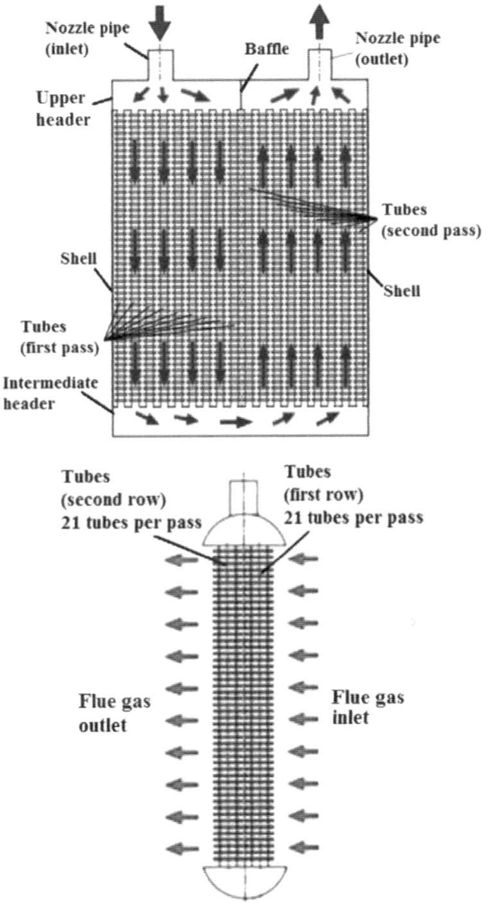

Fig. 5. Scheme of the two-pass cross-flow heat exchanger

The considerable variations in the water flow rate result in increased temperature of water in heat exchanger tubes (Fig. 7). Figure 7 reveals that the water temperature differences are up to 45 °C and the saturation temperature is reached in tubes no 10 (first pass of the heat exchanger) and 22 (second pass of the heat exchanger).

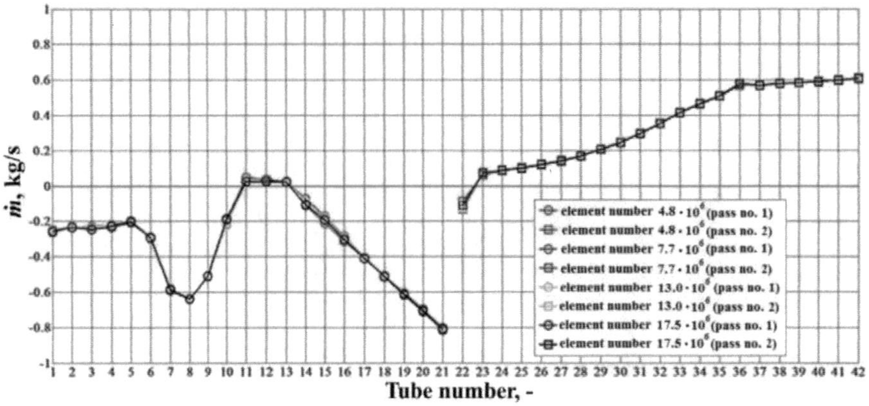

Fig. 6. Water mass flow rate in elliptical tubes of the heat exchanger of the first pass – tubes no. 1÷21, of the second pass – tubes no. 22÷42 (the legend shows number of finite element used)

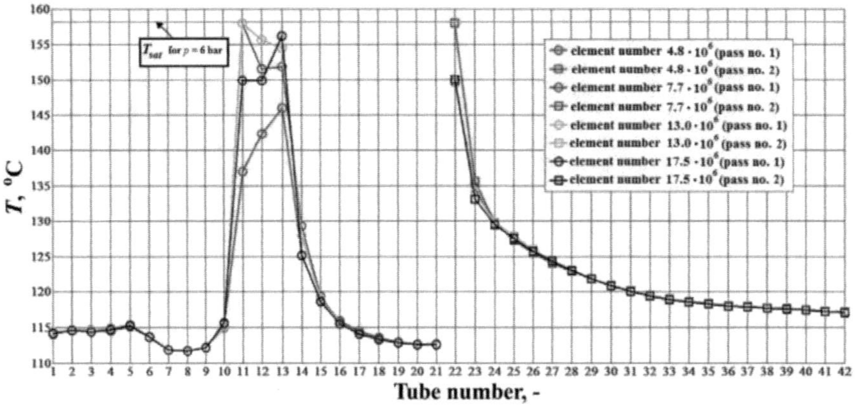

Fig. 7. Average water temperature in elliptical tubes of the heat exchanger of the first pass – tubes no. 1÷21, of the second pass – tubes no. 22÷42 (the legend shows number of finite element used)

The analysis of Fig. 6 reveals that the flow of coolant (in this case water) in the heat exchanger tubes is very diverse, with the largest value is approx. 0.8 kg / s (for tube 21) and the smallest, approx. 0, 02 kg / s (tube 11, 12 and 13). In both cases, the differences exist in the tubes of the first pass. In the second pass, the highest value of flow rate is approx. 0.6 kg / s, and applies to tubes 36 ÷ 42, and the lowest, approx. 0.1 kg / s, to the tube 22. From Fig. 6 one can observe the different sign of mass flow rate in tubes no. 11 to 13 of first pass and tube no 22 of the second pass (compared to adjacent tubes). This situation indicated the reverse flow of water. This situation leads to a significant increase in water temperature (may cause the flow boiling [17, 18] in the case of low flow rate) since the path of the flow is repeated several times. Figure 6 shows that the water flow rate in the pipes is very low. The flow rate is one of the most important factors affecting the intensity of the heat transfer process. Hence, the conditions of heat removal in the individual pipes are different. In this regard very unfavorable conditions apply to the tubes with low flow rates of water, especially characterized by backflow. In this case, one can expect high thermal stresses, which can contribute significantly to the heat exchanger failure. It should be noted that the previously described damage underwent the tubes corresponding to the unfavorable conditions of the heat exchanger (tubes no. 11-13 in the first pass, tube no. 22 in the second pass). The observed improper flow distribution in cross-flow heat exchanger (Fig. 6) leads to the significant temperature differences in the tubes located in both passes. As it can be seen in Fig. 7, in the tubes with the reverse-flow, the water temperature is significantly higher compared to other tubes. Maximum reach approx. 158 ° C (this corresponds to a saturation temperature at a pressure of 6 bars, for which calculations were performed [12]) and the lowest is approx. 112 ° C. Those differences significantly affect the temperature of the tube wall and result in the thermal stresses. Due to a considerable stiffness of sieve plates, the tubes with the lower temperature exhibit tensile stresses, while those with the higher the compressible stresses [10, 11].

The flow boiling and local evaporation of water [18] accelerate the process of limescale formation (in the case of raw or improperly treated water). In such a situation, the limescale will cause a further increase in the

* Corresponding author: lopata@mech.pk.edu.pl

temperature of the tube wall and may result in even higher thermal stresses. This disadvantageous effect (lower heat transfer rate) will also be caused by contaminants transported through a heating medium (e.g., the exhaust gas) and accumulated on the outer surface of tubes.

The possible solution to improve the working conditions of the analyzed heat exchanger is the change the shape of headers. The new solution (the patent procedure is in progress) is shown in Fig. 8.

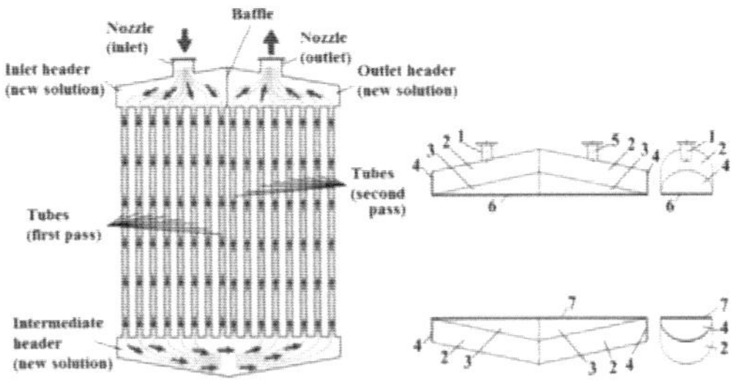

Fig. 8. A scheme of the two-row/two-pass high performance heat exchanger (of water/flue gas type) with the new shape of headers: 1 – inlet nozzle, 2, 3 – header combined from the tube sector (2) and two wedges (3), 4 – bottom, 5 – outlet nozzle, 6 – sieve plate (connected with outlet and inlet headers separated by the baffle), 7 – sieve plate (connected with intermediate header)

Because a new inlet, intermediate and outlet headers with increased flow space are proposed, a uniform flow distribution to heat exchanger tubes is expected. However, to confirm the expectation, the extensive experimental investigations are needed.

3 Experimental stand

Figure 9 presents a schematic and general view of the test stand (for investigation of flow distribution in the heat exchanger with elliptical tubes). Various interchangeable inlet and outlet chambers are depicted in Figs. 10 and 11.

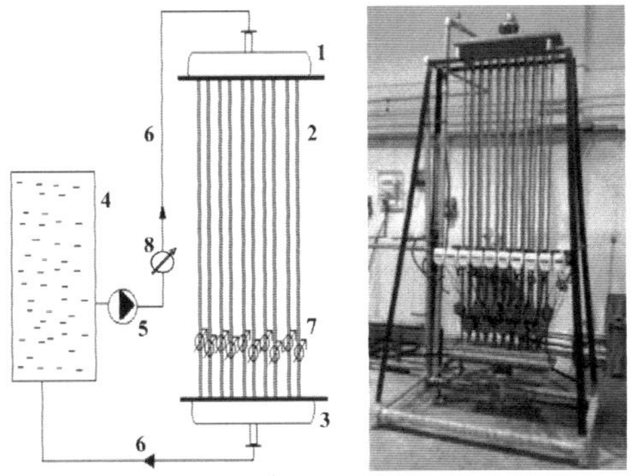

Fig. 9. Schematic of the stand for investigation of flow distribution in heat exchanger with elliptical tubes: 1 – inlet header with nozzle pipe, 2 – tubular space of heat exchanger (elliptical tubes), 3 – outlet header with nozzle pipe, 4 – water tank, 5 – feed pump, 6 – connections, 7 –flowmeters located in each tube, 8 – flow meter of total flow rate of working fluid

Fig. 10. Examples of inlet headers with various location of nozzle pipe

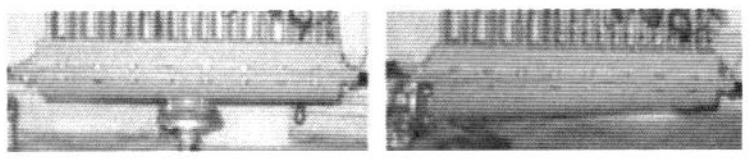

Fig. 11. Examples of outlet headers with various location of nozzle pipe

The stand presented in Figs. 10 and 11, enables to install two different shapes of the heat exchanger header. The solution used by heat exchanger manufacturers and a proposed shape of the header. Also, for each solution (the new and existing) it is possible to test various location of inlet/outlet nozzle pipes. Therefore, the investigation of various flow configuration is possible.

A view of sieve plate and its flange connection with heat exchanger header is shown in Fig. 12, the assembly of outlet header to the flange with sieve plate is presented in Fig. 13.

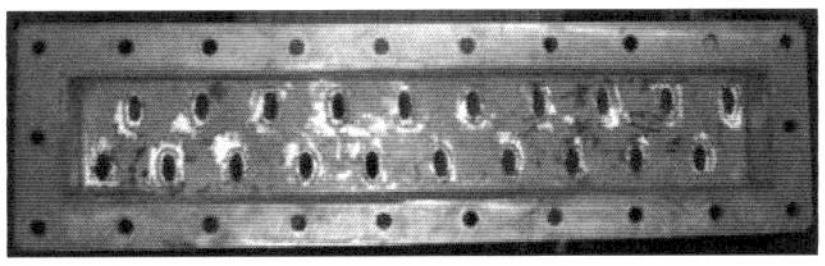

Fig. 12. A view of sieve plate and its connection with heat exchanger header

Fig. 13. The assembly of outlet header to sieve plate

The Sontex Superstatic 749 flowmeters are used for the measurement of the instantaneous flow rate in individual heat exchanger tubes [19]. The another flowmeter is installed before the heat exchanger to control the total flow rate and the indications of the flowmeters located within the tubes.

4 Closing remarks

The reason for a failure of the high-performance heat exchanger used for heat recovery from exhaust gasses may be the irregular distribution of working fluid (e.g. water) flowing in the tubular space. It can be related to the not appropriate design of the heat exchanger headers: the inlet, the intermediate and the outlet. As a result, in the tubes of the heat exchanger, flow rate and temperature of the medium may vary considerably. Numerical calculations confirm this fact and allow to determine the zones of heat exchanger tube, where the flow is significantly lower than in other ones. The low flow rate of coolant causes the ineffective heat reception and leads to the increased tube wall temperature. Consequently, the excessive thermal stresses occur and may result in the heat exchanger failure.

The new construction of the heat exchanger headers were proposed. Changing the design of the headers may improve the coolant flow distribution in heat exchangers tubular space. Both the coolant flow rate and temperature are more uniformly distributed within the tubes of heat exchanger pass.

The experimental verification of the proposed design of heat exchanger header shall be performed. Therefore, the experimental stand to the analysis of flow distribution in the heat exchanger with elliptical tubes was made. The stand allows measuring the working fluid flow rates separately within each tube of the heat exchanger. The measurements can be performed for various types of heat exchanger headers, and location of inlet and outlet nozzle pipe. In future, the authors plan to validate the CFD simulation results, through the investigation performed at the newly created stand.

References

1. GEA company website (http://www.gea-tc.com.pl/High-performance-heat-exchange)
2. Websites of fin-and-tube heat exchangers producers (http://gzgele.en.made-in-china.com/productimage/; http://www.made-in-china.com/showroom/; http://www.weiku.com/products-image/)
3. N.V. Challenger, R. Phaal, S.J. Garwood, Fracture mechanics assessment of industrial pressure vessel failures, Int. J. of Pressure Vessels and Piping, **61** (2-3), pp. 433-456 (1995)
4. S.H. Dai, Failure analysis and life prediction of a set of extra high pressure heat exchangers containing defects, Int. J. of Pressure Vessels and Piping, **70** (3), pp. 197-200 (1997)
5. M.S. Liu, Q.W. Dong, D.B. Wang, X. Ling, Numerical simulation of thermal stress in tube-sheet of heat transfer equipment, Int. J. of Pressure Vessels and Piping, **76** (10), pp. 671-675 (1999)
6. D. Butrymowicz, M. Trela, Effects of fouling and inert gases on performance of recuperative feed-water heaters, Archives of Thermodynamics, No. 1-2, **23** (2001)
7. S. Łopata, W. Krzanowski, S. Pytel, P. Ocłoń, *The heat exchanger of flue gas/water type failure analysis, and the possibility of design changes for the improvement of device operation* (in polish, unpublished expertise), Cracow University of Technology, PK M-5/429/2009 (2009)
8. S. Łopata, P. Ocłoń, Investigation of the flow conditions in a high-performance heat exchanger, Archives of Thermodynamics, **31**, No. 3, pp. 37-53 (2010)
9. P. Ocłoń, S. Łopata, Modeling of the flow distribution inside the collectors of the high-performance heat exchanger, Technical Transactions, Mechanics, Cracow University of Technology, z. 4-M/2011/B, Issue 7, Year 108, pp. 391-400 (2011)
10. S. Łopata, P. Ocłoń, Analysis of operating conditions for high-performance heat exchanger with the finned elliptical tube, Rynek Energii, **102**, No 5, pp. 112-124 (2012)
11. S. Łopata, P. Ocłoń, *Modelling and Optimizing Operating Conditions of Heat Exchanger with Finned Elliptical Tubes*. Edited by L. H. Juarez, *Fluid Dynamics, Computational Modeling and Applications*, Part 3, chapter 14, InTech, ISBN: 978-953-51-0052-2, pp. 327 – 356, Croatia (2012)
12. P. Ocłoń, *Coupled flow-thermal and structural analysis of operating conditions for heat exchanger with finned elliptical tubes*, PhD Thesis (in polish), Kraków (2013)
13. S. Łopata, P. Ocłoń, The stress state analysis of the high performance heat exchanger, Technical Transactions, Mechanics, Cracow University of Technology, z. 4-M/2011/B, Issue 7, Year 108, pp. 359-368 (2011)
14. P. Ocłoń, S. Łopata, M. Nowak, A.C. Benim, Numerical study on the effect of inner tube fouling on the thermal performance of high-temperature fin-and-tube heat exchanger, Progress in CFD, **15**, Issue 5, pp. 290–306 (2014)
15. S. Łopata, P. Ocłoń, Numerical study of the effect of fouling on local heat transfer conditions in a high-temperature fin-and-tube heat exchanger, Energy, **92**, 1, pp. 100–116 (2015)
16. D. Taler, P. Ocłoń, Thermal contact resistance in plate fin-and-tube heat exchangers, determined by experimental data and CFD simulations, Int. J. of Thermal Sciences, **84**, pp. 309-322 (2014)
17. P. Ocłoń, S. Łopata, M. Nowak, A novel 1D/2D model for simulating conjugate heat transfer applied to flow boiling in tubes with external fins, Heat and Mass Transfer/Waerme- und Stoffuebertragung, **51** (4), pp. 553-566 (2015)
18. P. Ocłoń, M. Nowak, S. Łopata, Simplified numerical study of evaporation processes inside vertical tubes, J. of Thermal Science, **23** (2), pp. 177-186 (2014)
19. P. Ocłoń, S. Łopata, K. Chłosta, Experimental and Numerical Investigation of Flow Distribution within the Heat Exchanger with Elliptical Tubes, Procedia Engineering, **157**, pp. 428-435 (2016)

Modelling of flow boiling heat transfer in a cylindrical annulus mini gap

Magdalena Piasecka[1,*], *Sylwia* Hożejowska[2] and *Tomasz* Musiał[3]

[1,3]Kielce University of Technology, Faculty of Mechatronics and Mechanical Engineering ,Al. 1000-lecia P.P. 7, 25-314 Kielce, Poland
[2] Kielce University of Technology, Faculty of Management and Computer Modelling, Al. 1000-lecia P.P. 7, 25-314 Kielce, Poland

Abstract. This paper presents the results of investigations into flow boiling heat transfer in a mini gap 1 mm wide. The mathematical approaches of the heat transfer coefficient determination were also proposed. It was considered two ways of describing layers in a measurement module with a cylindrical annulus mini gap. In the first approach, the measurement module was treated as a planar multilayer wall and in the second approach - as a multilayer cylindrical wall. The values of the local heat transfer coefficients, determined from two approaches, were compared. Both approaches gave similar results.

1 Introduction

In recent years, many theoretical and experimental works have been published to address the issues of heat transfer in flow boiling in mini gaps of different geometry and spatial orientation. The major advantage of mini gaps technology in heat exchangers is improving the effectiveness of heat transfer processes. Effective heat transfer is vital for the cooling processes in miniature equipment, as it prevents overheating and permanent damage to the components.

Studies devoted to the cooling technology that uses flow boiling processes in small channels will contribute to further development of current trends regarding the miniaturisation of products and devices. In [1,2], the authors present their investigations into heat transfer bubbly boiling process, harnessing environmentally friendly refrigerating media in conventional channels [1] and minichannels [2]. Research on heat transfer coefficients is reported in [3], where pressure drop and 'dryout' for flow boiling of water in an oil heated minichannel was discussed. In [4], the authors show an improved semi-empirical method for determining heat transfer coefficient in flow boiling in conventional and small diameter tubes. The issue of flow boiling heat transfer of R134a in the multiport minichannel heat exchangers is discussed in [5]. In [6], the study of R134a flow boiling heat transfer and evaluation of existing correlations was reported. In [7], in order to determine the heat transfer coefficient, the researches perform experimental tests during nanofluid (based on water and copper oxide particles) flow in a cylindrical channel. Paper [8] shows the results of experimental investigation into heat transfer of helium-xenon mixtures in cylindrical channels. In [9], the authors determine the heat transfer coefficient in a mini channel evaporator with R-134a as a refrigerant.

Experimental data were the basis for the analysis of flow boiling heat transfer in rectangular minichannels performed by Piasecka in her previous works [10-14]. This research interest is focused on flow boiling heat transfer in cylindrical annulus mini gap.

2 Experimental stand

The experimental setup shown in Fig.1, consists of several systems: the test loop in which the working fluid (distilled water) circulates, the supply and control system, the data and image acquisition system, and the lighting system.

The test loop is composed of: a gear pump, a compensating tank, a heat exchanger, a filter, mass flow meter, and a deaerator. The data and image acquisition system consists of: a high speed camera, a data acquisition station, a computer with special software, and the lighting.

Figure 2 presents the testing module with an annular mini gap. The mini gap 1 mm wide was created between the external tempered glass pipe for viewing flow patterns (1) and the copper pipe (2) positioned along the same axis. A cartridge heater (4) was located inside the pipe (2), axially and symmetrically, powered by autotransformer with adjusted current intensity. To fix the thermoelement sensors with a spacing of approximately 1 cm in the flow line longitudinal slots was made in the inside surface of the pipe. Thermoelement wires run in the gap between the copper pipe and the heater and in the thermal conductive filler layer (thermal paste). The module header had pressure transducers and the sensors of successive thermoelements to measure the water temperature and positive pressure in the inlet and the outlet to and from the mini gap.

* Corresponding author: tmpmj@tu.kielce.pl

During experimental series, there was a laminar flow of distilled water along a mini gap. It was accompanied by a gradual increase in the electric power supplied to cartridge heater. The experiments were performed at the subcooled boiling region, in which the liquid flowing in the mini gap was superheated at the interface with the heater and subcooled at the core of the flow. The data from three experimental series performed at similar thermal and flow parameters, obtained for selected similar heat flux value, were used in the calculation and further analysis.

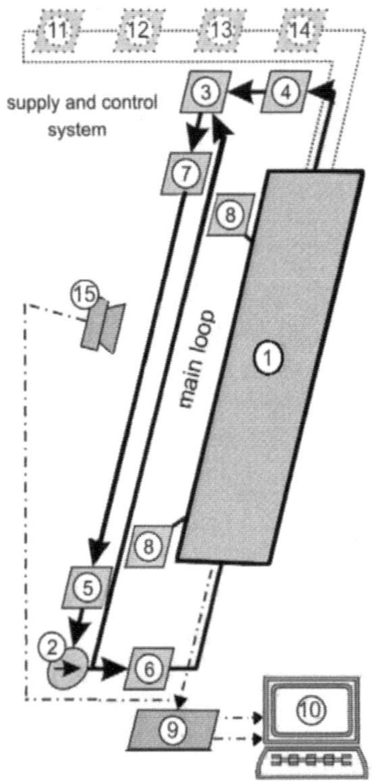

Fig. 1. The schematic diagram of the main loops at the experimental setup, 1-a measurement module with a mini gap; 2- a gear pump; 3- a compensating tank/pressure regulator; 4- a tube-type heat exchanger, 5- a filter, 6- a mass flow meter; 7- a deaerator; 8- a pressure transducer; 9- a data acquisition station; 10- a pc computer; 11- an inverter welder; 12- a voltmeter; 13- an ammeter; 14- a shunt; 15- a high-speed camera.

3 One-dimensional model

It is assumed that the steady state heat transfer process in the test module is stationary. Since the width of the gap is relatively small in relation to the channel radius (Fig. 3), the elements of the module can be treated as a system of planar layers with different thickness and thermal conductivity. In paper notation, the subscripts referring to: the cartridge heater, the thermal conductive layer (thermal paste) with thermocouples, the copper pipe and the fluid in the mini gap are denoted with the following letters: *h* for the cartridge heater, *s* for the thermal conductive layer (thermal paste) with thermocouples, *F* for the copper pipe and *f* for the fluid in the mini gap. The cartridge heater (with radius r_1) is the heat source with power output q_V.

For $r = r_1$ at each point $z_1, z_2, \ldots, z_M$, the condition below is satisfied

$$\tfrac{1}{2} r_1 q_V = \lambda_h \frac{\partial T_h(r_1, z_i)}{\partial r} = \lambda_s \frac{\partial T_s(r_1, z_i)}{\partial r} \tag{1}$$

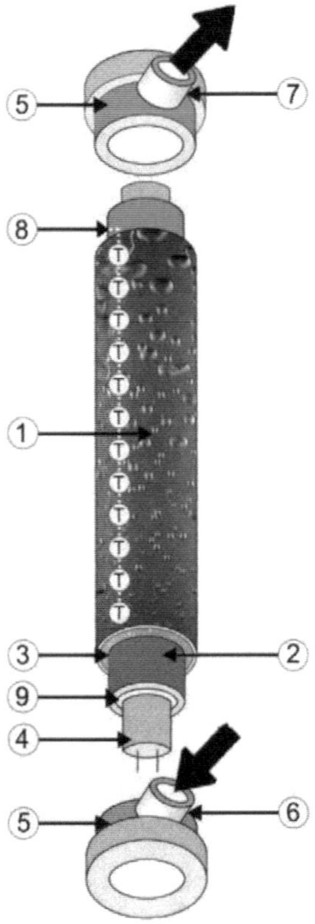

Fig. 2. The measurement module layout with an annular mini gap: 1- a glass pipe for viewing flow patterns, 2- a copper pipe, 3- an annular mini gap, 4- a cartridge heater, 5- a module header with sealing, 6- an inlet port for the medium, 7- an outlet port for the medium, 8-thermoelements between the heater (4) and the copper pipe (2) in the thermal conductive filler layer (thermal paste), 9- a thermal conductive layer (thermal paste).

Since difference $r_2 - r_1$ is very small, it is possible to replace the partial derivative with a finite difference:

$$\frac{\partial T_s(r_1, z_i)}{\partial r} \approx \frac{T_{s,i} - T_h(r_1, z_i)}{r_2 - r_1} \tag{2}$$

where $T_{s,i} = T_s(r_2, z_i)$ are the temperatures measured using the thermocouples at points (r_2, z_i), Fig. 3b.

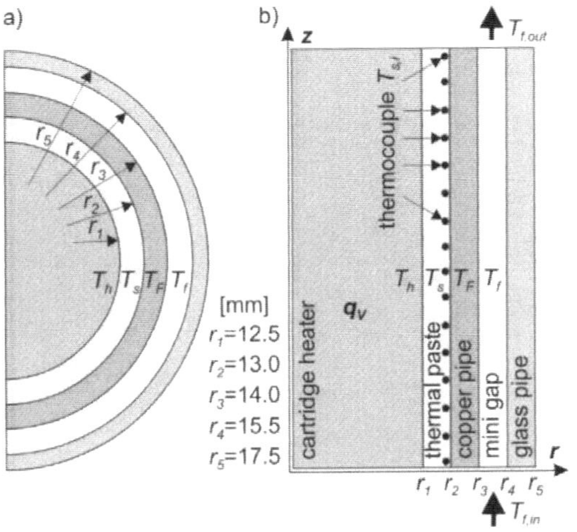

Fig. 3. Diagram illustrating the assumptions adopted for the planar multilayer wall: a) vertical cross-section, b) horizontal cross-section.

Combining equations (1) and (2) gives formula for the temperature $T_h(r_1, z_i)$:

$$T_h(r_1, z_i) = T_{s,i} + q_V \frac{(r_2 - r_1)r_1}{2\lambda_s} \quad (3)$$

The relationship between the cartridge heater temperature and the copper pipe surface temperature for the planar multilayer wall (Fig. 3b) can be written as:

$$T_h(r_1, z_i) - T_F(r_3, z_i) = \frac{q_V r_1}{2}\left(\frac{(r_2 - r_1)}{\lambda_s} + \frac{(r_3 - r_2)}{\lambda_F}\right) \quad (4)$$

Equations (3) and (4) lead to formula

$$T_F(r_3, z_i) = T_{s,i} - \frac{q_V r_1}{2}\frac{(r_3 - r_2)}{\lambda_F} \quad (5)$$

The formula for the local heat transfer coefficients at points $z_1, z_2, \ldots, z_M$, at the interface between the copper pipe surface and the fluid, derived from equation (5) and from the Robin boundary condition, has the form

$$\alpha_1(z_i) = \alpha_{1,i} = \frac{0.5 r_1 q_V}{T_{s,i} - 0.5 r_1 q_V \frac{(r_3 - r_2)}{\lambda_F} - T_f} \quad (6)$$

If it is assumed that the module elements create a cylindrical layer (Fig. 3a), formula (6) will have the following form

$$\alpha_2(z_i) = \alpha_{2,i} = \frac{0.5 r_1 q_V}{T_{s,i} - 0.5 r_1^2 q_V \frac{\ln(r_3/r_2)}{\lambda_F} - T_f} \quad (7)$$

In formulas (6) and (7) it was assumed that the fluid temperature T_f changes linearly in the mini gap, from the inlet temperature $T_{f,in}$ to the outlet temperature $T_{f,out}$. Both temperatures are known from the measurements.

4 Results

Results from the experiments performed at boiling heat transfer in the subcooled boiling region with distilled water as the working fluid flowing along a cylindrical annulus mini gap were presented. Selected results from three experimental series for the set similar heat flux values (approx. 62 MW/m^2 on average) were used in two mathematical approaches for determining the heat transfer coefficient. In the first approach, the measurement module was treated as a planar multilayer wall and in the second approach - as a multilayer cylindrical wall. The results from the three different experiments performed at the set volumetric heat fluxes of 57.6 [MW/m^3], 60.6 [MW/m^3] and 66.7 [MW/m^3] were analysed.

Figure 4 shows the heat transfer coefficient at the interface between the copper pipe and the water flowing in the mini gap, calculated from equation (6), i.e., using the calculation approach in which the measurement module was treated as a planar multilayer wall, vs. the mini gap length. It can be noticed that the heat transfer coefficient decreases, in the nearly monotonic mode, with the distance from the mini gap inlet. In all cases under analysis, local values of the heat transfer coefficient reached similar results, from the maximum value of 4 [$kW/(m^2K)$] at the inlet to the minimum value of 2.8 [$kW/(m^2K)$] at the outlet of the mini gap.

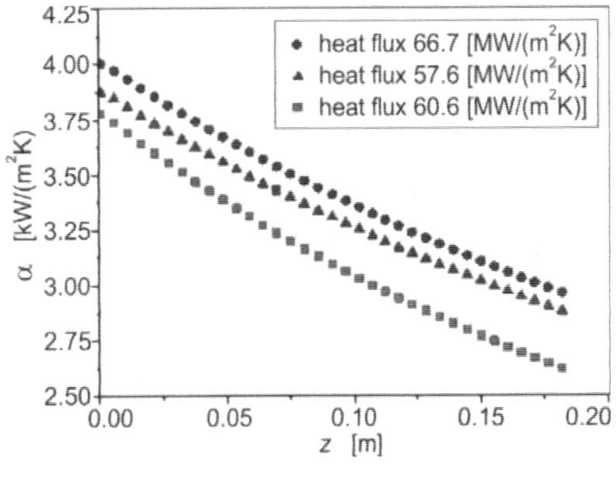

Fig. 4. Heat transfer coefficient vs. the mini gap length, calculated with the first approach, eq. 6 (the measurement module was treated as a planar multilayer wall).

The maximum relative error (MRE) describing the difference between the heat transfer coefficients obtained from equations (6) and (7), respectively, was calculated from the formula

DOI: 10.1051/e3sconf/20171302002

$$MRE = \max\left\{\left[\frac{\sum_{i=1}^{M}(\alpha_1(z_i)-\alpha_2(z_i))^2}{\sum_{i=1}^{M}(\alpha_1(z_i))^2}\right]^{0.5},\left[\frac{\sum_{i=1}^{M}(\alpha_1(z_i)-\alpha_2(z_i))^2}{\sum_{i=1}^{M}(\alpha_2(z_i))^2}\right]^{0.5}\right\} \tag{8}$$

In the three cases analysed (similar heat fluxes in the three experiments), the MRE was lower than 0.2 %. Local values of the heat transfer coefficient calculated from equation (7) (second calculation approach, in which the measurement module was treated as a multilayer cylindrical wall), take similar values, as that in Fig. 4, confirmed by very low MRE values.

Conclusions

The paper focuses on heat transfer modelling for the working fluid (distilled water) flowing in the symmetrically heated annular mini gap. In the mathematical approach presented here, two ways of describing the multilayer measurement module were used. In the first approach, measurement module was treated as a planar multilayer wall, and in the second approach - as a multilayer cylindrical wall. The calculation data for the heat transfer coefficient values were from three different experiments, with each of them performed with similar experimental thermal and flow parameters, including the most important one - heat flux.

The heat transfer coefficient calculated between the metal pipe and water in the mini gap took similar values in both approaches. The outcomes from the two methods allow concluding that both approaches give very similar results. This conclusion is confirmed analytically by a very low value of the maximum relative error (MRE), which in the cases under analysis was at most 0.2%.

Further work will be focused on constructing a two-dimensional heat transfer model in a mini annular gap and solving it with the use of CFD.

The research reported herein was supported in part by a grant from the National Science Centre in Poland (No. DEC-2013/09/B/ST8/02825).

References

1. T. Bohdal, Ann. Set Environ. Prot. **15**, 107–126. (2013)
2. T. Bohdal, Int. J. Heat Fluid Flow **21**, 449–455 (2000)
3. L.A. Campbell, S.G. Kandlikar, *Proc. Second Int. Conf. Microchannels Minichannels*, June 17-19, Rochester (2004)
4. D. Mikielewicz, J. Mikielewicz, J. Tesmar, Int. J. Heat Mass Tran. **50**, 3949–3956 (2007)
5. J. Kaew-On, K. Sakamatapan, S. Wongwises, Exp. Therm. Fluid Sci. **35**, 364–374 (2011)
6. Y. Xu, X. Fang, G. Li, A. Xu, Int. J. Heat Mass Transf. **98**, 150–163 (2016)
7. A.V. Minakov, V.Y. Rudyak, D.V. Guzei, A.S. Lobasov, High Temp. **53**, 246–253 (2015)
8. S.L. Elistratov, O.V. Vitovskii, E.Y. Slesareva, J. Eng. Thermophys. **24**, 33–35 (2015)
9. E.B. Dollera, E.P. Villanueva, *IOP Conf. Ser. Mater. Sci. Eng.* **88**, 12027 (2015)
10. M. Piasecka, Exp. Heat Tran. **27**, 231–255 (2014)
11. M. Piasecka, Heat Tran. Eng. **35** (10), 903–912 (2014)
12. M. Piasecka, Int. J. Heat Mass Tran. **81**, 114–121 (2015)
13. M. Piasecka, Int. J. Refrig. **56**, 198–212 (2015)
14. M. Piasecka, K. Strąk, B. Maciejewska, Heat Tran. Eng. **38**, 332–346 (2017)

Selected soil thermal conductivity models

Monika Rerak[1,*]

[1]Cracow University of Technology, Institute of Thermal Power Engineering, al. Jana Pawła II 37, 31-864 Krakow, Poland

Abstract. The paper presents collected from the literature models of soil thermal conductivity. This is a very important parameter, which allows one to assess how much heat can be transferred from the underground power cables through the soil. The models are presented in table form, thus when the properties of the soil are given, it is possible to select the most accurate method of calculating its thermal conductivity. Precise determination of this parameter results in designing the cable line in such a way that it does not occur the process of cable overheating.

1 Introduction

When designing the underground power cable system it is important to ensure its safe operation [1-5]. The cable line shall be designed in such a way to not exceed the safe operating temperature. What is also important is the economic aspect of the cable system installation. To best meet the safety conditions and the cost-effectiveness the heat dissipation from the power cables to the surrounding soil should be properly evaluated.

In the literature there are several methods that aim to provide the soil thermal conductivity values. Many of them are based primarily on the moisture content of the soil. Soils are divided into five types: gravel, loam, clay, peat and sand. This division corresponds to the categorization used by the Unified Soil Classification System. The soils exhibit two states: the frozen and unfrozen. The methods described below also take into account the effect of the soil dry density [6].

2 Van Rooyen and Winterkorn model

The first method presented in this work was developed by Van Rooyen and Winterkorn. This is an empirical relationship. It is used to determine the thermal conductivity of sand and gravel. It can be applied to the non-frozen soil, where the water content is between 1.5% and 10%.

$$\frac{1}{k} = A \cdot 10^{-BS_r} + s \tag{1}$$

Where:
k – soil thermal conductivity [W/m·K],
S_r – water content [%],
A – function of soil dry density,
B – mineral type,
s – granulometry.

Applying the above equation is not straightforward due to additional variables, which also one must determine (A, B and s). The selection of suitable parameters is complicated, and therefore the correct calculation of the thermal conductivity is difficult [7].

3 Johansen model

Another relationship was reported by Johansen. It can be used for soils both coarse and fine-grained in frozen or unfrozen state. The maximum water content for this method is 20% [8,9]. Johansen's equation becomes:

$$k = (k_{sat} - k_D) \cdot k_e + k_D \tag{2}$$

where:

k_{sat} – saturated soil thermal conductivity [W/m·K],
k_D – dry thermal conductivity of the soil [W/m·K],
k_e – function of soil saturation level [–].

The disadvantage of this method is fact that it does not account moisture migration. This attribute can have a significant impact on the effective thermal conductivity.

4 De Vries model

Next model was developed by De Vries. This method assumes that the soil is a two-phase material. It consists of a uniform, solid particles, which have an elliptical shape. These particles are dispersed in a liquid phase. This method can be used for the water content in the range of 10% to 20%. The soil must be coarse and not frozen [10,11]. De Vries formula on soil thermal conductivity is expressed by:

$$k = \frac{x_f \cdot k_f + F \cdot x_s \cdot k_s}{x_f + F \cdot x_s} \tag{3}$$

* Corresponding author: monika.rerak@mech.pk.edu.pl

where:

f – liquid phase,
s – solid phase,
x – volume fraction,

and:

$$F = \frac{1}{3}\sum_i \left[1 + \left[\frac{k_s}{k_f} - 1\right] g_i\right]^{-1} \quad (4)$$

where the values of g_i sum to unity. The parameter F was originally supposed to be the aspect ratio, but now its values are used to fit the empirical data [7,8,12,13].

This method assumes particles that are not in contact and this is contrary to the general condition in soils.

5 Kersten model

Another formula for soil thermal conductivity is Kersten's model. The model is successfully used for various soil types. It was created based on the relationships that were derived from empirical data. This model is widely used due to the derivation of equations for the frozen and unfrozen sandy soils and the soils formed of silt and clay. The Kersten's formula can be applied to sandy soils with water content higher than 1%, and for soils consisting of silt and clay with water content higher than 7%. It should be noted that Kersten's correlation gives the correct results only for frozen soils with a maximum water content of 90% [14,15].

Correlation for sandy soils:

- unfrozen soil

$$k = [0.101 \log w + 0.0577] 10^{0.000624\rho_d} \quad (5)$$

-frozen soil

$$k = 0.011(10)^{0.000812\rho_d} + 0.00462(10)^{0.000911\rho_d} \cdot w \quad (6)$$

The correlation for soils consisting of loam and clay:
- unfrozen soil

$$\mathrm{k} = [0.13 \log \mathrm{w} - 0.0288] 10^{0.000624\rho_\mathrm{d}} \quad (7)$$

- frozen soil

$$k = 0.00144(10)^{0.00137\rho_d} + 0.0123(10)^{0.000499\rho_d} \cdot w \quad (8)$$

where:

k – soil thermal conductivity [W/m·K],
w – water content [%],
ρ_d – dry density of the soil [kg/m^3].

Experiments conducted in order to verify the above method led to the conclusion that this method is valid and recommended for soils with low clay and loam content (20%). It works very well also for soil primarily consisting of quartz [7,8].

Figures 1 and 2 illustrate the results obtained by the Kersten's model. One can observe the impact on the soil thermal conductivity of water content and bulk density in the dry state. Using the charts presented above, for a given soil properties, one can determine the thermal conductivity value [16,17].

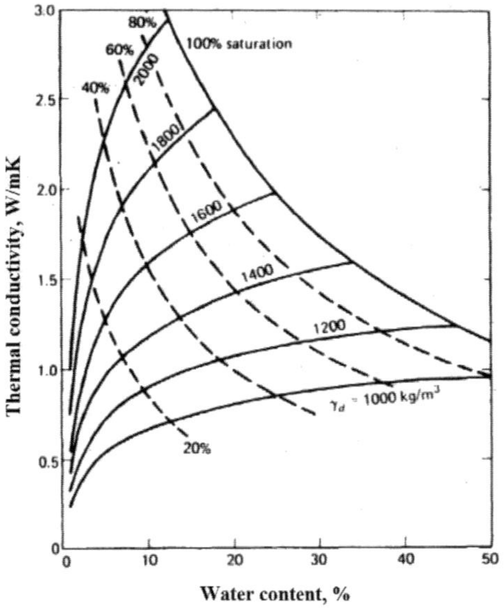

Fig. 1. The average thermal conductivity of the non-frozen sandy soil as a function of water content and the density in the dry state [7].

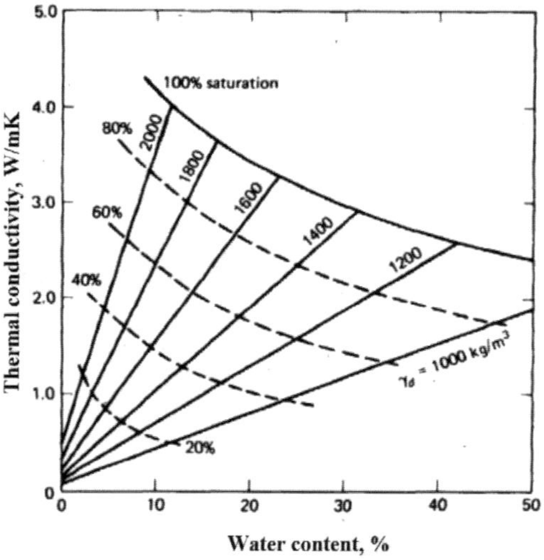

Fig. 2. The average thermal conductivity of the frozen sandy soil as a function of water content and the density in the dry state [7].

 DOI: 10.1051/e3sconf/20171302003

6 Gemant model

Another correlation for determining the soil thermal conductivity was proposed by Gemant. His model is based on the idealized shape of the soil particle. Gemant assumed that the particle has designated contact points around which the water is gathered. The contact points form thermal bridges. Based on the heat flow through thermal bridges the soil thermal conductivity is calculated [8,11,18]:

$$\frac{1}{k} = \frac{\left[\frac{(1-a)}{a}\right]^{\frac{4}{3}} \tan^{-1}\left[\frac{(k_s-k_w)}{k_w}\right]^{\frac{1}{2}}}{\left(\frac{h}{2}\right)^{\frac{1}{3}} [k_w(k_s-k_w)]^{\frac{1}{2}}} + \frac{(1-z)}{k_s a} f\left\{\frac{b^2}{a}\right\} \quad (9)$$

$$a = 0.078 s^{\frac{1}{2}} \quad (10)$$

$$h = 0.16 \cdot 10^{-3} s \cdot w - h_0 \quad (11)$$

$$z = \left(\frac{1-a}{a}\right)^{\frac{1}{2}} \left(\frac{h}{2}\right)^{\frac{1}{3}} \quad (12)$$

$$b^2 = \left(\frac{a}{1-a}\right)^{\frac{2}{3}} \left(\frac{h}{2}\right)^{\frac{2}{3}} \quad (13)$$

where :

k_s – the thermal conductivity of the solids [W/m·K],
k_w – the thermal conductivity of water [W/m·K],
h – apex water (water collected around the contact points),
h_0 – water absorbed as a film around the soil particles,
s – soil dry density [lb/ft^3],
w – water content [%],
a,b,z – shape functions.

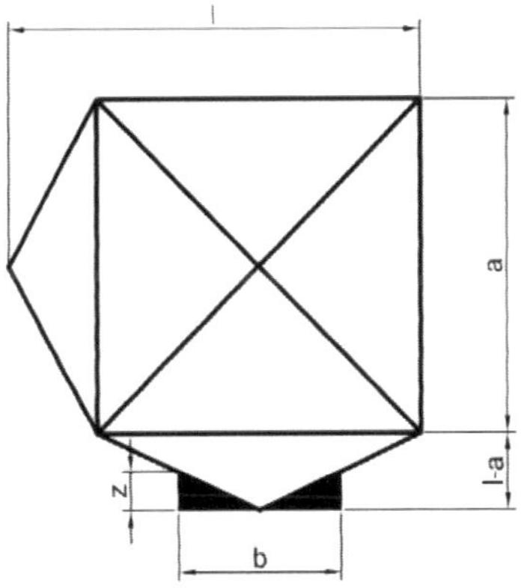

Fig.3. Idealized model of the soil particle [7].

7 Conclusions

Farouki [7] recommended the application ranges of the various soil thermal conductivity models. The validation ranges are given in Table 1. This table was created based on the experimental investigation for various soil state, textures, and water content [19-21]. As one can see, each method cannot be applied to all types of soils. So there is no single methodology that can be used. Therefore, the use of these methods in numerical algorithms related to the heat transfer needs to apply special conditions. The table allows you to choose the best method. Choosing the appropriate method allows to reduce the error.

Table 1. Application ranges for various soil thermal conductivity correlations [7].

State	Teksture	Water content	Correlation for thermal conductivity calculation
Unfrozen	Coarse soil	0,015-0,100	Van Rooyen and Winterkorn
		0,100-0,200	De Vries
		0,200-1,000	Johansen
		0,000-1,000	Gemant
		Saturated state	Johansen, De Vries, Gemant
	Fine soil	0,000-0,100	Johansen (error 15%)
		0,100-0,200	Johansen (error 5%)
		0,200-1,000	Johansen
		Saturated state	Johansen, De Vries, Gemant
Frozen	Coarse soil	0,100-1,000	Johansen
		Saturated state	Johansen, De Vries
	Fine soil	0,000-0,900	Kersten
		0,100-1,000	Johansen
		Saturated state	Johansen, De Vries

References

1. P. Ocłoń, M. Bittelli, P. Cisek, E. Kroener, M. Pilarczyk, D. Taler, R.V. Rao, A. Vallati, *The performance analysis of a new thermal backfill material for underground power cable system* Applied Thermal Engineering, **108**, pp. 233-250 (2016)
2. P. Ocłoń, P. Cisek, D. Taler, M. Pilarczyk, T. Szwarc, *Optimizing of the underground power cable bedding using momentum-type particle swarm optimization method* Energy, **92**, pp. 230-239 (2015)
3. P. Ocłoń, P. Cisek, M. Pilarczyk, D. Taler, *Numerical simulation of heat dissipation processes in underground power cable system situated in thermal backfill and buried in a multilayered soil* Energy Conversion and Management, **95**, pp. 352-370 (2015)

4. P. Ocłoń, D. Taler, P. Cisek, M. Pilarczyk, *Fem-Based Thermal Analysis of Underground Power Cables Located in Backfills Made of Different Materials* Strength of Materials, **47** (5), pp. 770-780 (2015)
5. P. Cisek, P. Ocłoń, M. Pilarczyk. *Thermal analysis of operating conditions for The 400 kV underground power cable transmission line as a power plant delivery system.* Rynek Energii; **114**:70–7 [in polish] (2014)
6. E. Kroener, A. Vallati, M. Bittelli, *Numerical simulation of coupled heat, liquid water and water vapor in soils for heat dissipation of underground electrical power cables.* App Therm Eng; **70**:510–23 (2014)
7. O.T. Farouki: *Thermal properties od soils*, Series Rock Soil Mech, (1986)
8. B.R. Becker, B.A. Fricke: *Effects of saturation and dry density on soil thermal conductivity*, (1997)
9. M. Bittelli, G. S. Campbell, F. Tomei; *Transport in the Soil-Plant-Atmosphere System,* Oxford University Press (2015)
10. G.S. Campbell, J.R. Jungbauer, W.R. Bidlake, R.D. Hungerford. *Predicting the effect of temperature on soil thermal conductivity.* Soil Sci; **5**:307–13 (1994)
11. B. Usowicz, J. Lipiec, J.B Usowicz, W. Marczewski. *Effects of aggregate size on soil thermal conductivity: Comparison of measured and model-predicted data.* Int J Heat Mass Tran, 2:536–**41** (2013)
12. D.A. De Vries, N.H. Afgan. *Heat and mass transfer in the biosphere.* John Wiley and Sons; (1975)
13. D.A. De Vries. *Thermal properties of soils.* In: Van Wijk WR, editor. Physics of plant environment. Amsterdam: North-Holland Publishing Company; p. 210–35. (1963)
14. ISO 14688-2, *Geotechnical investigation and testing – Identification and classification of soil – Part 2: Principles for a classification.*
15. M.S. Kersten. *Thermal properties of soil.* Bull Univ Minn Inst Technol **52**:1–225 (1949)
16. M. Bittelli, F. Ventura, G.S. Campbell, R.L. Snyder, F. Gallegati, P.R. Pisa. *Coupling of heat, water vapor, and liquid water fluxes to compute evaporation in bare soils.* J Hydrol 3–4:191–205 (2008)
17. F. Gori, S. Corasaniti . *New model to evaluate the effective thermal conductivity of three-phase soils.* Int Commun Heat Mass **47**:1–6 (2013)
18. A.R. Sepaskhah, L.Boersma. *Thermal conductivity of soils as a function of temperature and water content.* Soil Sci Soc Am J, **43**:439–44 (1979)
19. T. Kasubuchi. *Heat conduction model of saturated soil and estimation of thermal conductivity of soil solid phase.* Soil Sci, **138**:240–7 (1984)
20. V.R. Tarnawski, T. Mosome, W.H. Leong, G. Bovesecchi, P. Coppa. *Thermal conductivity of standard sands. Part I. Dry-state conditions.* Int J Thermophys, **30**:949–68 (2009)
21. C.D. Peters-Lidard, E. Blackburn, X. Liang, E.F. Wood. *The effect of soil thermal conductivity parameterization on surface energy fluxes and temperatures.* J Atmos Sci, **55**:1209–24 (1998)

The effect of soil and cable backfill thermal conductivity on the temperature distribution in underground cable system

Monika Rerak[1], and *Paweł* Ocłoń[1,*]

[1]Cracow University of Technology, Institute of Thermal Power Engineering, al. Jana Pawła II 37, 31-864 Krakow, Poland

Abstract. The paper presents a mathematical model of heat transfer in the underground cable system. The computations were performed for flat formation of power cables buried in the ground at a depth of 2 meters. The model allows determining the two-dimensional temperature distribution in the soil, thermal backfill and power cables. The simulations studied the effect of soil thermal conductivity on the maximum temperature of the cable conductor. Furthermore, the effect of thermal backfill soil conductivity on the cable conductor temperature was studied. Numerical analyses were performed based on a program written in MATLAB.

Nomenclature

Roman Symbols

A	cross-sectional area, m^2
b	the distance between the conductors' axes and the bottom of cable bedding layer, m
C	length of the sample, m
f	frequency, Hz
H	computational domain height, m
h	burial depth, m
I	current loading, A
k	thermal conductivity, W/(m K)
k_s, k_p	skin and proximity effect correction factors, -
l	the spacing between two consecutive cables, m
p	the distance between the conductors' axes and the top of cable bedding layer, m
q_v	volumetric heat source, W/m^3
$R_{e,AC}$	alternating current (AC) resistance, Ω/km
$R_{e,DC}$	direct current (DC) resistance, Ω/km
$R_{e,ref}$	reference cable conductor electrical resistance (calculated for 20°C), Ω/km
s	the spacing between the right edge of the bedding layer and the side cable axis, m
T	temperature, °C
W	computational domain width, m
x, y	Cartesian coordinates, -
y_s, y_p	skin and proximity coefficients, -

Subscripts

b	backfill,
c	cable conductor,
g	ground,
ins	cable insulation,
s	soil,

Greek Symbols

$\alpha_{e,ref}$	temperature coefficient, -
ρ_{20}	specific electrical resistance of the copper conductor in 20°C, Ω m
ΔQ	cable conductor heat losses, W

1 Introduction

When designing the underground electricity network, the thermal phenomena shall be considered. Due to the cable insulation meltdown occurrence, the cable conductor temperature shall not exceed 90 °C. Since the data on soil thermal resistance is mostly uncertain, the cable engineers design the underground power system in such a way that the cable core temperature does not exceed the optimum temperature of cable operation (65 °C). Therefore, the cable engineers in many cases use thermal backfill materials for locating the power cables. The cable backfill material exhibits higher thermal conductivity than the mother soil. Therefore, the heat dissipation from the buried cables to surroundings is enhanced, when compared to the cables situated in the native soil.

Thermal analysis of the underground power cables operation was performed in [1–5], among others. Both the

* Corresponding author: poclon@mech.pk.edu.pl

numerical and experimental investigations were conducted. Hwang and Jiang [1] presented a combined nonlinear magnetothermal analysis, including radiation effects, for calculating the thermal fields of an underground cable system. Al-Saud et al. [2] performed the numerical computations of temperature distribution in soil and underground power cables. A new perturbed finite-element analysis technique was developed. Al-Saud's approach involves the use of derived sensitivity coefficients associated with various cable parameters of interest, and use these coefficients to achieve optimal cable performance. De Lieto Vollaro et al. [3] performed a numerical study based on a Control-Volume Formulation of the Finite-Difference Method. That technique was used to determine the thermal resistance existing between an underground electrical power cable and the ground surface. The thermal behavior of the cables system was studied for various dimensions of the trench, cable burial depth, two backfill layers thicknesses, and the cable bedding. The authors developed a semi-empirical correlating equation using multiple regression procedures and verified proposed model experimentally in [4]. Kroener et al [5] presented the comprehensive approach to determining the cable ampacity. The authors developed the new numerical model of coupled liquid water, vapor and heat flow in a thermal system that consists of underground cable buried in the soil. The transient computations performed using the Finite Element Method were verified experimentally and demonstrated the strong relation of the cable temperature on soil water content. The thermal analysis of underground power cable systems were also presented in [6-10]. The optimization of thermal backfill dimensions was presented in [6,7], and the computational studies of multilayered soil effect [8] and various cable placement type [9,10] were performed .

This study presents a numerical model of underground power cable system. Three power cables arranged in the flat (in-line) formation and buried at a depth of 2 m are considered. The influence of soil thermal conductivity and thermal backfill conductivity on the maximum cable core temperature is studied. The numerical code developed in MATLAB software allows calculating the two-dimensional temperature distribution within the underground power cable system including cable core, thermal backfill, and soil. The presented mathematical model is efficient and straightforward in programming implementation.

2 Mathematical model of underground power cable system

The scheme of the computational domain is shown in Fig. 1:

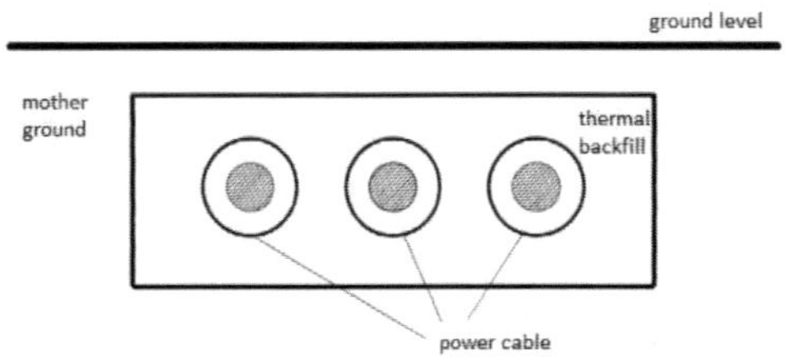

Fig. 1. Flat formation of underground power cable system.

Figure 2 depicts the characteristic dimensions of the system.

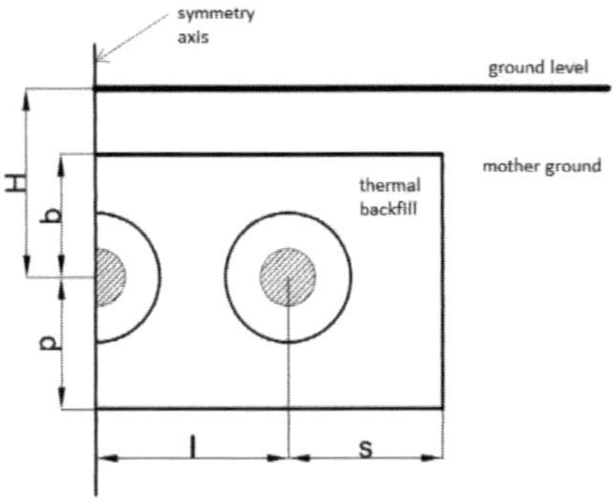

Fig. 2. Characteristic dimensions of the cable system.

The following dimensions are shown in Fig. 2
l – spacing between two consecutive cables (assumed as 0.4 m in this study),
s –distance from the right edge of the backfill layer to the side cable axis (assumed as 0.4 m in this study),
b – distance from the conductor's axis to the top of bedding layer (assumed as 0.4 m in this study),
p – distance between the conductor's axis and the bottom of bedding layer (assumed as 0.4 m in this study),
The cable conductor cross section A_c is selected from the XLPE HV cable design series provided by the cable producer, and equal to $A_c = 1400$ mm^2.
The computational scheme is given in Fig. 3a while the grid used in the analysis is shown in Fig. 3b.

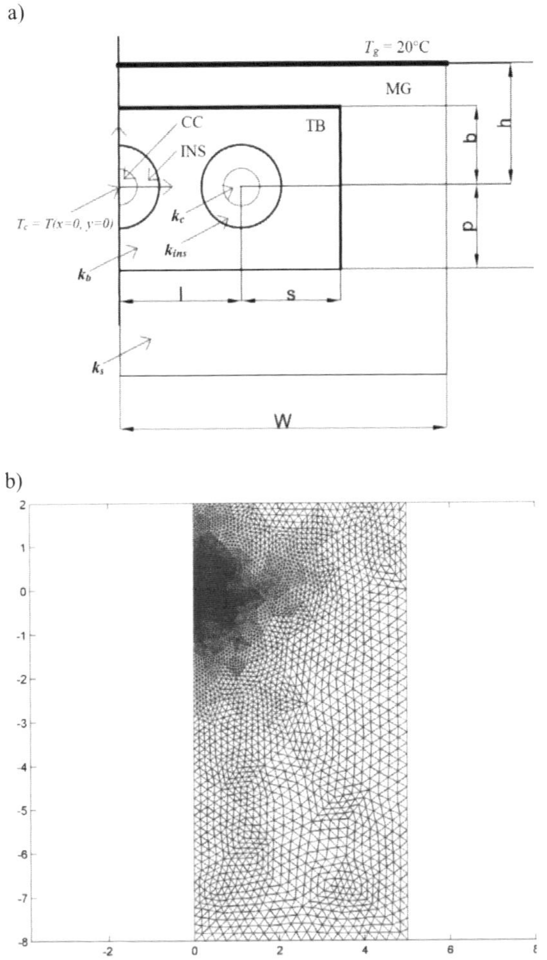

Fig. 3. a) the computational scheme b) computational grid.

The Underground Power Cable System (UPCS) computational domain consists of four subdomains:

- Mother ground (MG),
- Thermal Backfill (TB),
- Cable conductor (CC),
- Cable insulation (INS).

The temperature at certain point with (x,y) coordinates is calculated from the heat conduction equation [8]:

$$\frac{\partial}{\partial x}\left[k\frac{\partial T(x,y)}{\partial x}\right]+\frac{\partial}{\partial y}\left[k\frac{\partial T(x,y)}{\partial y}\right]=-q_v(T_c(x,y)) \qquad (1)$$

where:
T_c - cable conductor temperature, °C
q_v - volumetric heat source (valid only for cable conductor domain), W/m^3.
k - thermal conductivity, W/(m K),

The thermal conductivity values for power cable layers are assumed according to Table 1. The soil thermal conductivity is varied from 0.5 W/(m K) to 1 W/(m K) and depends on the computational case. The thermal backfill conductivity is ranged from 1 W/(m K) to 3 W/(m K)).

Table 1. Cable layer dimensions and thermal conductivities

Cable layer	Material	Thermal conductivity, W/(m·K)
Conductor	Copper	400.00
Insulation	XLPE	0.3232

The volumetric heat source is given as:

$$q_v(T(x,y))=\frac{\Delta Q}{A_c} \qquad (2)$$

Where the heat losses ΔQ are calculated based on the current rating I = 1000 A and the alternating current (AC) electric resistance $R_{e,AC}$:

$$\Delta Q=I^2R_{e,AC} \qquad (3)$$

The altering current electrical resistance is a function of direct current electrical resistance $R_{e,DC}$ and skin and proximity coefficients denoted as y_s and y_p, respectively:

$$R_{e,AC}=R_{e,DC}(1+y_s+y_p) \qquad (4)$$

with

$$R_{e,DC}=R_{e,ref}\left[1+\alpha_{ref}(T(x,y)-T_{ref}\right] \qquad (5)$$

where, $R_{e,ref}$ and $\alpha_{e,ref}$ are the reference cable conductor's electric resistance, and the temperature coefficient for the conductor material, both given at the reference temperature T_{ref} = 20°C. The reference electric resistance of the cable conductor is given by:

$$R_{e,ref}=\frac{\rho_{20}C}{A_c} \qquad (6)$$

where,
ρ_{20} - specific electrical resistance of copper conductor in 20°C, Ω ·m; for the following case study ρ_{20} = 1.7241 10-8 Ω ·m,
C - length of the sample, m; C = 1.0 m,
Ac - cable conductor area, m^2,
the temperature coefficient for copper conductor $\alpha_{e,ref}$ is equal to 0.00393.
The skin and proximity effect coefficients y_s and y_p are determined from the following relationships:

$$y_s=\frac{x_s^2}{192+0.8x_s^2},\ x_s=\frac{8\pi f}{R_{e,DC}}\cdot 10^{-7}k_s \qquad (7)$$

$$y_p=\frac{x_p^2}{192+0.8x_p^2}\left(\frac{d_c}{l}\right)^2\left[0.312\left(\frac{d_c}{l}\right)^2+\frac{1.18}{\frac{x_p^2}{192+0.8x_p^2}+0.27}\right], \qquad (8)$$

$$x_p = \frac{8\pi f}{R_{e,DC}} \cdot 10^{-7} k_p$$

Where, k_s and k_p are skin and proximity effect correction factors equal to 0.435 and 0.37, respectively, for the case of segmented conductor type[8]. An alternating current frequency f = 50 Hz and l is the distance between adjacent conductor axes.

2.1 Boundary condition

The symmetry pattern of the heat conduction equation solution is expected. Therefore only the half of the computational domain is considered. Hence,
the square domain with a height of 10.0 m and width of 5.0 m is used in the computations (Fig. 3). It is assumed that the right, left and bottom edges of the boundary region are perfectly insulated. At the top edge (the ground level), the temperature T_g is set to 20°C. This temperature value is specified by the standards [11,12]. The mathematical description of the applied boundary conditions is given by Eq. (9).

$$\begin{aligned} & k\left.\frac{\partial T}{\partial x}\right|_{x=W} = 0, \text{ for the boundary region right edge,} \\ & k\left.\frac{\partial T}{\partial x}\right|_{x=0} = 0, \text{ for the symmetry plane,} \qquad (9) \\ & k\left.\frac{\partial T}{\partial y}\right|_{y=h-H} = 0, \text{ for the boundary region bottom edge,} \\ & T(x, y = h) = 20°\text{C, for the boundary region top edge.} \end{aligned}$$

The Finite Element Method is applied to solve the heat conduction equation Eq. (1). The linear triangular elements are used to discretize the computational domain. Since the volumetric heat source depends on the temperature, the system of nonlinear equation, comprising from FEM discretization, needs to be solved. Therefore, the Jacobi iteration method is used. The computational domain mesh was created using a PDE toolbox of MATLAB software. The FEM solver was developed by the authors, and implement to the discretized model. The computations performed in MATLAB allow to determine the temperature distribution within entire underground power cable system.

3. Soil thermal conductivity effect on temperature distribution in underground power cable system

The variation in the mother ground thermal conductivity changes the intensity of the heat transfer from power cables. The larger the conductivity, the faster the soil receives the heat, and thus also lowers the temperature of the cable conductor. The soil thermal conductivity changes due to the decrease/increase of the water content (e.g. caused by rains or droughts). The lower the moisture content, the lower the soil thermal conductivity. Therefore, it may happen that the thermal calculations performed for moist soil that meets the security requirements do not satisfy them when the soil is dry.

Figure 4 presents the variation of cable core temperature with the increase in the mother ground thermal conductivity from 0.5 W/(m K) to 1 W/(m K). It is assumed that the cables are placed in thermal backfill (Fluidized Thermal Backfill) with a thermal conductivity of 1.54 W/(m K).

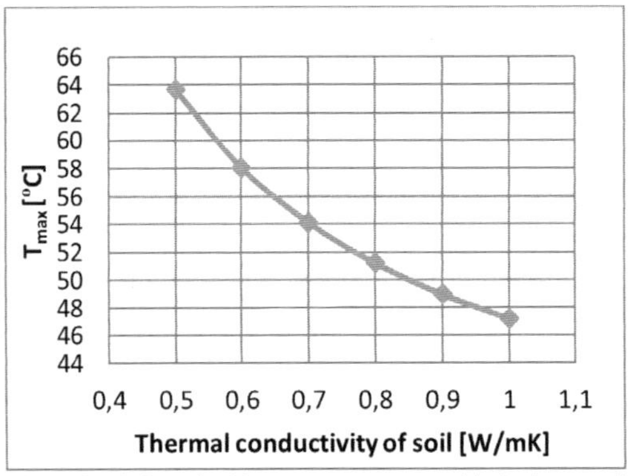

Fig. 4. Variation of cable core temperature with the increase in soil thermal conductivity.

Figure 4 shows the effect of soil thermal conductivity on cable core temperature (central cable conductor). It can be seen that for the analyzed computational case, the two times decrease in the soil thermal conductivity results in the 15°C increase of cable core temperature.

The temperature distribution within the cable system (Fig. 5) illustrates well the major impact of soil thermal conductivity on heat transfer.

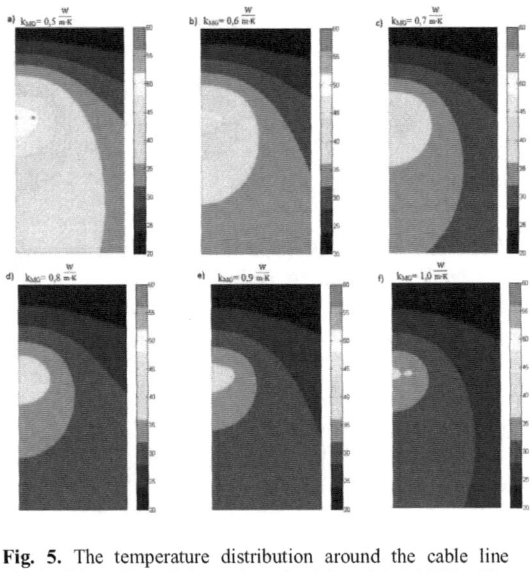

Fig. 5. The temperature distribution around the cable line obtained for cable soil thermal conductivity of a) 0,5 W/(m·K) , b) 0,6 W/(m·K) , c) 0,7 W/(m·K) , d) 0,8 W/(m·K) , e) 0,9 W/(m·K) , f) 1,0 W/(m·K) .

It can be observed that a properly designed cable system for the specified location (with high thermal conductivity), may not perform well in the region where the soil thermal conductivity is low. Even small fluctuations in the thermal conductivity value result in different temperature distributions.

4. Cable backfill thermal conductivity effect on temperature distribution in underground power cable system

Variation in the thermal conductivity of cable backfill has similar effects (as the soil thermal conductivity changes) on the cable core temperature. Cable backfill has a higher thermal conductivity than the mother ground, and therefore enhances the heat dissipation from buried cables. The study was performed for the soil thermal conductivity of 0.5 W/(m K) and cable backfill thermal conductivity varied in a range of 1 W/(m K) to 3 W/(m K). The first value (1 W/(m K)) refers to the conventional thermal backfill i.e. sand-cement mixture in proportions of 12:1. The last value (3 W/(m K)) relates to the POWERCRETE™ thermal backfill produced by Heidelberg Cement Group.

It should be noted that the higher the backfill thermal conductivity, the increased costs of the cable backfill. Therefore, during the design of underground power cable system, the economic conditions shall be considered.

Figure 6 shows that the increase in the cable backfill thermal conductivity from 1 W/(m K) to 3 W/(m K) causes the decrease in cable conductor temperature of 7°C.

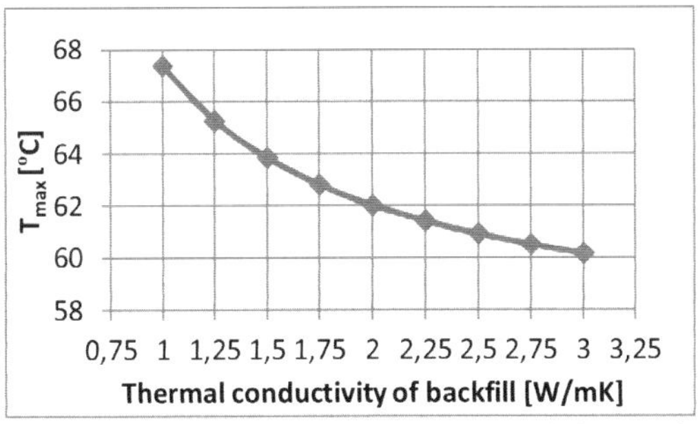

Fig. 6. Variation of cable core temperature with the increase in cable backfill thermal conductivity (the soil thermal conductivity is equal to 0.5 W/(m K).

Figure 7 presents the effect of cable backfill thermal conductivity on temperature distribution within the underground power cable system. The value of soil thermal conductivity is assumed as 0.5 W/(m K). It can be seen, that for the analyzed system, when the thermal backfill with high thermal conductivity is applied, the temperature gradients within the soil and cable backfill are lower. Hence, the heat is dissipated in a more efficient way from power cables to the surroundings.

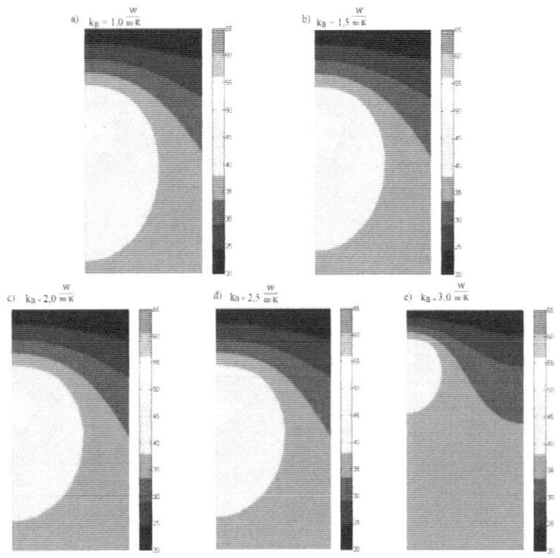

Fig. 7. The temperature distribution around the cable line obtained for cable backfill thermal conductivity of a) 1.0 W/(m·K) , b) 1.5 W/(m·K), c) 2.0 W/(m·K), d) 2.5 W/(m·K), e) 3.0 W/(m·K) and soil thermal conductivity of 1 W/(m·K).

Figure 8 compares two computational cases:

- The cables buried in thermal backfill with a thermal conductivity of 3 W/(m K) (Fig. 8a)
- the cables buried directly within the soil with a thermal conductivity of 0.5 W/(m K) (Fig. 8b)

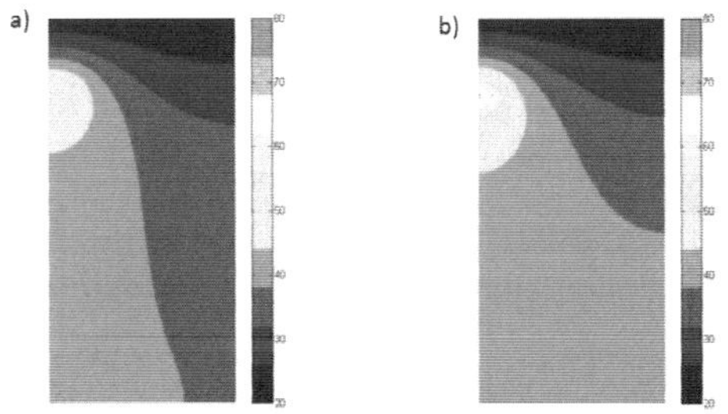

Fig. 8. Temperature distribution in underground power cable system for two cases a) with the use of thermal backfill layer b) without the use of thermal backfill layer.

This example shows that if the cable backfill is not used, but only the mother ground, the cable conductor would reach the temperature of 77,53°C. It is much higher than the cable design temperature of 65°C. This example shows how important is to use the thermal backfill layer. When the cable backfill is used instead of mother ground, the cable core temperature is 14°C lower.

5 Conclusions

This paper presents the thermal analysis of underground power cable system. The case when the cables are arranged in flat formation (in-line) and buried at a depth of 2 m underground was studied. The Finite Element model that allowed determining the temperature distribution within the system was developed. The computations were performed for the thermal conductivity of:

- Thermal Backfill from 1 W/(m K) to 3 W/(m K)
- Mother ground from 0.5 W/(m K) to 1 W/(m K)

Based on the performed analysis it can be concluded that:
- soil and thermal backfill conductivity play an important role in cable line design
- the higher the soil thermal conductivity, the lower the cable conductor temperature
- the replacement of mother ground with the thermal backfill material may reduce the cable core temperature up to 14°C.

References

1. C.C Hwang, Y.H. Jiang, *Extensions to the finite element method for thermal analysis of underground cable systems.* Electr Pow Syst Res 64:159–**64** (2003)
2. M.S. Al-Saud, M.A. El-Kady, R.D. Findlay, *A new approach to underground cable performance assessment.* Electr Pow Syst Res 78:907–**18** (2008)
3. V.R. De Lieto, L. Fontana, A. Vallati, *Thermal analysis of underground electrical power cables buried in non-homogeneous soils.* Appl Therm Eng 31:772–**8** (2011)
4. V.R. De Lieto, L. Fontana, A. Vallati, *Experimental study of the thermal field deriving from an underground electrical power cable buried in nonhomogeneous soils.* Appl Therm Eng; 62:390–**7** (2014)
5. E. Kroener, A. Vallati, M. Bittelli, *Numerical simulation of coupled heat, liquid water and water vapor in soils for heat dissipation of underground electrical power cables.* App Therm Eng; 70:510–**23** (2014)
6. P. Ocłoń, M. Bittelli, P. Cisek, E. Kroener, M. Pilarczyk, D. Taler, R.V. Rao, A. Vallati, *The performance analysis of a new thermal backfill material for underground power cable system* Applied Thermal Engineering, **108**, pp. 233-250 (2016)
7. P. Ocłoń, P. Cisek, D. Taler, M. Pilarczyk, T. Szwarc, *Optimizing of the underground power cable bedding using momentum-type particle swarm optimization method Energy*, **92**, pp. 230-239 (2015)
8. P. Ocłoń, P. Cisek, M. Pilarczyk, D. Taler, *Numerical simulation of heat dissipation processes in underground power cable system situated in thermal backfill and buried in a multilayered soil* Energy Conversion and Management, **95**, pp. 352-370 (2015)
9. P. Ocłoń, D. Taler, P. Cisek, M. Pilarczyk, *Fem-Based Thermal Analysis of Underground Power Cables Located in Backfills Made of Different Materials* Strength of Materials, **47** (5), pp. 770-780 (2015)
10. P. Cisek, P. Ocłoń, M. Pilarczyk. *Thermal analysis of operating conditions for The 400 kV underground power cable transmission line as a power plant delivery system.* Rynek Energii, **114**:70–7 [in polish]. (2014)
11. IEC 60287-2-1: Electric Cables-Calculation of the current rating – Part 2: Thermal resistance – Section 1: calculation of the thermal resistance (1995)
12. IEEE 835-1994: IEEE Standard Power Cable Ampacity Tables (1994)

Experimental and theoretical investigations of special type coil heat exchanger with the nanofluid buffer layer

Robert Smusz[1], and *Joanna* Wilk[1,*]

[1]Rzeszów University of Technology, Thermodynamics Department, Faculty of Engineering and Aeronautics, Al. Powstańców Warszawy 12, 35-959 Rzeszów, Poland

Abstract. The paper presents the results of experimental and theoretical investigations of special type of coil heat exchanger. The tested device is equipped with three vertical coils and the temperature stratification system. Water is a heating medium in two coils. The refrigerant transferring the waste heat from air conditioning system is the heating medium in the third coil. The finned pipe of this coil has a double wall in which the annular buffer layer with nanofluid is mounted. Thermophysical properties of the applied water based Cu nanofluid cause the enhancement of heat transfer through the buffer layer. The paper presents thermal characteristics of the exchanger received on the basis of measurements performed on the industrial test stand. Measurements were conducted during the operation of the coil with refrigerant. Heat loss to the surroundings, distributions of water temperature in the storage tank, changes of water temperature in time and thermal power of the coil heat exchanger were obtained. The measurement results were compared with those received on the basis of theoretical analysis of the exchanger.

1 Introduction

Heat exchangers with helical coils have widespread domestic and industrial applications. They may be used in domestic hot water systems, air conditioning systems, refrigerators, chemical and nuclear reactors, steam generators, etc. Applied in heat exchangers the helical coils are often used since they can accommodate a large heat transfer area in compact space and heat transfer coefficients have high values. In recent time coil heat exchangers were the subject of many researchers. The exemplary works on the considered exchangers are [1–5]. Prabhanjan et al. presents in [1] the comparison of heat transfer between coil heat exchanger and straight tube heat exchanger. The work [2] by Zachar deals to improvement of heat transfer processes in coil heat exchanger with spirally corrugated wall. In the paper [3] Genic et al. presents the results of investigations of the influence of geometry parameters of coil heat exchangers on heat transfer coefficients. Urbanowicz and Wojtkowiak presents in [4] a comparative analysis of correlations for heat transfer at surfaces of coil and suggests an original method of calculation of heat exchanger with helical coil. In turn, characteristics of agitated helical coil heat exchanger operating with the nanofluid are described by Srinivas and Vinod in [5]. Tested in the present work the coil heat exchanger works with the nanofluid too. The investigated device is designed to warm domestic hot water. It is equipped with three vertical coils immersed in storage tank filled with heated water. The orientation of the coil system is shown in Fig.1. Additionally the temperature stratification system is applied in the storage tank. Water is a heating medium in two coils. The refrigerant transferring the waste heat from air conditioning system is the heating medium in the third coil. The finned pipe of this coil has a double wall in which the annular buffer layer with nanofluid is mounted. According to the rules the layer have to protect against the possible refrigerant leakage to the hot domestic water tank. The gap is filled by the nanofluid consisting of distilled water and coper nanoparticles. Thermophysical properties of the applied water based Cu nanofluid cause the enhancement of heat transfer through the buffer layer. Thus, the buffer layer with nanofluid acts as a protection, and in the other hand it improves thermal characteristics of the exchanger.

A detailed description of the idea and the analytical model of the proposed coil heat exchanger is contained in [6]. In turn the results of calculations of thermal power of water coil of the exchanger considered are included [7]. This paper presents also the experimental results of water coil in comparison with received theoretical results. The present work contains the results of thermal measurements which were conducted during the operation of the coil with refrigerant. Heat loss to the surroundings, distributions of water temperature in the storage tank, changes of water temperature in time and thermal power of the coil heat exchanger were obtained. The measurement results were compared with those received on the basis of theoretical analysis of the exchanger.

* Corresponding author: joanwilk@prz.edu.pl

2 Experimental investigations

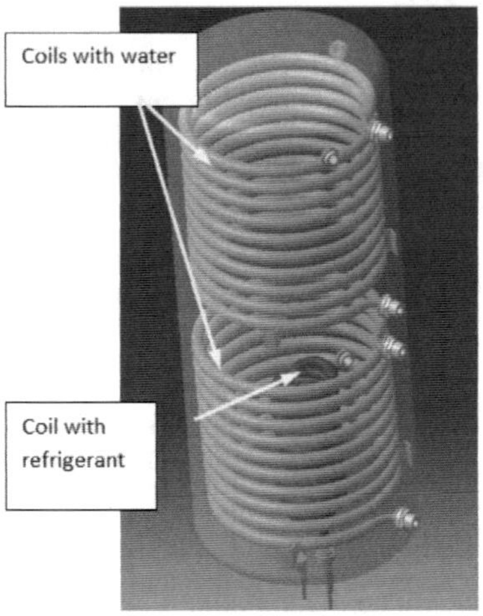

Fig. 1. System of coils applied in the heat exchanger

Experimental investigations on considered heat exchanger with the coil that uses the waste heat from air conditioning system were performed on industrial test stand. The detailed description including the technological scheme of the stand, the specification of applied elements and the research possibilities of the stand were presented in [8]. The simplified scheme of the experimental stand is shown in Fig.2.

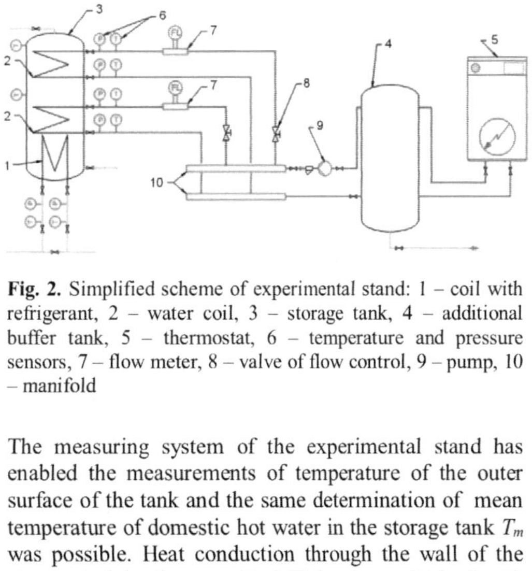

Fig. 2. Simplified scheme of experimental stand: 1 – coil with refrigerant, 2 – water coil, 3 – storage tank, 4 – additional buffer tank, 5 – thermostat, 6 – temperature and pressure sensors, 7 – flow meter, 8 – valve of flow control, 9 – pump, 10 – manifold

The measuring system of the experimental stand has enabled the measurements of temperature of the outer surface of the tank and the same determination of mean temperature of domestic hot water in the storage tank T_m was possible. Heat conduction through the wall of the storage tank has been omitted. T_m has been obtained with the use of eight resistance temperature detectors T_i [8] located on the surface of the tank (see Fig. 3) according to the formula

$$T_m = \sum_{i=1}^{8} T_i \cdot \frac{V_i}{V}, \qquad (1)$$

where V_i – control volume assumed for the temperature measured T_i.

The basic geometrical parameters necessary to calculate T_m is summarised in Fig.3.

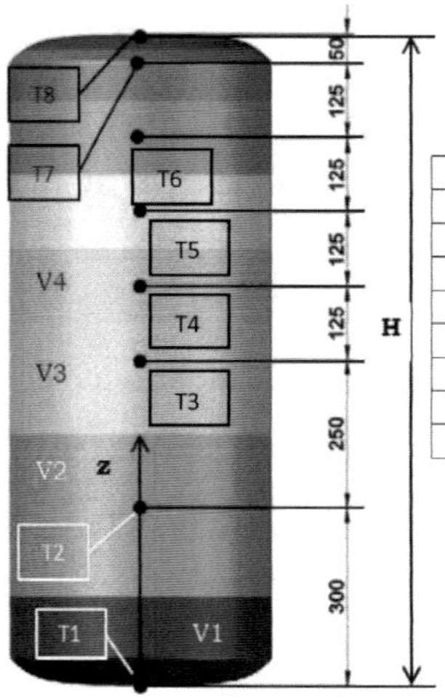

T_i	z/H	V_i/V
T1	0	0,124
T2	0,27	0,260
T3	0,5	0,175
T4	0,61	0,118
T5	0,73	0,118
T6	0,84	0,118
T7	0,95	0,072
T8	1	0,015

Fig. 3. Location of temperature detectors and relative control volumes of the storage tank

The results of investigations of the heating process of water in the storage tank are presented in Fig. 4. There are changes of water temperatures T1 – T8 during the thermal start-up of the exchanger operating with the use of coil with refrigerant – the exchanger of heat recovery system. The refrigerant R407C has been applied.

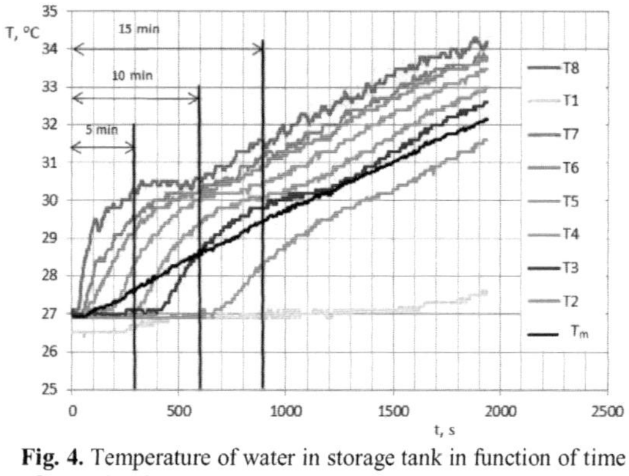

Fig. 4. Temperature of water in storage tank in function of time of water heating

The applied measuring system also enabled the measurement of the ambient temperature and supply and return temperatures of the coil. The experiment was performed at measured mean ambient temperature 26.4°C. The mean supply temperature during the water heating with the use of the coil with refrigerant was equal 44.74°C. The return temperature was respectively 40.2°C. Calculated according to eq.(1) the mean water temperature T_m is shown in Fig.5.

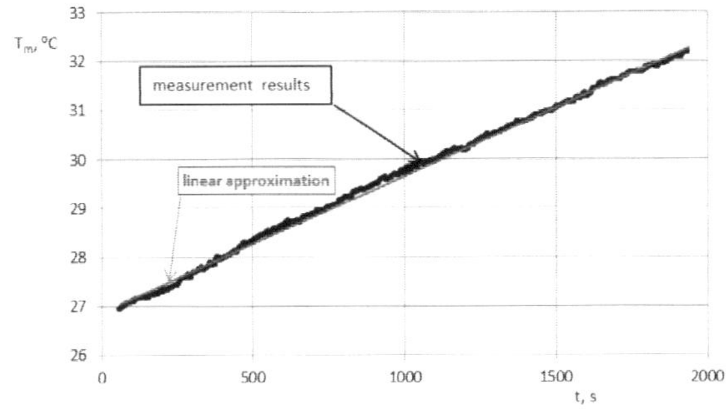

Fig. 5. Mean water temperature in function of heating time

The linear approximation of T_m in function of time has been made. It was necessary to obtain the rate of heating water in storage tank dT_m/dt.
Energy balance of the exchanger – heater – has a form

$$V\rho_w c_w \frac{dT_m}{dt} = \dot{Q} - \dot{Q}_{loss}, \qquad (2)$$

where: V– volume of water in the tank, ρ_w, c_w – density and heat capacity of water, $\dot{Q}$- thermal power of the exchanger, $\dot{Q}_{loss}$ - heat loss to the surroundings.

On the basis of eq.(2) the thermal power of coil exchanger of heat recovery system has been calculated. Heat losses from the storage tank have been measured with the use of three film heat flux sensors [8]. The results of the measurements and the linear approximation are presented in Fig.6. The product $\rho_w c_w$ in eq.(2) has been calculated on the basis of approximation of literature data [9] according to the formula

$$\rho_w c_w = 4211.7 - 1.6796\ T_m. \qquad (3)$$

The obtained mean value of thermal power of coil heat exchanger operating with refrigerant as a heating medium is equal to 2.452 kW.

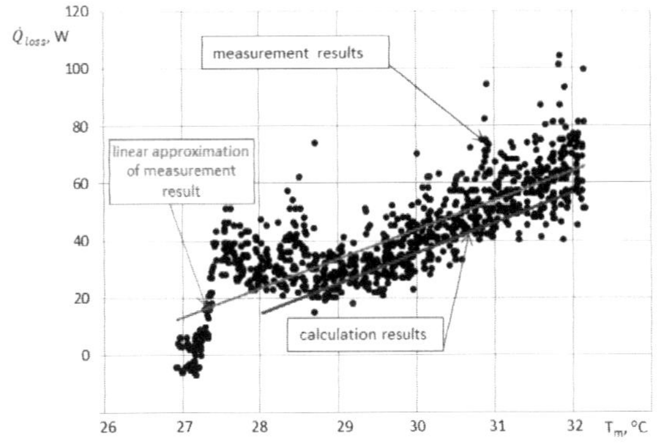

Fig. 6. Heat losses from the outer surface of the storage tank

3 Theoretical calculations

Theoretical calculations of the coil heat exchanger were based on heat transfer processes occurring in the exchanger during operation. The analysis has related to: convective and radiative heat transfer from the outer surface of the storage tank, convective heat transfer between water and the inner surface of the tank, natural convection from the outer surface of finned coil pipe, heat conduction through the coil wall with taking into consideration the additional thermal resistance of the nanofluid buffer layer and convective turbulent flow of the refrigerant inside the coil pipe. The main formulas used to calculations are summarised in Table 1. Characteristic dimensions occurring in dimensionless numbers were described in [6].

Table 1. Equations for the calculation of heat transfer process in the coil heat exchanger [10 – 12]

Heat transfer process	Formula
Radiative heat transfer from the outer surface of the tank	$h_r = \dfrac{\sigma \cdot \varepsilon \cdot \left(T_{wall}^4 - T_a^4\right)}{T_{wall} - T_a}$
Convective heat transfer between air or water and the storage tank walls	$Nu = 0.59 \cdot Ra^{0.25}$ $\quad 10^4 < Ra < 10^9$ $Nu = 0.1 \cdot Ra^{0.333}$ $\quad 10^9 < Ra < 10^{13}$
Natural convection from the surface of the finned pipe	$Nu = 0.028 \cdot Ra^{0.4}$ $1.2 \times 10^7 < Ra < 1.4 \times 10^8$
Heat transfer in annular buffer layer with nanofluid	$\dfrac{k_{eq}}{k} = 0.11 \cdot Ra^{0.29}$ $6000 < Ra < 10^6$
Heat transfer during the turbulent flow of refrigerant through the coil	$Nu = 0.023\left[1 + 3.6\left(1 - \dfrac{d_{in}}{D}\right)\left(\dfrac{d_{in}}{D}\right)^{0.8}\right] \times$ $\times \Pr^{1/3} \mathrm{Re}^{0.8}$ $2 \times 10^4 < \mathrm{Re} < 1.5 \times 10^5$

d_{in} – inner diameter of the coil pipe
D – characteristic dimension of the coil
h_r – radiative heat transfer coefficient
k, k_{eq} – thermal conductivity and equivalent thermal conductivity
Nu – Nusselt number
Pr – Prandtl number
Ra – Rayleigh number
Re – Reynolds number
T_{wall}, T_a– wall and ambient temperature
ε – surface emissivity
σ – Boltzmann constant

The presented procedure has enabled obtain heat losses from the storage tank and the thermal power of the coil heat exchanger. The thermal power of coil exchanger of heat recovery system – the coil with refrigerant – has been calculated with the use of the general formula

$$\dot{Q} = A \cdot U \cdot \frac{(T_1 - T_m) - (T_2 - T_m)}{\ln \frac{T_1 - T_m}{T_2 - T_m}}, \quad (4)$$

where: A – overall coil pipe surface; U – overall heat transfer coefficient; T_1, T_2 – temperature of refrigerant at the inlet and the outlet of the coil, respectively, T_m – assumed mean temperature of water. The overall heat transfer coefficient takes into account the following components: convective heat transfer inside coil, conduction through the inner wall of the coil pipe, heat transfer in the nanofluid buffer layer, conduction through the outer wall, convective heat transfer from finned wall of the coil. It is given by

$$U \cdot A = \quad (5)$$

$$= \frac{1}{\frac{1}{h_i \cdot A_{ip}} + R_{k1} + R_b + R_{k2} + \frac{1}{h_{eq} \cdot A_{in\_f} + h_{eq}\eta_f N_f A_f}},$$

where: h_i – heat transfer coefficient at the inner surface of the coil pipe A_{ip}, h_{eq} – equivalent heat transfer coefficient of the finned surface, A_{in_f} and A_f – intercostal surface and the surface of the fin respectively, η_f – efficiency of the fin, N_f – the number of fins. R_{kl} – conduction thermal resistance of the inner coil pipe; R_{k2} – conduction thermal resistance of the outer finned coil pipe; R_b – conduction thermal resistance of the buffer layer taking into account equivalent thermal conductivity of the nanofluid.

Calculations were performed with taking into account the temperature dependence of thermophysical properties of the applied medium. The latent heat of the refrigerant was omitted because of temperature range of the working medium. Since heat transfer coefficients were implicit temperature functions, the system of nonlinear equations was built. It has been solved using the secant method. Geometrical parameters of the tank and the coil were assumed as the same as in the real object.

Table 2. Geometry of the finned copper pipe of the coil with refrigerant and calculated thermal power of the exchanger

Outer and inner diameter of finned coil pipe	16mm, 12mm
Thickness of the inner wall of the double-wall coil pipe δ_i	0.8 mm
Thickness of the outer wall of coil pipe δ_o	0.8 mm
Thickness of buffer layer filled with nanofluid δ_b	0.4 mm
Nanofluid – Cu particles of purity 97% and spherical form of diameter about 40 nm dispersed in water of 1% mass concentration	
Mean coil diameter	130 mm
Coil pitch	31 mm
Number of coils	14
Coil pipe length	5.7 m
Fin height	3.5mm
Fin width	0.5 mm
Number of fins per inch of coil pipe length	11

Table 3. The results of thermal power calculations

Supply temperature of refrigerant	Thermal power of coil heat exchanger
60 °C	4.5 kW
55 °C	3.9 kW
55 °C	3.3 kW
45 °C	2.7 kW

The main data are summarised in Table 2 according to the presented in Fig. 7 the scheme of the double wall finned pipe. Table 2 also presents the main characteristic of the nanofluid applied in the buffer layer. The results of calculations of thermal power of the exchanger at assumed temperatures of the refrigerant in the coil are summarized in Table 3.

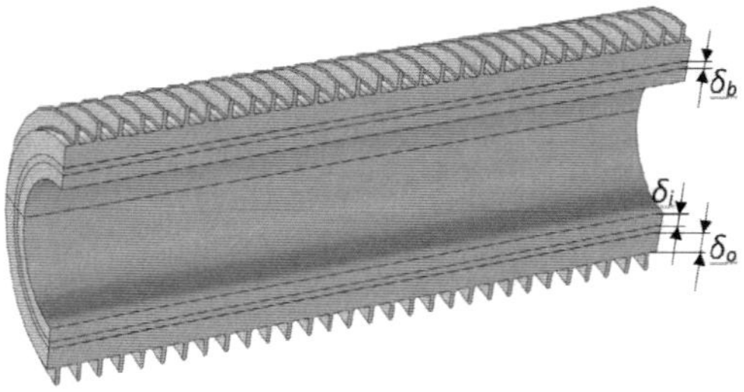

Fig. 7. Scheme of the double wall finned pipe

Table 4. Heat losses from the storage tank at T_a = 26.4°C, V = 200dm^3

Mean temperature of water in the storage tank	Heat loss from the tank
28 °C	14.1 W
29 °C	24.2 W
30 °C	34.8 W
31 °C	45.9 W
32 °C	57.3 W

In turn Table 4 includes the results of calculations of heat loss to the surrounding from the outer surface of the storage tank. Calculations were made for uninsulated tank (real conditions) at assumption ambient temperature 26.4°C. The linear approximation of the calculation results is shown in Fig.5.

4 Conclusions

Experimental and theoretical investigations on a special type of coil heat exchanger have allowed to formulate some conclusions from the obtained results.

The fast process of water heating in the storage tank was observed. By the time 5 minutes after the start of the exchanger the temperature of water is not rising to the 60% of the relative height of the tank. After 1900 second from the start of water heating with the use of coil with refrigerant the difference between T8 and T3 (see Fig. 4) is only 1.6 °C. Hot water is transported to the upper zone of the tank by a temperature stratification system. The obtained results indicate good operating of the system and may be helpful in optimization of the stratification system.

The calculation results of heat loss from the storage tank are in good agreement with measured heat loss. The linear approximation of the calculated heat loss is in the area of the scatter of measurement results. Mean value of heat loss measured at maximum water temperature 32°C is equal to 64.1W while the calculated value is 57.3W. The relative difference is about 10%. Good compatibility is also between the measured and calculated thermal power of the exchanger. Experimentally obtained thermal power is 2.452 kW at measured supply temperature of refrigerant 44.74 °C. Calculated values are respectively equal: 2.7 kW at 45 °C. The results obtained indicate a valid model for theoretical calculations. It can be applied to appropriate selection of insulation of the storage tank to reduce heat losses. Also it can be used for modeling of heat transfer in the buffer layer with nanofluid.

References

1. D.G. Prabhanjan, G.S.V. Raghavan, T.J. Rennie, Int. Comm. Heat Mass Transfer, **29**, 185-191 (2002)
2. A. Zachár, Int. J. of Heat Mass Transfer, **53**, 3928-3939 (2010)
3. S.B. Genić, B.M. Jaćimović, M.S. Jarić, N.J. Budimir, M.M. Dobrnjac, Int. J. of Heat and Mass Transfer, **55**, 4295-4300 (2012)
4. A.E. Urbanowicz, J. Wojtkowiak, *Metoda obliczeń wymienników ciepła z pionową wężownicą* (Wydawnictwo Politechniki Poznańskiej, Poznań 2013)
5. T. Srinivas, A.V. Vinod, Exp. Thermal and Fluid Science, **51,** 77-83 (2013)
6. R. Smusz, J. Wilk, Applied Mechanics and Materials, **831**, 223-231 (2016)
7. R. Smusz, Journal of Physics (to be published)
8. R. Smusz, J. Wilk, *Proceedings of 16th Int. Symposium on Heat Transfer and Renewable Sources of Energy*, 397-403 (2016)
9. Main IAPWS Thermodynamics Property Formulations, Updated July 6, http://www.iapws.org/newform.html (2015)
10. J. P. Holman, *Heat transfer* (McGraw-Hill Inc. 1992)
11. B. Niezgoda-Żelasko, W. Zalewski, *Chłodnicze i klimatyzacyjne wymienniki ciepła* (Wydawnictwo Politechniki Krakowskiej, Kraków 2013)
12. H. Berkel, N.K. Mitra, M. Fiebig, Wärme und Stoffübertragung, **18**, 167-173 (1984)

Generalization of Martinelli-Nelson method of pressure drop calculation in two-phase flows

Marian Trela*, *Roman* Kwidzinski, and *Marcin* Lackowski

The Szewalski Institute of Fluid-Flow Machinery, Polish Academy of Sciences, Fiszera 14, 80-231 Gdansk, Poland

Abstract. A simple method of pressure drop calculation for two-phase flows of different fluids during convective boiling in channels is presented. It is based on experimental data of pressure drop multiplier R and void fraction φ obtained by Martinelli and Nelson for boiling of water in vertical tubes. The data cover the whole two-phase domain from ambient to critical pressure. Unfortunately, they have been presented in graphical forms. The first step in the procedure proposed in the paper was a transformation of the graphical data into analytical formulas which contain such dimensionless quantities as steam quality x, Martinelli parameter X, multiplier Φ_l^2 and dimensionless coefficients D, m, E and k. In the second step, simple analytical formulas were determined to express the dimensionless coefficients as a function of physical property parameter K. In this way two simple analytical expressions for the multiplier R and void fraction φ were obtained. They are in analytical dimensionless form so they may be used directly for different fluids, not only for water. This is the main advantage of the proposed method.

Notation

A – channel cross section area, m^2
g – gravity, kg/s^2
G – mass velocity, $G = \dot{m}/A$, $kg/(s \cdot m^2)$
K – physical property parameter
L – length, m
$\dot{m}$ – mass flow rate, kg/s
p – pressure, Pa
R – friction factor or multiplier, $R = \Delta p_{TPf}/\Delta p_{lo}$
v – specific volume, m^3/kg
x – steam quality, $x = \dot{m}_g/(\dot{m}_g+\dot{m}_l)$
X_1, X_2 – Martinelli parameters
Greek symbols:
φ – void fraction, $\varphi = A_g/A$
Φ – parameter proposed by Martinelli
μ – dynamic viscosity, Pa·s
ρ – density, kg/m^3
Θ – angle to horizontal plane, rad
subscripts:
g – gas (steam)
go – gas only flow rate equal to that of two-phase flow
l – liquid
lo – liquid only flow at a rate equal to that of two-phase flow
TPf – two-phase friction term

1 Introduction

In a number of engineering systems it is important to be able to predict the pressure drop during boiling of two-phase medium, such as water or other fluids. Generally, total two-phase pressure drop for flows in vertical channels consists of three components – frictional loss Δp_{TPf}, momentum change Δp_a and elevation pressure drop Δp_h arising from the effect of the gravitational force field. Thus

$$\Delta p_{TP} = \Delta p_{TPf} + \Delta p_a + \Delta p_h \,. \tag{1}$$

Generally, the current methods for modelling of two-phase pressure drop fall into two categories: homogenous model and separated model. The latter is chosen in the paper for particular consideration. According to Collier and Thome [1], the total two-phase pressure drop may be written as

$$\Delta p_{TP} = \frac{2 f_{lo} G^2 v_l L}{D}\left[\frac{1}{x}\int_0^x R\,dx\right] + G^2 v_l\left[\frac{x^2}{\varphi}\left(\frac{v_g}{v_l}\right) + \frac{(1-x)^2}{(1-\varphi)^2} - 1\right] + \frac{Lg\sin\Theta}{x}\int_0^x\left[\rho_g\varphi + \rho_l(1-\varphi)\right]dx \tag{2}$$

It order to apply the above equation it is necessary to develop two expressions for multiplier R and void fraction φ. Two different approaches are shown in the paper. The first one is the well-known Lockhart–Martinelli (LM) correlation for calculation of both quantities, the second one is developed by the present authors.

* Corresponding author: marian.trela@imp.gda.pl

2 The development of analytical formula for pressure multiplier *R* and void fraction φ

2.1 The Lockhart–Martinelli Correlation

This method was basically proposed by Lockhart and Martinelli in 1949 [2]. They introduced a few pertinent parameters. The first one is known as Φ_l^2 parameter, defined as

$$\Phi_l^2 = \frac{(dp/dz)_{TPf}}{(dp/dz)_l} \tag{3}$$

that is as a ratio of two-phase pressure gradient due to friction in a flow at mass rate equal to $\dot{m} = \dot{m}_o$ to the pressure gradient for a liquid stream flowing alone in the same channel at the mass flow rate $\dot{m}_l = (1-x)\dot{m}_o$.

The second parameter is known as Martinelli one and is defined as

$$X = \left[\frac{(dp/dz)_l}{(dp/dz)_g}\right]^{1/2} = X_1 \tag{4}$$

what means that it is a ratio of pressure gradient due to friction of liquid phase flowing at mass rate equal to $\dot{m} = \dot{m}_l$ to the pressure gradient for a gas (vapour) stream flowing alone in the same channel at the mass flow rate $\dot{m}_g = x\dot{m}_o$. For each phase flowing separately in the channel at flow rate $\dot{m}_l$ or $\dot{m}_g$, the Martinelli parameter X_1 can be easily calculated as

$$X_1 = \left(\frac{\rho_g}{\rho_l}\right)^{0.5}\left(\frac{\mu_l}{\mu_g}\right)^{0.1}\left(\frac{1-x}{x}\right)^{0.9}. \tag{5}$$

It follows from (3) that

$$\left(\frac{dp}{dz}\right)_{TPf} = \Phi_l^2\left(\frac{dp}{dz}\right)_l. \tag{6}$$

The pressure gradient $(dp/dz)_l$ may by written as

$$\left(\frac{dp}{dz}\right)_l = \left(\frac{dp}{dz}\right)_{lo}(1-x)^{2-n} \tag{7}$$

where n is the exponent in Blasius friction factor expression for pressure drop in single phase flow. Taking into account equations (6), (7) and definition of the multiplier R one can obtain relation

$$R = \frac{(dp/dz)_{TPf}}{(dp/dz)_{lo}} = (1-x)^{2-n}\,\Phi_l^2. \tag{8}$$

Lockhart and Martinelli (LM) [2] have obtained experimental correlations for the parameter Φ_l and void fraction φ in graphical forms as $\Phi_l = f(X_1)$ and $(1-\varphi) = f(X_1)$. The investigations were carried out for adiabatic horizontal two-phase flows of two-component systems (air-water, air-oil) at low pressure close to atmospheric one. Therefore, application of their graphs to other than the above mentioned conditions is not recommended. Another shortcoming of Lockhart–Martinelli correlations comes from their solely graphical form.

Two substantial improvements of LM correlations were done by Chisholm [3]. First, using theoretical approach, he proved that the usage of Φ_l and X_l parameters is justified for the separate flow model of two-phase flow. Secondly, because the graphical form of LM correlation is cumbersome in practice, he proposed to replace the LM graphs by analytical correlation in the form

$$\Phi_l^2 = 1 + \frac{C}{X_1} + \frac{1}{X_1^2} \tag{9}$$

where C is a constant which depends on flow mechanisms that can exists during the simultaneous flow of a liquid and a gas (vapour) in the channel [3]. It is known that four types of flow mechanisms appear since the flow may be viscous (laminar) or turbulent. The particular values of C are listed below for the given flow regime:

liquid		gas	C
turbulent	–	turbulent (tt)	20
viscous	–	turbulent (vt)	12
turbulent	–	viscous (vt)	10
viscous	–	viscous (vv)	5

Using $n = 0.25$ in (8) and $C = 20$ in (9), for turbulent–turbulent flow regime the multiplier R takes the form

$$R = (1-x)^{1.75}\left(1 + \frac{20}{X_1} + \frac{1}{X_1^2}\right) \tag{10}$$

which reproduces quite well the LM data for low pressure two-phase flow. Other, more recent modifications of LM correlation for micro- and macrochannels may be found in [4,5].

For high pressure and diabatic (with heat addition, e.g. at boiling) flows, the use of LM correlation (10) leads to large errors. Therefore the attempt was made by one of the present authors [6], to obtain analytical formulas for two-phase multiplier R and void fraction φ. The present version of this approach differs significantly from that original one.

2.2 Derivation of analytical formula for *R* multiplier

Experimental investigation of pressure drop during boiling of water in vertical tubes have been carried out by Martinelli and Nelson [7] using also Davidson *et. al.* [8] data. The data for multiplier R and void fraction φ, Fig. 1 and Fig. 2, cover the whole two-phase domain from ambient pressure to critical one. The two-phase multiplier R was presented versus a mass quality x with pressure as a parameter, unfortunately in graphical form and in tables. Martinelli and Nelson [7] assumed that

DOI: 10.1051/e3sconf/20171302006

during flow boiling, the flow regime would always be "turbulent–turbulent".

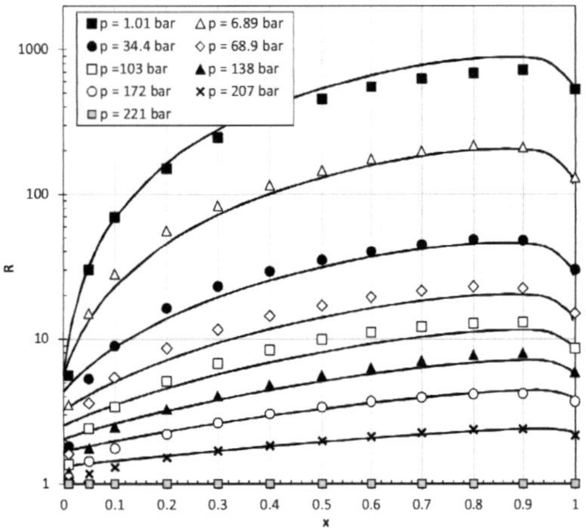

Fig. 1. Multiplier R versus mass quality x and pressure p: symbols – experimental data [7], solid lines – correlation (20).

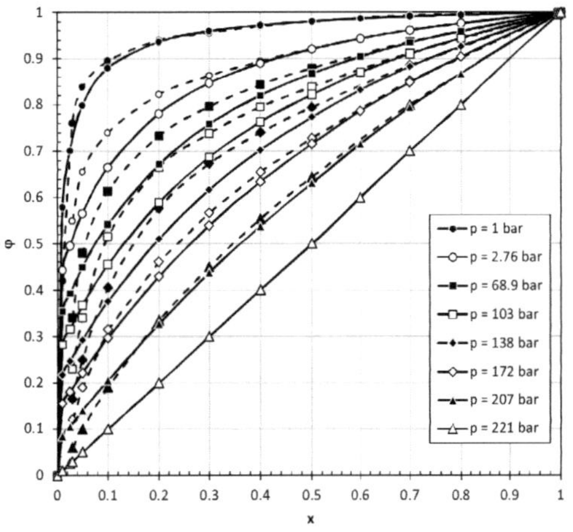

Fig. 2. Comparison of original (broken lines) and correlated (solid lines) values of void fraction φ as a function of dryness fraction x for selected values of pressure p. Original values are taken from Martinelli-Nelson experiments [7], approximated values are calculated according to (28).

The first aim of the present paper is therefore to develop mathematical formula for the multiplier R. The procedure starts from evaluation of the parameter Φ_l values using Fig. 1 for given pressure and steam quality x. Then from (8) it follows that

$$\Phi_l = \sqrt{\frac{R}{(1-x)^{2-n}}}. \tag{11}$$

For critical pressure $R = 1$ so with $n = 0.25$ the critical values of Φ_l equal to

$$\Phi_{lcr} = \sqrt{\frac{1}{(1-x)^{1.75}}}. \tag{12}$$

Moreover, for the critical pressure the Martinelli parameter (5) reduces to

$$X_{1cr} = \left(\frac{1-x}{x}\right)^{0.9} \tag{13}$$

since the physical properties of steam and liquid phases are the same. It may be shown that after extracting steam quality x from (13) and substituting it into (12) one obtains critical value of parameter Φ_l

$$\Phi_{lcr} = \sqrt{\left(1+\frac{1}{X_{1cr}^{1.111}}\right)^{1.75}}. \tag{14}$$

It was found earlier by Trela [6] that there is a relation between Φ_l and its critical value Φ_{lcr} at the same value of the Martinelli parameter, i.e. at condition that $X_1 = idem$ for the critical pressure p_{cr} and a pressure p. This relation was proposed in the form

$$\Phi_l = D\left(\Phi_{lcr}\right)^m \tag{15}$$

where m is the slope of the $\log(\Phi_l)$ line with the reference to the abscissa and D is a coefficient described by the relation

$$D = \frac{\Phi_l}{\left(\Phi_{lcr}\right)^m}. \tag{16}$$

It was found in [6] that formula (15) is valid within the range of steam quality $0.03 \le x \le 0.99$.

Using formula (14), the relation (15) can be rewritten in the following form

$$\Phi_l = D\left(1+\frac{1}{X_1^{1.111}}\right)^{0.875m}. \tag{17}$$

Relation (17) may be simplified if new Martinelli parameter X_2 is introduced, defined as

$$X_2 = X_1^{1.111} = \left(\frac{\rho_g}{\rho_l}\right)^{0.555}\left(\frac{\mu_l}{\mu_g}\right)^{0.111}\left(\frac{1-x}{x}\right). \tag{18}$$

Then

$$\Phi_l = D\left(1+\frac{1}{X_2}\right)^{0.875m} \tag{19}$$

and accordingly to (8), for $n = 0.25$, the multiplier R takes the form

$$R = (1-x)^{2-n}\Phi_l^2 = (1-x)^{1.75}D^2\left(1+\frac{1}{X_2}\right)^{1.75\cdot m}. \tag{20}$$

The above formula replaces the empirical diagram $R = f(x,p)$ given for boiling water by Martinelli Nelson [7]. Two parameters, D and m, appear in the equation (20). In the former paper [6], the parameters D and m were expressed versus the property parameter K in the graphical form. The parameter K is defined by Eq. (21) which also means that K is equal to the parameter R when $x \rightarrow 1$,

$$K = \frac{\left(\frac{dp}{dz}\right)_{go}}{\left(\frac{dp}{dz}\right)_{lo}} = \left(\frac{\mu_g}{\mu_l}\right)^{0.25} \frac{\rho_l}{\rho_g}. \tag{21}$$

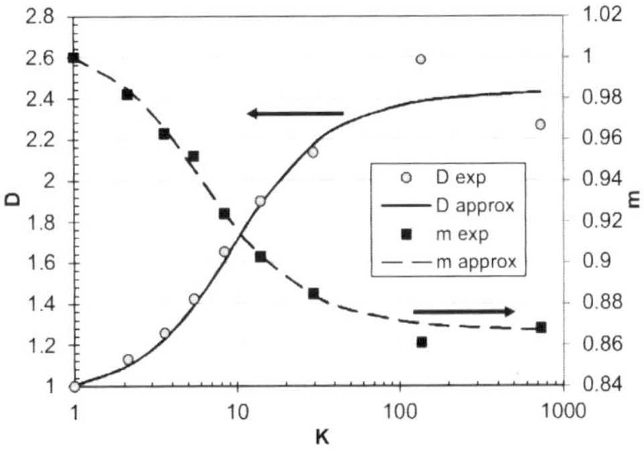

Fig. 3. Fitting of coefficients D, m in multiplier R correlation (20).

Figure 3 presents the variation of the parameters D and m versus the property parameter K. The former two parameters may be quite well approximated by the inverse trigonometric functions:

$$D = 1.18 + 0.8\,\mathrm{arctg}\left(\frac{K-3}{9}\right) \tag{22}$$

for D parameter and

$$m = 1 - 0.085\,\mathrm{arctg}\left(\frac{K-1}{6}\right) \tag{23}$$

for the m parameter.

2.3 Derivation of analytical formula for void fraction φ

Experimental investigations of the void fraction during flow boiling of water in vertical tubes have also been carried out by Martinelli and Nelson [7]. The data cover the whole two-phase domain from ambient pressure to critical one. The void fraction φ was presented versus mass quality x with pressure as a parameter in a graphical form, Figure 2. Procedure to develop an analytical formula for void fraction φ is analogous to the previous one for the multiplier R. Under critical pressure, the physical properties of vapour and liquid phases are identical and the value of φ becomes equivalent to quality x. Taking this into account one can write down that

$$(1-\varphi)_{cr} = (1-x)_{cr}. \tag{24}$$

Martinelli parameter for the critical pressure is given by the formula (13). Extracting steam quality x from (13) and substituting it into (24), one gets the expression

$$(1-\varphi)_{cr} = \left(1 + \frac{1}{X_1^{1.111}}\right)^{-1} = \left(1 + \frac{1}{X_2}\right)^{-1}. \tag{25}$$

For the pressures lower than the critical, it was found in [6] that an analytical formula may be proposed for φ

$$1 - \varphi = E(1-\varphi)_{cr}^{k} \tag{26}$$

where k is the slope of straight lines in logarithmic coordinates and E is a coefficient given by relation

$$E = \frac{1-\varphi}{(1-\varphi)_{cr}^{k}}. \tag{27}$$

Taking into account (25) and (26), it is easy to show that the void fraction formula takes the following form

$$1 - \varphi = E\left(1 + \frac{1}{X_2}\right)^{-k}. \tag{28}$$

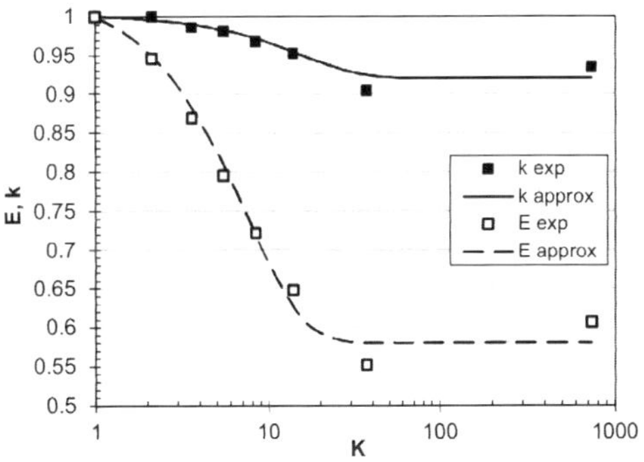

Fig. 4. Fitting of coefficients k, E in void fraction approximation formula (28).

The values of the two coefficients E and k appearing in the above equation may be made dependent on the property parameter K (21) and may be quite well approximated by the hyperbolic functions

$$E = 1 - 0.42\tanh\left[0.11(K-1)\right] \tag{29}$$

and

$$k = 1 - 0.08\tanh\left[0.05(K-1)\right] \tag{30}$$

as shown in Fig. 4.

2.4 Pressure drop similarity in two-phase flows

Similarity criteria for the pressure drop in a flow can be based on the general theory of physical phenomena similarity [6]. It is required that geometrical similarity and the similarity of physical processes (described by relevant equations in dimensionless form together with boundary conditions) will be preserved.

Taking this into account, similarity criteria of two (geometrically similar) systems comprising two-phase flows include the equation describing the variation of the friction coefficient R in the flow domain,

$$R = (1-x)^{1.75} D^2 \left(1+\frac{1}{X_2}\right)^{1.75\cdot m}, \tag{31}$$

and the boundary conditions:

$$R = 1 \quad \text{for } x = 0,$$

$$R = K = \frac{(dp/dz)_{go}}{(dp/dz)_{lo}} = \left(\frac{\mu_g}{\mu_l}\right)^{0.25} \frac{\rho_l}{\rho_g} \quad \text{for } x = 1,$$

$$R = 1 \quad \text{for } p = p_{cr}.$$

If two two-phase fluid flows (I and II), one of which is water, are compared then it follows from the above relations that the two flows will be similar if the boundary conditions will be the same, as the Eq. (31) is valid for both flows, i.e. $R_I = R_{II}$. For the boundary conditions to be identical, the equality

$$K_I = K_{II} \tag{32}$$

must be true for both flows considered. This means that the property parameter K is a similarity number which must be the same for the two systems. The conclusion from these considerations is such that by making the coefficients D and m in Eq. (31) functions of the physical parameter K, that equation can be applied to calculate the frictional pressure drop in two-phase flows of various working fluids, not only for water.

3 Validation of the proposed method

The validation of the above method may be done in two ways. Firstly, as illustrated in Figure 1, the experimental values of multiplier R (symbols) can be compared with their approximated values (solid lines) calculated according to (20). It is seen that analytical approximation fits quite well the Martinelli–Nelson data. Similarly, Figure 2 shows the comparison of Martinelli–Nelson data with analytical approximation of void fraction φ (28). In this case, there is also good agreement between original Martinelli–Nelson data and the analytical approximation. It has to be noted that the approximation applied is valid for a steam quality in the range $0.03 \leq x \leq 0.99$.

Second validation of the method comes from the comparison of the experimental data of the multiplier R for Freon 21 with theoretical values calculated from Eq. (20) with coefficients D and m given by (22) and (23), Fig. 5 and Fig. 6. The experimental investigation was carried out by Trela [9] in Laboratory of IF-FM in Gdańsk. The influence of mass velocity G on the multiplier R was also investigated.

Figure 5 shows also the prediction of the R multiplier according to well-known formulas of Lottes [10] and Levy [11]:

- Lottes formula $$R = \left(\frac{1-x}{1-\varphi}\right)^2, \tag{33}$$

- Levy's formula $$R = \frac{(1-x)^{1.75}}{(1-\varphi)^2}. \tag{34}$$

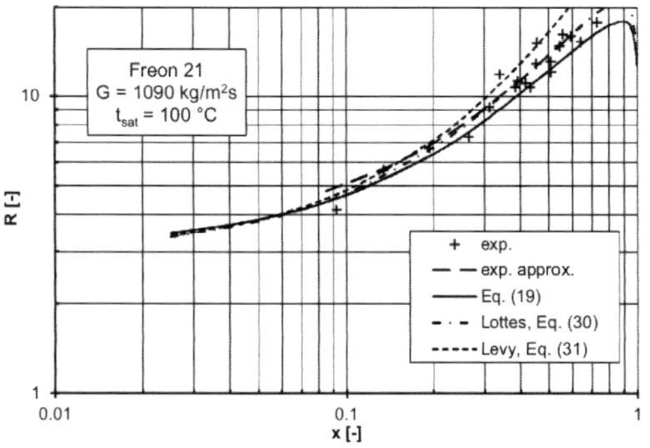

Fig. 5. Calculated (solid line) and experimental values (cross marks) of multiplier R versus quality x for Freon 21 at saturation temperature $t_{sat} = 100$ °C and mean mass velocity $G = 1090$ kg/m²s.

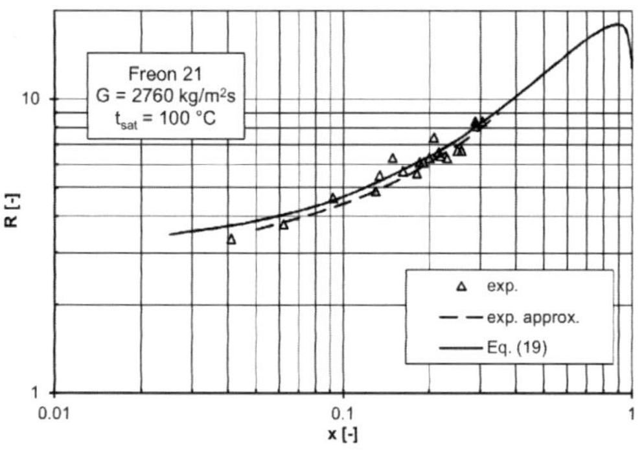

Fig. 6. Calculated (solid line) and experimental values (triangles) of multiplier R versus quality x for Freon 21 at saturation temperature $t_{sat} = 100$ °C and mean mass velocity $G = 2760$ kg/m²s.

Figure 6 presents experimental results of R multiplier for higher mass velocity $G = 2760$ kg/m²s. The dashed lines in Figures 5 and 6 show average values of R evaluated from the experiments. It is seen that R is decreasing when the mass velocity increases. This tendency is confirmed by literature, e.g. [1]. The experimental data of multiplier R obtained for Freon 21 [9] suggest that effect of mass velocity on R may be written as

$$R_{\text{mod}} = R\Omega \tag{35}$$

where the correction factor Ω is expressed as

$$\Omega = \left(\frac{1400}{G}\right)^{0.25} \tag{36}$$

in the range $G = 1000 \div 2500$ kg/m^{2}s. A systematic investigation of mass velocity influence on pressure drop in two-phase flow was conducted by Baroczy [12] for a variety of fluids. His results show very complex dependence of the correction factor Ω not only on the mass velocity G but also on property index and steam quality. The Baroczy correction factor Ω_B was presented in the form of graphs as a function of property index K_B defined as

$$K_B = \left(\frac{\mu_l}{\mu_g}\right)^{0.2} \frac{\rho_g}{\rho_l}. \tag{37}$$

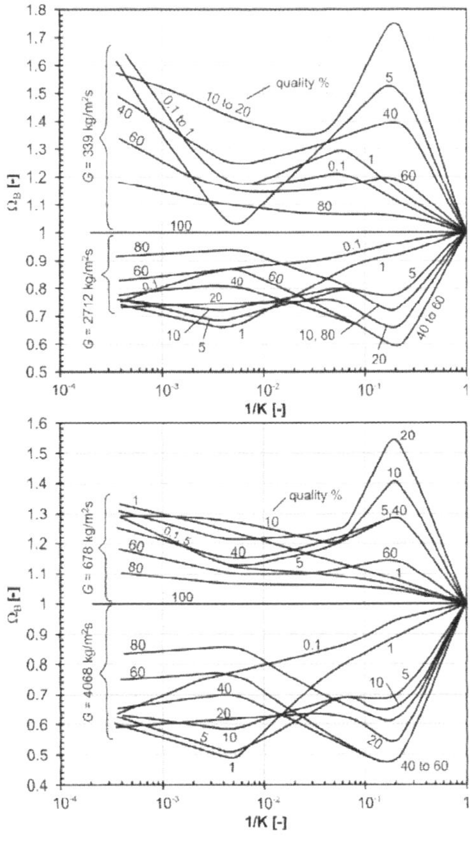

Fig. 7. Mass velocity correction factor Ω_B as a function of property parameter $1/K$ and steam quality x for selected values of mass velocity G, adapted from Baroczy [12]

It is seen that Baroczy property index (37) is similar to the property parameter K (20), namely

$$K_B \approx 1/K. \tag{38}$$

Therefore it is proposed to adapt Baroczy correction factor Ω_B to use with the multiplier R in (35) as a more general formula than (36). Due to complex representation of Ω_B, it is impossible to find a simple approximation function allowing direct numerical calculations so its values should be read from the plots shown in Fig. 7.

Table 1. Initial conditions for pressure drop measurements in R134a two-phase flow [13].

Case No.	G [kg/m^{2}s]	p [bar]	p/p_{crit}
1	1200	10	0.25
2	1200	25	0.62
3	1200	36	0.89
4	1800	10	0.25
5	1800	25	0.62
6	1800	36	0.89

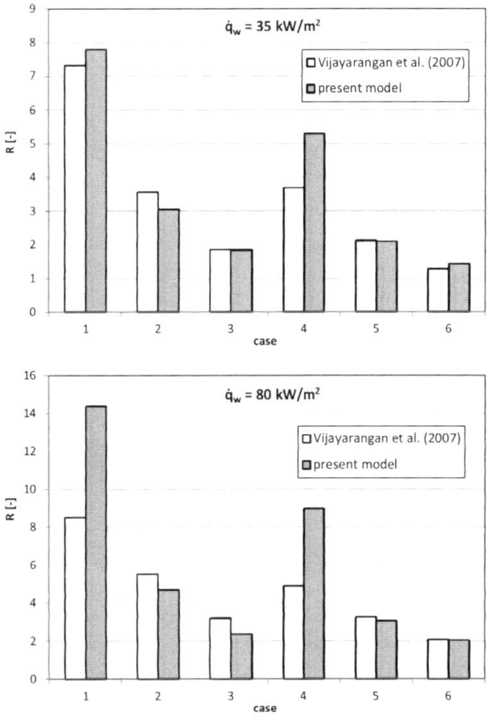

Fig. 8. Comparison of multiplier R calculated with present model, i.e. Eq. (20) with Baroczy correction factor Ω_B, and derived from Vijayarangan *et al.* [13] measurements for R134a.

The predictions of the proposed method of multiplier R calculation (20) with the correction factor Ω_B were compared with experimental data for R134a refrigerant reported by Vijayarangan *et al.* [13]. For this purpose, a total of 12 Vijayarangan *et al.* measurements of frictional pressure drop in a vertical tube of 12.7 mm internal diameter were selected. Initial flow conditions for the comparison, which are a combination of two mass velocity values and three pressure values, are named as Case 1 to 6 and defined in the Table 1. In each case, two heat flux $\dot{q}_w$ values (35 or 80 kW/m^2) were supplied to the R134a flow from the uniformly heated tube wall.

The comparison results are shown in Fig. 8. Good agreement is achieved between the Vijayarangan *et al.* data and the present model predictions, except for the cases with p = 10 bar and $\dot{q}_w$ = 80 kW/m^2, where the model clearly overestimates the R value.

4 Conclusions

The paper presents a simple methodology to transform Martinelli–Nelson graphical forms of two-phase friction multiplier R and void fraction φ into their equivalent analytical expressions.

Comparison of Martinelli–Nelson data of R and φ with their analytical approximations, Eq. (20) and (28), shows very good agreement, Fig.1 and Fig 2.

Non–dimensional property parameter K, Eq. (21), was introduced to allow the use of the above method to calculate multiplier R and void fraction φ for other fluids than water, at the condition that K for water has to be equal to K for that other fluid.

Experimental investigation of pressure drop of Freon 21 in vertical tube (d_{in} = 11.6 mm) during boiling proved the validity of Eqs. (20) and (28), as well as the usefulness of the property parameter K.

The above investigation also showed the effect of mass velocity on the multiplier R. Formula (35) was proposed to take this effect into account for Freon 21. For other fluids Baroczy [12] correction factor was adapted in the form of graphs shown in Fig. 7 and checked against the experimental data for R134a [13].

This research has been supported by National Science Centre within the Project No. 2012/05/B/ST8/02742

References

[1] J.G. Collier, J.R. Thome, *Convective Boling and Condensation*, Clarendon Press, Oxford (1994)

[2] P.A. Lockhart, R.C. Martinelli, *Proposed correlation of data for isothermal two-phase two-component flow in pipes*, Chemical Engineering Progress, **45**, No. 1, 39–48 (1949)

[3] D. Chisholm, *A theoretical basis for the Lockhart-Martinelli correlation for two-phase flow*, Int. J. Heat Mass Transfer, **10**, 1767–1778 (1967)

[4] Y.S. Muzychka, M.M. Awad, *Asymptotic generalizations of the Lockhart-Martinelli method for two phase flows*, Transactions of the ASME Journal of Fluids Engineering, **132**, Issue 3, 031302-1 – 031302-12 (2010)

[5] S.K. Saha, G.P. Celata, *Heat transfer and pressure drop in flow boiling in microchannels*, Springer, Cham, Switzerland (2016)

[6] M. Trela, *A new approach to predicting pressure drop in two-phase flows*, Trans. of the Institute of Fluid Flow Machinery (IMP PAN), **64**, 41–57 (1974)

[7] R.C. Martinelli, D.B. Nelson, *Prediction of pressure drop during forced circulation boiling of water*, Trans. ASME, **70**, 695–702 (1948)

[8] W.F. Davidson, P.H. Hardie, C.R. Humphrey, *Studies of heat transmission through boiler tubing at pressures from 500 to 3300 pounds*, Trans. ASME, **65**, 553–591 (1943)

[9] M. Trela, *Experimental investigations of the pressure drop of freon 21 during flow boiling in vertical pipes*, Trans. of the Institute of Fluid Flow Machinery (IMP PAN), **64**, 59–78 (1974)

[10] P.A. Lottes, M. Petrick, F. Marchattere, *Lecture notes on heat extraction from boiling water power reactors*, USA EC Report ANL-6063 (1959)

[11] S. Levy, *Steam slip theoretical prediction from momentum model*, J. Heat Transfer, Trans. ASME series C, **82** (1960)

[12] C.J. Baroczy, *A systematic correlation for two-phase pressure drop*, Chemical Engineering Progress Symposium Series, **62**, No. 64, 232–249 (1966)

[13] B.R. Vijayarangan, S. Jayanti, A.R. Balakrishnan, *Pressure drop studies on two-phase flow in a uniformly heated vertical tube at pressures up to the critical point*, Int. J. Heat Mass Transfer, **50**, 1879–1891 (2007)

 DOI: 10.1051/e3sconf/20171302007

Determination of transient temperature fields in thick-walled elements using the inverse method

Magdalena Jaremkiewicz*, and *Artur* Cebula

Institute of Thermal Power Engineering, Cracow University of Technology, Al. Jana Pawła II 37, 31-864 Cracow, Poland

Abstract. The purpose of this paper is to propose a method of determining the transient temperature of the inner surface of thick-walled elements. The method can be used to determine thermal stresses in pressure elements. An inverse marching method is proposed to determine the transient temperature of the thick-walled element inner surface with high accuracy. The inverse method was validated computationally. The comparison between the temperatures obtained from the direct heat conduction problem solution and the results obtained by means of the proposed inverse method is very satisfactory. The advantage of the method is the possibility of determining the heat transfer coefficient at a point on the exposed surface based on the local temperature distribution measured on the insulated outer surface. The heat transfer coefficient determined experimentally can be used to calculate thermal stresses in elements with a complex shape. The proposed method can be used in online computer systems to monitor temperature and thermal stresses in thick-walled pressure components because the computing time is very short.

1. Introduction

Unsteady-state thermal stresses arising in pressure elements cannot be calculated correctly without knowing the temperature of the fluid and the heat transfer coefficient on the inner surface of these elements. In order to determine the heat transfer coefficient, very precise measurements of the transient temperature are required due to the small difference between the temperature of the fluid under high pressure and the temperature of the element inner surface. The heat transfer coefficient has a significant effect on the optimum rate of changes in the fluid temperature, which is determined from the condition of not exceeding the thermal stress value on the inner surface of the pressure element [1].

The temperature measurement of the pressure element surface which is in contact with the fluid is difficult, especially if the fluid temperature, pressure and velocity parameters are high. It is not always possible to mount a temperature sensor on the element surface, and temperature measurements can be disturbed by significant errors due to contact resistance.

The most common method for determining time-dependent changes in the element surface temperature described in reference literature is based on solving the direct heat conduction problem analytically or numerically. This method is based on the heat transfer coefficient, determined from the correlation for the Nusselt number. The analysis usually concerns one-dimensional cases. Two-dimensional or three-dimensional issues are discussed in [2-4]. Using the direct method to determine the inner surface transient temperature can be problematic. The heat transfer coefficients on the inner surfaces of elements are determined from correlations known from reference literature. These are usually correlations found for hydrodynamically and thermally developed flows under steady-state conditions.

In recent years, extensive research has been conducted on the development of methods based on solving inverse heat conduction problems [5-13]. These methods enable determination of the inner surface temperature on the basis of two known boundary conditions on the outer surface. In the case of elements with regular shapes, the two- or three-dimensional transient temperature field can be determined by solving the inverse heat conduction problem using marching methods [7, 8, 11, 12]. In order to calculate the inner surface transient-state temperature, the inverse region is divided into control volumes. The method marches in space towards the element inner surface by using the energy balance equations for the finite volumes placed on the boundary to find the temperature values in adjacent nodes.

A modified space marching method for transient-state nonlinear inverse heat conduction problems is presented in [9]. It is based on the classical space marching method and reduces numerical oscillation by using future temperatures in the least square manner.

* Corresponding author: mjaremkiewicz@pk.edu.pl

However, for elements with complex shapes and with holes, identification of thermal stresses and temperature fields by means of the marching inverse method is difficult. In this article, a new approach to temperature and thermal stress determination is proposed. Using the heat transfer coefficient determined on the element inner surface and the fluid temperature [12], commercial programs based on the finite element method can be applied to calculate thermal stresses. In this way, thermal stresses on the edges of holes or other places of stress concentration can be found.

2. Identification of the temperature distribution in the flat thick-walled plate

An inverse space marching method is used to determine the temperature distribution inside the plate (cf. Figure 1). The transient heat conduction equation for a flat plate is

$$c(T)\rho(T)\frac{\partial T}{\partial t} = \\ = \frac{\partial}{\partial x}\left[k(T)\frac{\partial T}{\partial x}\right] + \frac{\partial}{\partial y}\left[k(T)\frac{\partial T}{\partial y}\right] + \frac{\partial}{\partial z}\left[k\frac{\partial T}{\partial z}\right] + \dot{q}_v \tag{1}$$

where T denotes temperature, c, ρ, k – specific heat, density and thermal conductivity of the plate material, respectively, and $\dot{q}_v$ – energy generation rate per unit volume.

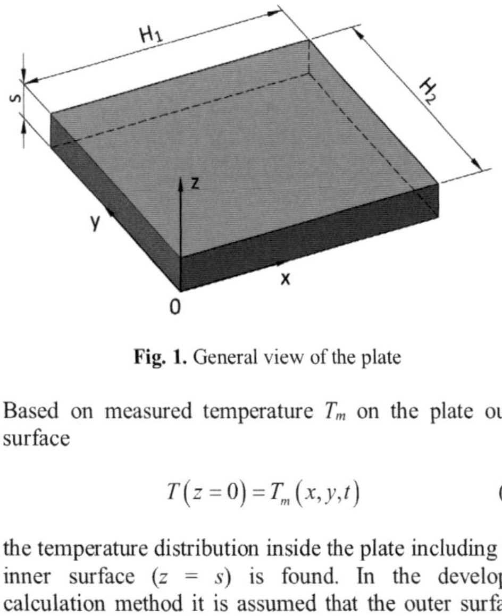

Fig. 1. General view of the plate

Based on measured temperature T_m on the plate outer surface

$$T(z=0) = T_m(x,y,t) \tag{2}$$

the temperature distribution inside the plate including the inner surface ($z = s$) is found. In the developed calculation method it is assumed that the outer surface, where the temperature is measured, is thermally insulated. Therefore, the boundary condition for $z = 0$ is as follows:

$$k\left.\frac{\partial T}{\partial z}\right|_{z=0} = 0 \tag{3}$$

There are two known boundary conditions on the surface $z = 0$, i.e. conditions (2) and (3), while the boundary condition on the opposite surface of the plate $z = s$ is unknown.

The side surfaces of the plate are thermally insulated, so the boundary conditions of the second type are as follows:

$$k\left.\frac{\partial T}{\partial x}\right|_{x=0} = 0, \quad k\left.\frac{\partial T}{\partial x}\right|_{x=H_1} = 0 \tag{4}$$

$$k\left.\frac{\partial T}{\partial y}\right|_{y=0} = 0, \quad k\left.\frac{\partial T}{\partial y}\right|_{y=H_2} = 0 \tag{5}$$

The solution of the inverse problem (1)-(5) is obtained using the finite volume method (FVM).

The plate division into finite volumes is shown in Figure 2. The energy balance equations can be written using the control volume method for all nodes located in the centres of the control volumes.

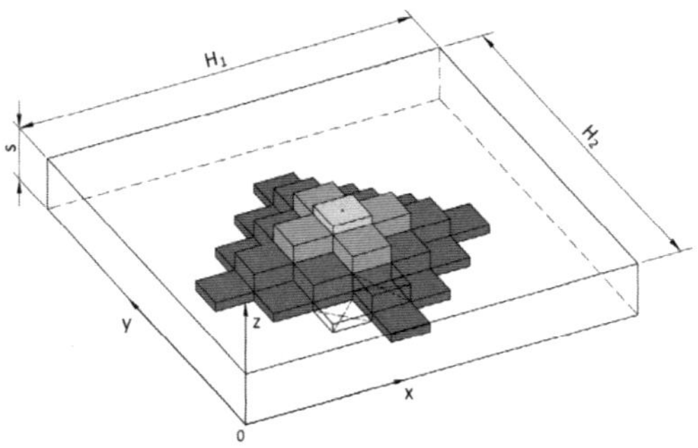

Fig. 2. Plate division into control volumes

The energy balance equation for node (i, j, k) inside the analysed finite volume is:

$$\begin{aligned} &\Delta x\,\Delta y\,\Delta z\,c\left(T_{i,j,k}\right)\rho\left(T_{i,j,k}\right)\frac{\mathrm{d}T_{i,j,k}}{\mathrm{d}t} = \\ &= \Delta y\,\Delta z\frac{k\left(T_{i,j,k}\right)+k\left(T_{i+1,j,k}\right)}{2}\frac{T_{i+1,j,k}-T_{i,j,k}}{\Delta x} + \\ &+ \Delta y\,\Delta z\frac{k\left(T_{i,j,k}\right)+k\left(T_{i-1,j,k}\right)}{2}\frac{T_{i-1,j,k}-T_{i,j,k}}{\Delta x} + \\ &+ \Delta x\,\Delta z\frac{k\left(T_{i,j,k}\right)+k\left(T_{i,j+1,k}\right)}{2}\frac{T_{i,j+1,k}-T_{i,j,k}}{\Delta y} + \\ &+ \Delta x\,\Delta z\frac{k\left(T_{i,j,k}\right)+k\left(T_{i,j-1,k}\right)}{2}\frac{T_{i,j-1,k}-T_{i,j,k}}{\Delta y} + \\ &+ \Delta x\,\Delta y\frac{k\left(T_{i,j,k}\right)+k\left(T_{i,j,k+1}\right)}{2}\frac{T_{i,j,k+1}-T_{i,j,k}}{\Delta z} + \\ &+ \Delta x\,\Delta y\frac{k\left(T_{i,j,k}\right)+k\left(T_{i,j,k-1}\right)}{2}\frac{T_{i,j,k-1}-T_{i,j,k}}{\Delta z} + \\ &+ \dot{q}_v\left(x_i,y_j,z_k\right)\Delta x\,\Delta y\,\Delta z \end{aligned} \tag{6}$$

DOI: 10.1051/e3sconf/20171302007

In Eq. (6) the symbols Δx, Δy, Δz denote the control volume dimensions, $c(T_{i,j,k})$, $\rho(T_{i,j,k})$, $k(T_{i,j,k})$ – specific heat, density and thermal conductivity in temperature $T_{i,j,k}$, respectively, and $\dot{q}_v(x_i, y_j, z_k)$ – energy generation rate per unit volume at points (x_i, y_j, z_k).

Figure 3 shows examples of nodes together with the surrounding cells, for which the heat balance equations are written. Figure 3(a) illustrates all finite volumes taken into account when the heat balance equation is written for a node situated on an insulated surface (z = 0). The thickness of control volumes adherent directly to the insulated surface is $\Delta z/2$. Figure 3(b) shows the cell around the internal node; such a case occurs in the plate distribution into a larger number of cells in the z-axis. All the cells have full dimensions: Δx, Δy, and Δz. The balance equation is written for the central cell. In the case of the heat balance compilation for a node situated at distance Δz from the cooled surface, the top cell has thickness $\Delta z/2$, and the rest are full size (Figure 3(c)).

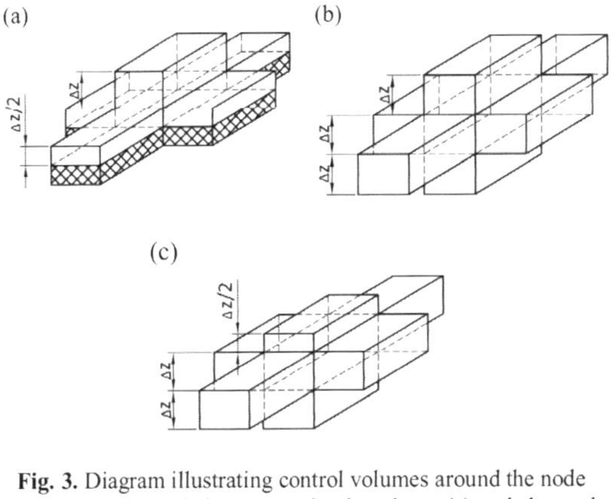

Fig. 3. Diagram illustrating control volumes around the node for which the heat balance equation is written: (a) node located on the plate insulated surface; (b) node located within the space between the plate insulated and cooled surfaces; (c) node located on the cooled surface

The heat balance equations are written for all nodes $P_{i,j,k} = P(x_i, y_j, z_k)$ located in the centres of gravity of each control volume. The coordinates of nodes in the centre of the control volumes are defined by the following formulae:

$$x_i = \frac{\Delta x}{2} + (i-1)\Delta x,\ i = 1,\ldots,7 \tag{7}$$

$$y_j = \frac{\Delta y}{2} + (j-1)\Delta y,\ j = 1,\ldots,7 \tag{8}$$

$$z_k = (k-1)\Delta z,\ k = 1,\ldots,4 \tag{9}$$

In order to solve the inverse problem, the analysed thick-walled plate is divided into finite volumes (Figures 2-5). First, to solve the inverse problem, the heat balance equations are written for nodes located on the insulated surface z = 0 (Figures 3(a) and 4). Then time-dependent temperatures in the nodes shown in Figures 3(b) and 4 lying in plane z = Δz are determined from the heat balance equations which are written for the nodes shown in Figures 3(a) and 4. In a similar way, time-dependent temperatures in the nodes in plane $z = 2\Delta z$ (Figures 3(c) and 4) are determined from the heat balance equations for the nodes lying in plane $z = \Delta z$. Next, the temperature in node (4,4,4), lying on surface $z = 3\Delta z = s$ is determined from the heat balance equations for the nodes lying in plane $z = 2\Delta z$. The finite volumes, in the middle of which temperature is measured or determined, lying in different planes z are marked with different colours (Figures 2, 4 and 5).

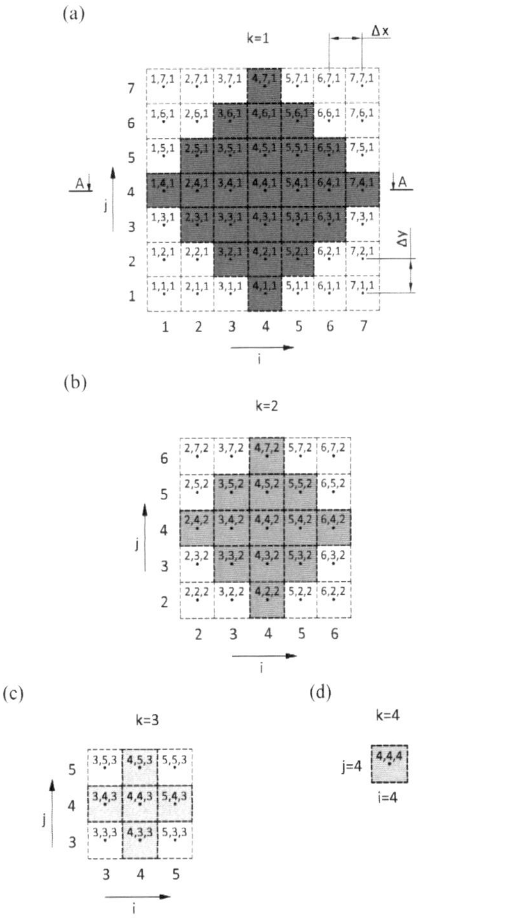

Fig. 4. Location of nodes in individual planes where temperature is measured or calculated: (a) plane in which temperature measurement points are situated; (b) plane $z = \Delta z$; (c) plane $z = 2\Delta z$; (d) plane $z = 3\Delta z = s$

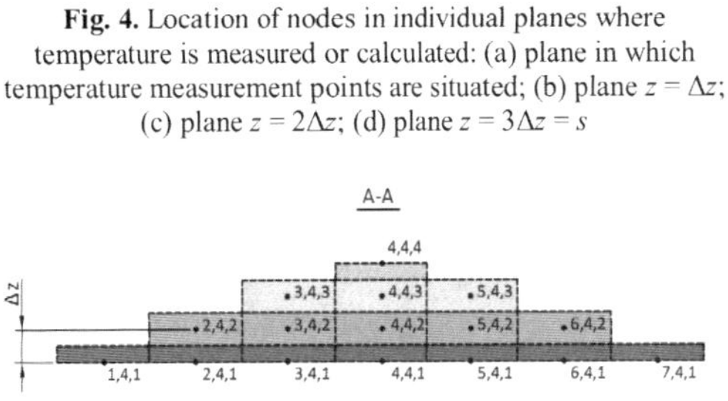

Fig. 5. Cross-section of finite volumes where temperature is measured or determined

Using the calculation procedure described above, only the temperature at point (4,4,4) is found.

In a similar way, the calculation procedure (calculation template) can be used to determine the temperature of additional nodes lying in plane $z = 3\Delta z = s$. In order to determine the heat flux density at point (4, 4, 4), it is necessary to measure the temperature in 16 additional nodes lying on the insulated surface ($z = 0$). In this case, it is possible to determined temperature values in five nodes on the plate surface $z = 3\Delta z = s$: (4,4,4), (3,4,4), (5,4,4), (4,3,4), and (4,5,4). The flux density of the heat transferred to the fluid from the plate surface in node (4,4,4) can then be found from the heat balance equation for node (4,4,4). The heat transfer coefficient h on surface $z = 3\Delta z = s$ is determined from the following formula:

$$h = \frac{\dot{q}(x_4, y_4, z_4)}{T(x_4, y_4, z_4) - T_f} \tag{10}$$

where T_f denotes the fluid temperature.

3. Computational validation of the inverse method

The calculations are carried out for a flat plate. It is assumed that the nodes on both surfaces ($z = 0$ and $z = s = 0.045$ m) and inside the plate are arranged as in Figures 3 and 4. The distances between the nodes are as follows: $\Delta x = 0.085$ m, $\Delta y = 0.085$ m, $\Delta z = s/3 = 0.015$ m, and the time step is $\Delta t = 1$ s. Calculations are made for constant values of the steel physical properties: $k = 57$ W/(mK), $\rho = 7850$ kg/m^3 and $c = 460$ J/(kgK). The outer surface of the plate ($z = 0$) is thermally insulated. In the calculations, it is assumed that the density of the heat flux transferred through the plate is $\dot{q} = 20$ kW/m^2, and the initial temperature T_0 is 20°C.

At first, to validate the developed method, "measured data" are generated using the direct method, assuming uniform heating of the plate inner surface ($z = s$).

Description of the direct method

The "measurement data" are obtained using an analytical method. Because the plate inner surface heating is uniform, the heat conduction equation for the flat wall with independent physical properties of the plate material is

$$\kappa \frac{\partial^2 T}{\partial z^2} = \frac{\partial T}{\partial t} \tag{11}$$

where $0 \le z \le s$. Symbol $\kappa = k/(c\rho)$ in Eq. (11) denotes thermal diffusivity.

The solution of Eq. (11) is obtained for the following boundary conditions

$$k \left. \frac{\partial T}{\partial z} \right|_{z=0} = 0 \tag{12}$$

$$k \left. \frac{\partial T}{\partial z} \right|_{z=s} = \dot{q} \tag{13}$$

and the initial condition

$$T|_{t=0} = T_0 \tag{14}$$

The problem of (11)-(13) with the initial condition expressed in (14) above is solved using the Laplace transform. The result is

$$T(z,t) = T_0 + \frac{\dot{q}s}{k}\left[\mathrm{Fo} + \frac{1}{2}\left(\frac{z}{s}\right)^2 - \frac{1}{6} + \sum_{n=1}^{\infty} (-1)^{n+1} \frac{2}{n^2\pi^2} \cos\left(n\pi \frac{z}{s}\right) e^{-n^2\pi^2 \mathrm{Fo}} \right] \tag{15}$$

where $n = 1, 2, \ldots$ and $\mathrm{Fo} = \kappa t/s^2$ designates the Fourier number.

All calculations are performed using the FORTRAN programming language. The calculation results of time-dependent temperature changes on both surfaces of the plate obtained using the analytical method are presented in Figure 6(a). Because the plate is heated evenly, the calculated temperature is the same in 25 nodes located on the insulated outer surface.

Then, on the basis of the obtained temperatures on the insulated surface, the temperature inside the wall at $z = \Delta z$ and $z = 2\Delta z$ and on the plate inner surface are determined using the inverse marching method. The calculation results are presented in Figure 6(b). The comparison between the temperature values on the plate inner surface obtained by means of the direct method and the inverse marching method is shown in Figure 7.

Due to high thermal inertia of the thick-walled plate, the inner surface temperature calculated using the inverse method, with a small error compared to the temperature calculated using the direct method, is obtained with some time lag.

(a)

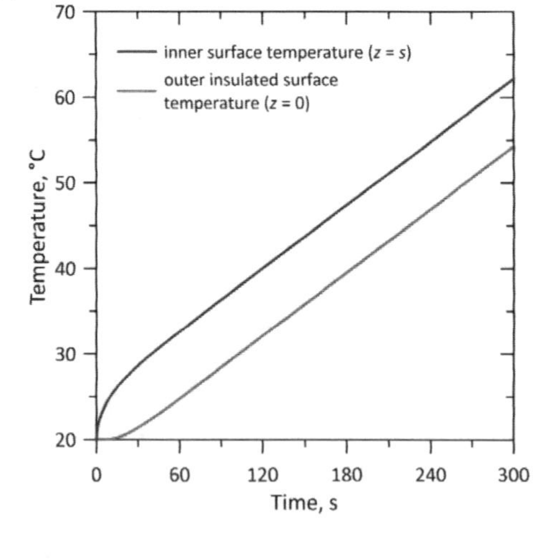

(b)

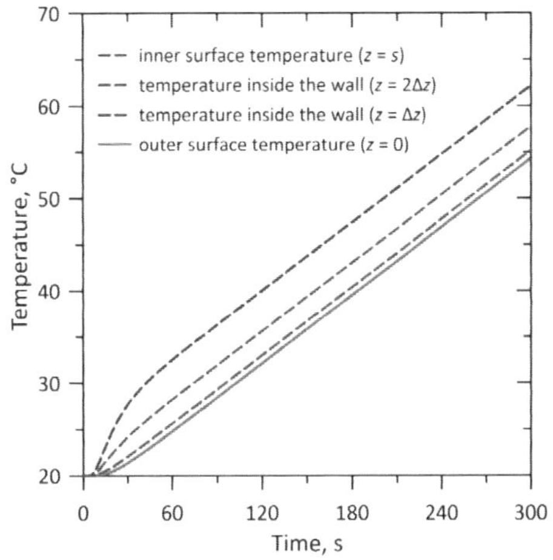

Fig. 6. Time-dependent changes in temperature: (a) data generated by the direct method; (b) results of calculations using the inverse marching method

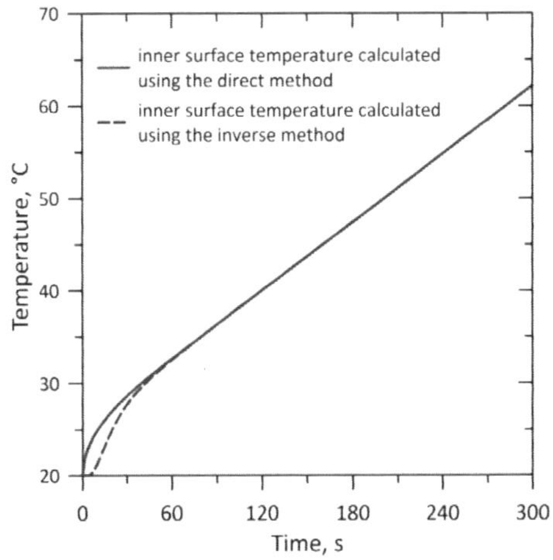

Fig. 7. Comparison between temperature values on the plate inner surface $z = s$ obtained using the direct method and the inverse marching method

However, when a small increase in the outer plate surface temperature is recorded, the inverse method enables rapid determination of the inner surface temperature. Better results can be achieved by carrying out the temperature measurements inside the plate wall, drilling holes in it, but this is not always allowed for components operating under high pressure.

4. Conclusions

This paper presents a general method for determination of temperature, the heat flux density and the heat transfer coefficient on the inner surface of a thick-walled element. Based on the described procedure, a calculation program was compiled and validated for determining the plate inner surface temperature.

The inverse method was validated computationally. The comparison between the temperature values obtained from the direct heat conduction problem solution and the results achieved by means of the proposed inverse method is very satisfactory. Stable and accurate measurement results are obtained using the proposed method based on the solution of the inverse heat conduction problem due to partial elimination of random errors in the input signal by using a 9-point moving digital filter [13].

After some modifications, the developed method can be used to monitor thermal stresses and evaluate the degree of wear and estimate residual life of the power plant pressure components such as boilers or turbines. By measuring the pressure element outer surface temperature in a small area, it is possible to determine the temperature of the inner surface. The presented method combined with the method of measuring the fluid unsteady-state temperature described in [12] enables calculation of the heat transfer coefficient on the inner surface. Owing to precise determination of the heat transfer coefficient on the inner surface, thermal stresses arising in the pressure element can be calculated by means of the finite element method.

The presented method can be used in online applications. Another advantage of the method is the ease of practical application.

References

1. J. Taler, P. Dzierwa, D. Taler, M. Jaremkiewicz, and M. Trojan, *Monitoring of Thermal Stresses and Heating Optimization Including Industrial Applications* (Nova Science Publishers, New York 2016)
2. A.O. Adelaja, J. Dirker, J.P. Meyer, Appl Energ **130**, 838-845 (2014)
3. S. Darici, S. Bilir, A. Ates, Int J Heat Mass Tran **84** 1040-1048 (2015)
4. C. Marugan-Cruz, O. Flores, D. Santana, M. Garcia-Villalba, Int J Heat Mass Tran **96** 256-266 (2016)
5. H. Ahmadi-Noubari, A. Pourshaghaghy, F. Kowsary, A. Hakkaki-Fard, Int J Numer Method H **18** 217-236 (2008)
6. X. Li, Int J Numer Method H **25** 190-198 (2015)
7. J. Taler, W. Zima, Int J Heat Mass Tran **42** 1123-1140 (1999)
8. A. Cebula, J. Taler, Appl Therm Eng **63** 158-169 (2014)
9. T. Lu, B. Liu, P.X. Jiang, Appl Therm Eng **31** 1976-1982 (2011)
10. E. Sun, B. Li, L. Shi, F. Chen, J Therm Sci Tech-JPN **11** 1-9 (2016)
11. J. Taler, B. Węglowski, T. Sobota, M. Jaremkiewicz, and D. Taler, *Inverse Space Marching Method for Determining Temperature and Stress Distributions in Pressure Components*, chapter in *Developments in Heat Transfer* edited by Marco Aurélio dos Santos Bernardes (InTech, Rijeka, Croatia, 2011)
12. M. Jaremkiewicz, D. Taler, T. Sobota, Int J Therm Sci **87** 241-250 (2015)

13. J. Taler, *Theory and Practice of Identification of Heat Transfer Processes* (Zakład Narodowy imienia Ossolińskich, Wrocław, 1995 (in Polish))

E3S Web of Conferences **13**, 02008 (2017) DOI: 10.1051/e3sconf/20171302008
WTiUE 2016

Simple heat transfer correlations for turbulent tube flow

Dawid Taler[1,*] and *Jan* Taler[2]

[1] Institute of Heat Engineering and Air Protection, Cracow University of Technology, 31-155 Cracow, Poland
[2] Institute of Thermal Power Engineering, Cracow University of Technology, 31-864 Cracow, Poland

Abstract. The paper presents three power-type correlations of a simple form, which are valid for Reynolds numbers range from $3 \cdot 10^3 \leq \mathrm{Re} \leq 10^6$, and for three different ranges of Prandtl number: $0.1 \leq \mathrm{Pr} \leq 1.0$, $1.0 < \mathrm{Pr} \leq 3.0$, and $3.0 < \mathrm{Pr} \leq 10^3$. Heat transfer correlations developed in the paper were compared with experimental results available in the literature. The comparisons performed in the paper confirm the good accuracy of the proposed correlations. They are also much simpler compared with the relationship of Gnielinski, which is also widely used in the heat transfer calculations.

1 Introduction

Heat transfer correlations for turbulent fluid flow in the tubes are commonly used in the design and performance calculations of heat exchangers [1-4]. The value of heat transfer coefficient significantly affects the value of the thermal stress [5]. Time changes in optimum fluid temperature determined from the condition of not exceeding the stress at a point lying on the inner surface of the pressure component, very strong depend on the heat transfer coefficient [6-9]. Therefore, continuous experimental research is carried out to find a straightforward and accurate heat transfer correlations.
The empirical correlation of Dittus-Boelter [10-12] has gained widespread acceptance for prediction of the Nusselt number with turbulent flow in the smooth-surface tubes

$$\mathrm{Nu} = 0.023\,\mathrm{Re}^{0.8}\,\mathrm{Pr}^{n} \tag{1}$$

$$0.6 \leq \mathrm{Pr} \leq 160, \qquad \mathrm{Re} \geq 10^4, \qquad L/d_w \geq 60$$

where : $\mathrm{Nu} = h d_w / k$, $\mathrm{Re} = u_m d_w / \nu$, $\mathrm{Pr} = c_p \mu / k$,

$d_w = 2r_w$ - inner diameter of the tube, L – tube length, h – heat transfer coefficient, k – thermal conductivity, u_m – mean velocity of the fluid, c_p - specific heat at constant pressure, $\nu = \mu / \rho$ - kinematic viscosity, μ-dynamic viscosity, ρ - fluid density.

The exponent of the Prandtl number is $n = 0.4$ for heating of the fluid and $n = 0.3$ if the fluid is being cooled. The similar power-type relationship of Colburn is based on the Chilton-Colburn analogy for heat and momentum transfer [13]

$$\mathrm{Nu} = 0.023\,\mathrm{Re}^{0.8}\,\mathrm{Pr}^{1/3} \tag{2}$$

$$0.6 \leq \mathrm{Pr} \leq 160, \qquad \mathrm{Re} \geq 10^4, \qquad L/d_w \geq 60$$

The correlations (1) and (2) are valid for moderate temperature differences $|T_w - T_m|$ under which properties may be evaluated at the mean bulk temperature $\overline{T}_m$. When variations of physical properties are significant, the Sieder and Tate relationship [14] can be used

$$\mathrm{Nu} = 0.027\,\mathrm{Re}^{0.8}\,\mathrm{Pr}^{1/3}\left(\frac{\mu_m}{\mu_w}\right)^{0.14} \tag{3}$$

$$0.7 \leq \mathrm{Pr} \leq 16{,}700, \qquad \mathrm{Re} \geq 10^4, \qquad L/d_w \geq 60$$

where all properties are evaluated at the local bulk temperature T_m except the viscosity μ_w that is estimated at the wall surface temperature T_w.
However, the power-type correlations like Dittus-Boelter, Colburn, and Colburn relationships do not provide good approximations within larger ranges of the experimental data because of its simple form. The power type correlation like that of Dittus-Boelter is not able to approximate the experimental results over a broad range of the Prandtl number. The maximum deviation between experimental data and predictions using Eq. (1) is about 20% [15-17]. When the working medium is water, the Dittus-Boelter relationship (1) overpredicts the heat transfer coefficient for smaller Reynolds numbers when Re < 10,000 and underpredicts for higher Reynolds numbers [18]. In the last forty years, the power-type equations (1-3) were often being replaced by the

* Corresponding author: dtaler@pk.edu.pl

Gnielinski correlation [19-20]

$$\mathrm{Nu} = \frac{\frac{\xi}{8}(\mathrm{Re}-1000)\mathrm{Pr}}{1+12.7\sqrt{\frac{\xi}{8}}\left(\mathrm{Pr}^{2/3}-1\right)}\left[1+\left(\frac{d_w}{L}\right)^{2/3}\right]\left(\frac{\mathrm{Pr}_m}{\mathrm{Pr}_w}\right)^{0.11} \tag{4}$$

$$4\cdot 10^3 \le \mathrm{Re} \le 10^6, \quad 0.5 \le \mathrm{Pr} \le 200$$

where the symbol ξ denotes the friction factor for smooth tubes calculated from the Filonienko relationship [21]

$$\xi = \left(1.82\log \mathrm{Re} - 1.64\right)^{-2} \tag{5}$$

The symbols Pr_m and Pr_w designates the Prandtl number at the bulk and wall temperature, respectively.

The Gnielinski relationship (4) is a modification of the Petukhov formula [22]

$$\mathrm{Nu} = \frac{\frac{\xi}{8}\mathrm{Re}\,\mathrm{Pr}}{1.07+12.7\sqrt{\frac{\xi}{8}}\left(\mathrm{Pr}^{2/3}-1\right)}\left(\frac{\mu_m}{\mu_w}\right)^n \tag{6}$$

$$10^4 \le \mathrm{Re} \le 10^6, \quad 0.5 \le \mathrm{Pr} \le 200$$

to increase the accuracy of Eq. (6) for smaller Reynolds numbers. In Eqs. (4) and (6), the subscript m refers to evaluating the fluid properties at the mean (bulk) average temperature $\overline{T}_m$ that is usually defined as the average of the inlet and exit bulk temperature. The subscript w refers to the evaluation of the fluid properties at the wall surface temperature T_w. The exponent n in Eq. (6) is equal to 0.11 for $T_w > \overline{T}_m$ and $n = 0.25$ for $T_w < \overline{T}_m$. The constant n is equal zero for gases. The relationship (4) of Gnielinski is very widely used, as well approximates the experimental data [17, 23-25].

In recent years, however, a new rise in popularity of power-type correlation to an approximation of the measurement data is observed [24-26].

The thermal-hydraulic performance of a dimpled enhanced tube was studied experimentally and numerically by Li et al. [24]. The power type correlation based on the experimental results was proposed. The Wilson method was used to adjust unknown constants in the searched relationship. The Prandtl number changed from 5.2 to 30.7 and the Reynolds number from 500 to 8,000.

Simple power correlations are also proposed to fit experimental data for turbulent flows of nanofluids in tubes. Li and Xuan [25] and Azmi and al. [26] also used electrically heat tubes to study hydrodynamically fully developed flow. Before starting the actual experiments with nanofluids, tests were conducted in which the working fluid was water [25] or a mixture of water and ethylene glycol [26]. The Reynolds number varied in the study of Azmi et al. from 3,000 to 25,000. The Nusselt numbers determined experimentally showed good agreement with the Nusselt numbers obtained from the Dittus-Boelter equation.

Li and Xuan [25] also found a heat transfer correlation for internal flow of Cu-water nanofluid that for water reduces to

$$\mathrm{Nu} = 0.0059\,\mathrm{Re}^{0.9238}\,\mathrm{Pr}^{0.4} \tag{7}$$

$$2.5\cdot 10^3 \le \mathrm{Re} \le 2.5\cdot 10^4$$

Previous studies indicate that power-type correlations are still used successfully to approximate the experimental results when the Reynolds and Prandtl numbers vary in a narrow range.

Unknown constants occurring in the power-type relationships can be readily determined by standard [27] or modified [28] Wilson method while maintaining reasonable effort.

Easy determination of searched constants based on the experimental results is also the reason for the use of power type functions to find heat transfer correlations for the turbulent flow of nanofluids or molten salts in tubes.

2 Nusselt numbers for turbulent tube flow

Energy conservation equation for turbulent tube flow averaged by Reynolds has the following form [23, 29]

$$\rho c_p \overline{u}\frac{\partial \overline{T}}{\partial x} = -\frac{1}{r}\frac{\partial}{\partial r}(rq) \tag{8}$$

where: q – heat flux density, x – Cartesian coordinate, r – radius, $\overline{u}$ - time averaged velocity.

The heat flux q consists of the molecular q_m and turbulent q_t component

$$q = q_m + q_t \tag{9}$$

where

$$q_m = k\frac{\partial \overline{T}}{\partial r}, \quad q_t = \rho\, c_p\, \varepsilon_q \frac{\partial \overline{T}}{\partial r} = k\frac{\mathrm{Pr}}{\mathrm{Pr}_t}\frac{\varepsilon_\tau}{\nu}\frac{\partial \overline{T}}{\partial r} \tag{10}$$

The symbols in Eq. (10) stand for: $\overline{T}$ - time averaged temperature, ε_q - eddy diffusivity for heat transfer, ε_τ eddy diffusivity for momentum transfer (turbulent kinematic viscosity), $\mathrm{Pr}_t = \varepsilon_\tau / \varepsilon_q$ - turbulent Prandtl number. Equation (8) is subject to the following boundary conditions

$$k\left.\frac{\partial \overline{T}}{\partial r}\right|_{r=r_w} = q_w \tag{11}$$

$$\left[\frac{\partial \bar{T}}{\partial r}\right]\Bigg|_{r=0} = 0 \quad (12)$$

$$\bar{T}\Big|_{x=0} = T_m\Big|_{x=0} \quad (13)$$

where q_w is the wall heat flux. The mass averaged (bulk) temperature $T_m(x)$ is defined as follows

$$T_m(x) = \frac{2}{r_w^2 u_m}\int_0^{r_w} \bar{u}(r)\bar{T}(x,r)\,r\,\mathrm{d}r \quad (14)$$

The time averaged velocity profile $\bar{u}(r)$ in tube cross-section is necessary to determine the temperature distribution in the fluid. The eddy diffusivity for momentum transfer ε_τ and time averaged fluid velocity $\bar{u}(r)$ were calculated using Reichardt's [30-31] empirical relationships, which are straightforward and accurate.

Equation (8) with the boundary conditions (11)-(13) was solved using the finite difference method [29]. The Nusselt number was determined based on the temperature distribution determined numerically using the following formula

$$\mathrm{Nu} = \frac{2hr_w}{k} = \frac{2q_w r_w}{k\left(T\big|_{r=r_w} - T_m\right)} \quad (15)$$

The Nusselt number was evaluated for various Reynolds and Prandtl numbers, and the results are listed in Table 1 [29].

Table 1. Nusselt Nu number as a function of the Reynolds and Prandtl numbers obtained from the solution of the energy conservation equation for fully developed turbulent flow in tubes with constant wall heat flux [29].

Pr \ Re	$3\cdot10^3$	$5\cdot10^3$	$7.5\cdot10^3$	10^4
0.1	7.86	9.42	11.13	12.69
0.2	9.41	11.86	14.58	17.08
0.5	12.65	16.96	21.81	26.31
0.71	14.32	19.60	25.57	31.12
1	16.22	22.61	29.86	36.61
3	24.43	35.57	48.39	60.47
5	29.56	43.64	59.94	75.36
7.5	34.34	51.15	70.68	89.21
10	38.16	57.14	79.24	100.25
12.5	41.40	62.20	86.47	109.56
15	44.23	66.62	92.78	117.70
30	56.73	86.09	120.53	153.40
50	67.98	103.56	145.36	185.32
100	86.64	132.44	186.35	237.96
200	110.11	168.69	237.71	303.83
1000	190.70	292.75	413.17	528.61

Pr \ Re	$3\cdot10^4$	$5\cdot10^4$	$7.5\cdot10^4$
0.1	22.76	31.02	40.20
0.2	33.53	47.23	62.59
0.5	56.55	82.31	111.63
0.71	68.78	101.16	138.18
1	82.91	123.04	169.15
3	145.31	220.65	308.45
5	184.68	282.64	397.43
7.5	221.38	340.53	480.68
10	250.64	386.73	547.14
12.5	275.31	425.71	603.25
15	296.85	459.72	652.22
30	391.25	608.73	866.75
50	475.39	741.46	1,057.70
100	613.70	959.44	1,371.11
200	786.27	1,231.01	1,761.37
1000	1,373.01	2,153.24	3,085.03

Pr \ Re	10^5	$3\cdot10^5$	10^6
0.1	48.62	104.68	256.61
0.2	76.78	172.57	437.11
0.5	138.97	327.20	860.91
0.71	172.84	413.35	1,102,29
1	212.50	515.41	1,391.88
3	391.85	987.68	2,764.63
5	506.89	1,295.71	3,677.45
7.5	614.66	1,586.18	4,544.73
10	700.78	1,818.99	5,242.89
12.5	773.51	2,015.88	5,834.61
15	836.98	2,187.81	6,352.24
30	1,115.04	2,941.80	8,624.88
50	1,362.53	3,612.61	10,647.89
100	1,768.51	4,711.28	13,959.35
200	2,273.65	6,076.39	18,067.10
1000	3,986.35	10,692.06	31,968.41

The Nusselt numbers shown in Table 1 with Prandtl numbers ranging from 0.1 to 1,000 and Reynolds numbers ranging from 3,000 to 10^6 were approximated by a product of two power functions of the Reynolds and Prandtl numbers

$$\mathrm{Nu} = c_1\,\mathrm{Re}^{c_2}\,\mathrm{Pr}^{c_3} \quad (16)$$

Unknown constants c_1, c_2, and c_3 occurring in the approximating function (16) were determined using the least squares method

$$S(c_1,c_2,c_3) = \sum_{i=1}^{n_{\mathrm{Re}}}\sum_{j=1}^{n_{\mathrm{Pr}}}\left(\mathrm{Nu}_{ij}^{m} - \mathrm{Nu}_{ij}^{c}\right)^2 = \min \quad (17)$$

where: $\mathrm{Nu}_{ij}^{m} = \mathrm{Nu}^{m}\left(\mathrm{Re}_i, \mathrm{Pr}_j\right)$ and $\mathrm{Nu}_{ij}^{c} = \mathrm{Nu}(\mathrm{Re}_i, \mathrm{Pr}_j), i = 1,\ldots,n_{\mathrm{Re}}, j = 1,\ldots,n_{\mathrm{Pr}}$ - given and approximated values of the Nusselt number, respectively, n_{Re} and n_{Pr} - the number of the Reynolds and Prandtl numbers which are considered in the sum (17). As Nu_{ij}^{m} the Nusselt numbers shown in Table 1 were taken. The values of Nu_{ij}^{c} were evaluated using Eq. (17).

Substituting the constants c_1, c_2, and c_3 determined from the condition (17) into the correlation (16), the following power-type formulas for turbulent flow in round tubes were obtained

$$\mathrm{Nu} = 0.02155\,\mathrm{Re}^{0.8018}\,\mathrm{Pr}^{0.7095} \quad (18)$$
$$3 \cdot 10^3 \le \mathrm{Re} \le 10^6,\ 0.1 \le \mathrm{Pr} \le 1$$

$$\mathrm{Nu} = 0.01253\,\mathrm{Re}^{0.8413}\,\mathrm{Pr}^{0.6179} \quad (19)$$
$$3 \cdot 10^3 \le \mathrm{Re} \le 10^6,\ 1 < \mathrm{Pr} \le 3$$

$$\mathrm{Nu} = 0.00881\,\mathrm{Re}^{0.8991}\,\mathrm{Pr}^{0.3911} \quad (20)$$
$$3 \cdot 10^3 \le \mathrm{Re} \le 10^6,\ 3 < \mathrm{Pr} \le 1000$$

The analysis of equations (18)-(20) shows that power exponents at the Reynolds and Prandtl numbers strongly depend on the Prandtl number. If the Prandtl number is smaller than one, power exponents with Reynolds and Prandtl numbers are almost equal to each other, as in the case of liquid metals when Pr < 0.1 [1, 17, 23]. The exponent with Reynolds number increases with the Prandtl number, while the exponent of Prandtl number behaves conversely. As the Prandtl number increases, the power with Prandtl number decreases.

Equation (16) can be generalized to a tube of finite length by introducing the same multiplier as in the Gnielinski formula (4)

$$\mathrm{Nu} = c_1\,\mathrm{Re}^{c_2}\,\mathrm{Pr}^{c_3}\left[1 + \left(\frac{d_w}{L}\right)^{2/3}\right] \quad (21)$$

The heat transfer correlations (18) - (20) were compared with the experimental data available in the literature taking into account the correction factor $\left[1 + \left(d_w / L\right)^{2/3}\right]$.

3 Comparison of the proposed correlations with experimental data

The proposed correlation (18) was compared with the experimental results of turbulent air flow in the pipes and the correlation (20) with experimental data for water, which are available in the literature. Comparison of the results of experimental studies of doctoral dissertations [32-35] for the air flowing inside the heating pipe is shown in Fig. 1. Abraham [36] was the first who presented the experimental data for smooth tubes with uniform wall heat flux reported in these theses and used by him to verify the CFD results. Equation (18) gives lower values of Nusselt number as compared to experimental results, whereas the correlation of Dittus-Boelter (1) overpredicts them.

The proposed relationship (18) was also compared in Fig. 2 with the experimental data of Eiamsa-ard and Promvonge [37] for turbulent air flow in a tube with constant wall heat flux.

Measurements were performed in an electrically heated tube. The outer and inner diameter of the copper tube were 47 mm and 50 mm, respectively. The tube was heated uniformly over the length of 1,250 mm. An unheated calming section preceded heated pipe section.

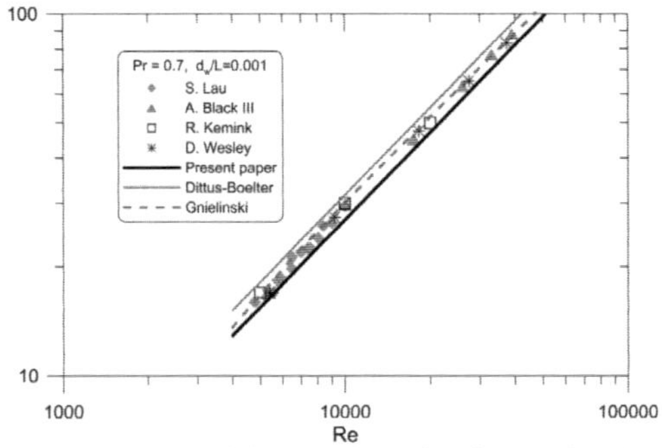

Fig. 1. Comparison of the Nusselt number for circular tubes obtained from the correlation (18) proposed in the paper with the experimental results of Lau, Black, Kemink, and Wesley [32-35] and the Gnielinski relationship (4) [19-20].

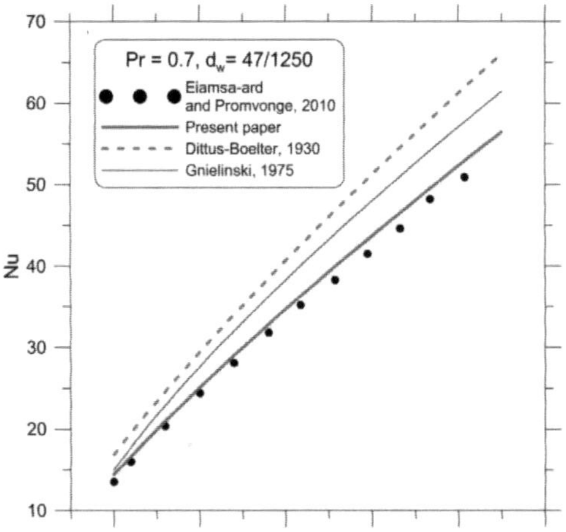

Fig. 2. Comparison of the correlation (18) proposed in the present paper and the correlation of Dittus-Boelter [10] and Gnielinski [19-20] with the experimental data of Eiamsa-ard and Promvonge [37] obtained for turbulent air flow in a tube with constant wall heat flux.

The hydrodynamically developed turbulent flow of water enters a uniformly heated tube. The Nusselt number determined experimentally was 10 to 20% higher than values predicted by the Dittus-Boelter equation. The discrepancy between the results of measurements and values of Nusselt number obtained from the Dittus-Boelter formula is greater for higher Reynolds numbers.

Then, Eq. (20) was compared with experimental results available of Allen and Eckert [16]. The smooth-tube heat transfer results reported by Allen and Eckert [16] were obtained for developed turbulent flow of water under the uniform wall heat flux boundary condition at Pr=7 and Pr=8, and $1.3\cdot10^4 \leq \mathrm{Re} \leq 1.11\cdot10^5$ [16]. Steel pipe with an inner diameter of 19.05 mm and a wall thickness of 1.5875 mm was heated electrically so that the heat flux at the internal surface of the tube was constant.

Allen and Eckert [16] found friction and heat transfer correlations for turbulent pipe flow of water at uniform wall heat flux for Pr=8 experimentally. The friction factor and Nusselt number determined experimentally were approximated by the following functions [16]

$$\xi = 0.00556 + \frac{0.432}{\mathrm{Re}^{0.308}} \tag{22}$$

$$1.3\cdot10^4 \leq \mathrm{Re} \leq 1.11\cdot10^5,\ \mathrm{Pr}=8$$

$$\mathrm{Nu} = 0.1576\frac{\xi}{8}\mathrm{Re}^{1.0685}\,\mathrm{Pr} \tag{23}$$

$$1.3\cdot10^4 \leq \mathrm{Re} \leq 1.11\cdot10^5,\ \mathrm{Pr}=8$$

The relationship (20) proposed in this paper is close to the correlation (23) of Allen and Eckert [16](Fig.3). The best agreement is observed between Eq. (4) of Gnielinski [19-20]. The correlations of Dittus-Boelter (1) and Li and Xuan (7) give Nusselt number smaller than the correlation of Allen and Eckert (23). Evaluating the relative difference from the relation

$$e = \frac{\mathrm{Nu}_{A-E} - \mathrm{Nu}}{\mathrm{Nu}_{A\text{-}E}}100 \tag{24}$$

the correlations (20) and (1) were compared with (23) for the following Reynolds numbers: 10,000, 50,000, and 110,000. The symbol $\mathrm{Nu}_{A\text{-}E}$ designates the Nusselt number obtained from the Allen and Eckert relation (23) while Nu is the Nusselt number calculated using Eq. (20) or (1). The relative differences for the Eq. (20) developed in the paper are: 14.2% , 3.9% , and 0.08% , respectively, whereas for the Dittus-Boelter equation (1) the differences e are: 8.4%, 12.5%, and 15.9% .

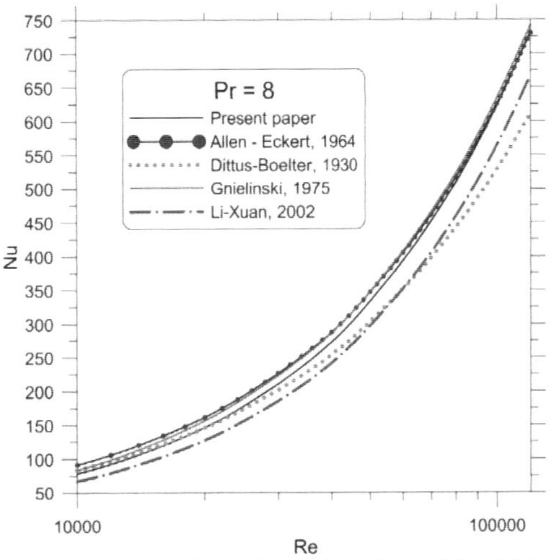

Fig. 3. Comparison of experimental correlation (23) of Allen and Eckert [16] with the correlation (20) proposed in the present paper, the correlation of Dittus-Boelter (1) [10], the Gnielinski relationship (4) [19-20], and the correlation (7) developed by Li and Xuan [25].

The correlation (20) found in the paper closely approximates the experimental equation (23) of Allen and Eckert for Reynolds numbers greater than 20,000 as compared to the Dittus-Boelter equation (1).

Comparisons of the correlations (8) and (20) to experimental data obtained by different researchers confirm quite well accuracy of the proposed relationships, especially when taking into account a variety of test conditions and different ways of experimental data processing.

4 Conclusions

The paper presents three power-type correlations of a simple form, which are valid for Reynolds numbers range from $3\cdot10^3 \leq \mathrm{Re} \leq 10^6$, and for three different ranges of Prandtl number: $0.1 \leq \mathrm{Pr} \leq 1.0$, $1.0 < \mathrm{Pr} \leq 3.0$, , and $3.0 < \mathrm{Pr} \leq 10^3$.

Formulas proposed in the paper have good theoretical basis, as they have been obtained by approximating Nusselt numbers obtained from the solution of the energy conservation equation for turbulent flow in a pipe by the method of least squares. Heat transfer correlations developed in the paper were compared with experimental results available in the literature. The performed comparisons confirm the good accuracy of the proposed correlations, better than the accuracy of the Dittus-Boelter correlation. Heat transfer correlations proposed in the paper can be used in a broader ranges of Reynolds and Prandtl numbers compared with a widely used correlation of Dittus-Boelter. They are also much simpler in comparison to the relationship of Gnielinski, which is also widely used in the heat transfer calculations.

References

1. S. Kakaç, H. Liu, A. Pramuanjaroenkij, *Heat exchangers. Selection, rating, and thermal design. 3rd edition*, CRC Press -Taylor & Francis Group, Boca Raton (2012)
2. S. G. Penoncello, *Thermal energy systems*. CRC Press-Taylor and Francis Group, Boca Raton (2015).
3. D. Taler, *Mathematical modeling and control of plate fin and tube heat. Energy Conversion and Management* **96** (2015), 452-462.
4. M. Trojan, D. Taler, *Thermal simulation of superheaters taking into account the processes occurring on the side of the steam and flue gas*. Fuel **150** (2015), 75-87.
5. D. Taler, K. Kaczmarski, *Mathematical modelling of the transient response of pipeline*. Journal of Thermal Science **25** (2016), 1-9.
6. J. Taler, P. Dzierwa, D. Taler, M. Jaremkiewicz, M. Trojan, *Monitoring of thermal stresses and heating optimization including industrial applications*. Nova Publishers, (New York 2016).
7. P. Dzierwa, *Optimum heating of pressure components of steam boilers with regard to thermal stresses*, Journal of Thermal Stresses **39** (2016), 874-886.
8. P. Dzierwa, M. Trojan, D. Taler, K. Kamińska, J. Taler, *Optimum heating of thick-walled pressure components assuming a quasi-steady state of temperature distribution*, Journal of Thermal Science **25** (2016), 380–388.
9. J. Taler, P. Dzierwa, D. Taler, P. Harchut, *Optimization of the boiler start-up taking into account thermal stresses*, Energy **92** (2015), 160-170.
10. F. W. Dittus, L. M. K. Boelter, *Heat transfer in automobile radiators of the tubular type*. The University of California Publications on Engineering **2** (1930) 443-461, Reprinted in Int. Commun. Heat Mass **12** (1985) 3-22.
11. R. H. S., Winterton, *Where did the Dittus and Boelter equation come from?*. Int. J. Heat Mass Tran. **41** (1998), 809-810.
12. W. H. McAdams, *Heat Transmission, 3rd edn.*, McGraw-Hill, (New York 1954).
13. A. P. Colburn, *A method of correlating forced convectin heat transfer data and a comparison with fluid friction*, Trans. AIChE **29**(1933) 174-210.
14. E. N. Sieder, E. C. Tate, *Heat transfer and pressure drop of liquids in tubes*. Ind. Eng. Chem. **28**(1936) 1429-1436.
15. R. L. Webb, *A critical evaluation of analytical solutions and Reynolds analogy equations for turbulent heat and mass transfer in smooth tubes*. Warme Stoffubertrag. **4** (1971), 197-204.
16. R. W. Allen, E. R. G. Eckert, *Friction and heat-transfer measurements to turbulent pipe flow of water (Pr = 7 and 8) at uniform wall heat flux*. J. Heat Trans-T. ASME **86**(1964), 301-310.
17. F. Kreith, R. M. Manglik, M. S. Bohn, *Principles of heat transfer. 7th edition*, Cengage Learning, (Stamford 2011).
18. D. R. Mirth, S. Ramadhyani, D. C. Hittle, *Thermal performance of chilled-water cooling coils operating at low water velocities*. ASHRAE Transactions, Part 1, **99**(1993), 43-53.
19. V. Gnielinski, *Neue Gleichungen für den Wärme- und den Stoffübergang in turbulent durchströmten Rohren und Kanälen*. Forsch. Ingenieurwes. (Engineering Research) **41** (1975),8-16.
20. V. Gnielinski, *New equations for heat and mass transfer in the turbulent pipe and channel flow*. Int. Chem. Eng. **16** (1976), 359-368.
21. G. K., Filonienko, *Friction factor for turbulent pipe flow*, Teploenergetika *1* (1954), no. **4**, 40-44 (in Russian).
22. B. S. Petukhov, *Heat transfer and friction in turbulent pipe flow with variable physical properties, in Advances in Heat Transfer*, Vol. **6**, (1970), 503-564, Edited by J. P. Hartnett and T. F. Irvine, Academic Press, (New York 1970).
23. S. Kakaç, Y. Yener, A. Pramuanjaroenkij, *Convective heat transfer. 3rd Edition*, CRC Press – Taylor & Francis, Boca Raton (2014).
24. M. Li, T.S. Khan, E. Al-Hajri, Z. H. Ayub, *Single phase heat transfer and pressure drop analysis of a dimpled enhanced tube*. Appl. Therm. Eng. **101** (2016), 38-46.
25. Q. Li, Y. Xuan, *Convective heat transfer and flow characteristics of Cu-water nanofluid*. Sci. China Ser. E: Technol. Sci. **45** (2002),408-416.
26. W. H. Azmi, K. Abdul Hamid, R. Mamat, K.V. Sharma, M. S. Mohamad, *Effects of working temperature on thermo-physical properties and forced convection heat transfer of TiO_2 nanofluids in water-Ethylene glycol mixture*. Appl. Therm. Eng. **106** (2016),1190-1199.
27. E. E. Wilson, *A basis for rational design of heat transfer apparatus*. Trans. ASME, **37** (1915), 47-82.
28. J. W. Rose, *Heat-transfer coefficients, Wilson plots and accuracy of thermal measurements*, Exp. Therm. Fluid Sci. **28** (2004), 77-86.
29. D. Taler, *A new heat transfer correlation for transition and turbulent fluid flow in tubes*. Int. J. Therm. Sci.**108** (2016), 108-122.
30. H. Reichardt, *Vollständige Darstellung der turbulenten Geschwindigkeitsverteilung in glatten Leitungen*, Z. Angew. Math. Mech. **31** (1951) no.**7**, 208-219.
31. H. Reichardt, *The principles of turbulent heat transfer, Transl. by P.A. Scheck in Recent advances in heat and mass transfer*, ed. by J. P. Hartnett, McGraw-Hill, (Boston 1961), 223-252.
32. S. Lau, *Effect of plenum length and diameter on turbulent heat transfer in a downstream tube and on plenum-related pressure loss*, Ph.D. Thesis, University of Minnesota, (1981).

33. A. Black III, The effect of circumferentially-varying boundary conditions on turbulent heat transfer in a tube, Ph.D. Thesis, University of Minnesota, 1966.
34. R. Kemink, *Heat transfer in a downstream tube of a fluid withdrawal branch*, Ph.D. Thesis, University of Minnesota, (1977).
35. D. Wesley, *Heat transfer in pipe downstream of a Tee*, Ph.D. Thesis, University of Minnesota, (1976).
36. J. P. Abraham, E. M. Sparrow, W. J. Minkowycz, *Internal-flow Nusselt numbers for the low-Reynolds-number end of the laminar-to-turbulent transition regime*, Int. J. Heat Mass Tran. **54** (2011), 584-588.
37. S. Eiamsa-ard, P. Promvonge, *Thermal characteristics in round tube fitted with serrated twisted tape*, Appl. Therm. Eng. **30** (2010),1673-1682.

DOI: 10.1051/e3sconf/20171303001

Variability of photovoltaic panels efficiency depending on the value of the angle of their inclination relative to the horizon

Marek Majdak[1,*]

[1]Cracow University of Technology, Institute of Thermal Power Engineering, al. Jana Pawła II 37, 31-864 Cracow, Poland

Abstract. The objective of this paper was to determine the relationship between the efficiency of photovoltaic panels and the value of the angle of their inclination relative to the horizon. For the purpose of experimental research have been done tests on the photovoltaic modules made of monocrystalline, polycrystalline and amorphous silicon. The experiment consisted of measurement of the voltage and current generated by photovoltaic panels at a known value of solar radiation and a specified resistance value determined by using resistor with variable value of resistance and known value of the angle of their inclination relative to the horizon.

Symbols used in formulas

P_{ideal} – power of ideal photovoltaic cell, W
V_{oc} – open circuit voltage, V
I_{sc} – short circuit current, A
FF – fill factor, -
η_{max} – efficiency of photovoltaic cell, -
P_{max} – maximum power of real photovoltaic cell, W
q – power of solar radiation, W
q_r – solar irradiance, W/m^2
A – active area of photovoltaic cell, m^2

1 Principle of operation of photovoltaic cells

Photovoltaic cells produce electricity through direct conversion of solar radiation energy. The conversion of solar radiation is caused by the occurrence of the internal photoelectric effect in the interior of a photovoltaic cell - changes in the electron energy due to absorption of energy of photon beam incident on the solar module [1-4]. The principle of operation of photovoltaic cell is shown in Fig. 1.

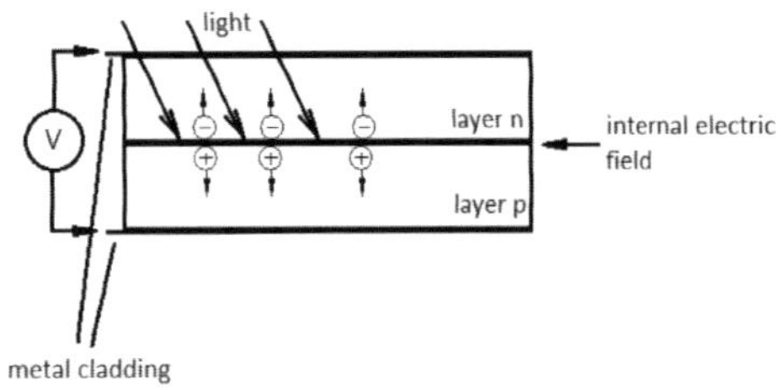

Fig. 1. Scheme and principle of operation of photovoltaic cell.

In PV module is used internal photoelectric effect involving the passage of electrons to the corresponding energy levels and the formation of excess charge carriers inside the solid - in case of silicon cells, this effect occurs within the crystal lattice of silicon at the interface "p"-"n", at the junction of layers with excess of electrons and holes (positive electrical charge carriers). Theory of bands is used to describe the phenomenon of internal photoelectric effect - according to this theory in material are energy levels which are "allowed" and "not allowed" for electrons. The photoelectric effect, described in accordance with the theory of bands, occurs in all semiconductors, but its intensity and wavelength range, which cause radiation, depends on the composition of the semiconductor.

Silicon, of which are made most of currently used photovoltaic cells, is intrinsic semiconductor and has a relatively high value of the conductivity at temperatures close to room temperature. In order to increase the conductivity of the crystal structure of the silicon are added small amounts of the appropriate elements (arsenic, phosphorus, aluminum, gallium). The "p"-type semiconductors, in which occurs a deficiency of electrons, are obtained by the use of acceptor, trivalent, dopant, "n"-type semiconductors, characterized by an excess of electrons, are obtained by applying a donor, pentavalent, dopants. At the interface between "n"-type semiconductor and the "p"-type semiconductor is created "p"-"n" junction [5]. By the diffusion of electrical charges is created barrier layer consisting of charge carriers with opposite sign, with a thickness of several mean free paths of electron. After reaching the "p"-"n" junction by a photon with sufficiently high energy, at least the energy of band gap, is formed in the potential difference - as a result of electron transfer to a segment of "n"-type semiconductor.

* Corresponding author: marek.majdak@mech.pk.edu.pl

If a "p"-"n" junction is connected to the resistor, the circuit is closed and occurs a flow of electric current.

2 Construction of silicon photovoltaic cells

The first monocrystalline and polycrystalline silicon solar cells were produced in thick layer technology - thickness of plates of silicon in modules of this type amounts to 200 ÷ 400 microns. Photovoltaic panel predominantly consists of the 36 or 40 interconnected cells as a single module allows to obtain a voltage of 0.5 ÷ 0.6 volts and a current of about 0.3 amperes. The cells, which are made of monocrystalline silicon, take the shape of a square with bevelled vertices, have uniform color, and their production relies on cutting them from a single crystal of silicon [6-8]. Polycrystalline cells take the shape of squares, and they do not have a uniform color. Silicon wafers in modules made in the thick layer technology are placed between two panes made from tempered glass with a thickness of 3 ÷ 5 mm and a reduced content of iron oxide. In order to create an electrical circuit individual links are joined using silver tape and adhered to the bottom of the glass. The whole structure is surrounded by the aluminum frame and rubber pads are mounted around pane as an additional sealing of the photovoltaic module [9]. Thin film cells are made of amorphous silicon dusted with a thickness of several micrometers onto a substrate made of e.g. glass. During the manufacturing process to the silicon layer are connected electrodes and the whole system is covered with a layer of solar glass with reduced content of iron oxide.

3 Characteristic parameters of photovoltaic cells

Photovoltaic cells are characterized by short circuit current I_{sc}, open circuit voltage V_{oc} and the values of voltage and current during the generation of the maximum power of cell, indicated by symbols V_{max} and I_{max} [10]. The values of these parameters can be read in the approximate manner by using the current-voltage and power-voltage characteristic for the photovoltaic module. The current-voltage and power-voltage characteristics created based on tests prepared for this paper are presented in Fig. 2 and Fig. 3.

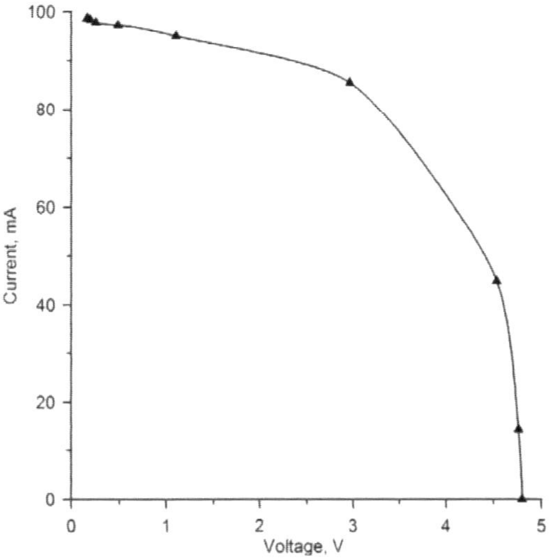

Fig. 2. Current-voltage characteristics of the photovoltaic cell.

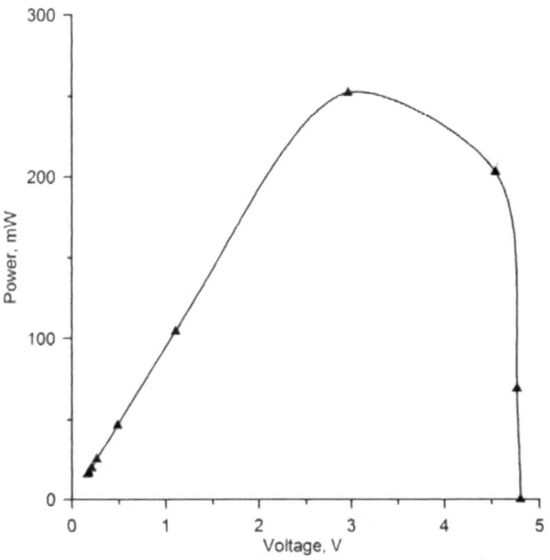

Fig. 3. Power-voltage characteristic of the photovoltaic cell.

Short circuit current I_{sc} is the value of the maximum current that will flow through tight contacts of photovoltaic cell, between which the voltage value is equal to zero. The value of short circuit current is related to the construction of the photovoltaic cell and the material from which it is made. Open circuit voltage V_{oc} is the maximum value of the potential difference that can be produced by a lighted photocell, which is not connected to the load. Maximum power point defines a pair of parameters (current and voltage) for which the real cell generates the maximum possible value of power at given conditions - it is equal to the largest rectangle designated under the current-voltage characteristics for the coordinates (I_{max}, V_{max}). The value of maximum power of real cell is defined by the formula:

$$P_{max}=V_{max}I_{max} \quad (1)$$

DOI: 10.1051/e3sconf/20171303001

and is less than the maximum power of the ideal cell, which is defined by the equation:

$$P_{ideal}=V_{oc}I_{sc}. \quad (2)$$

The relationship between the maximum power of the real cell and the maximum power of the ideal cell determines the fill factor *FF*. The coefficient can be expressed by using the formula:

$$FF=P_{max}/P_{ideal}=(V_{max}I_{max})/(V_{oc}I_{sc}) \quad (3)$$

The fill factor *FF* reaches the greater value, the more shape of the current-voltage characteristic of real photovoltaic cell is similar to the shape of the characteristic of the ideal cell. The value of fill factor of currently used silicon photovoltaic cells amounts in most cases 0.6 to 0.82.

Efficiency of the photovoltaic cell is a very important parameter because it largely depends on it how effectively will be processed solar energy. Efficiency of the module is defined as the ratio of the photovoltaic cell power to the sunlight radiation power operating on active area of the photovoltaic module. It can be expressed by using the formula:

$$\eta_{max}=P_{max}/q \quad (4)$$

4 Measuring system

In order to determine efficiency of tested photovoltaic panels and changes of its value depending on the angle of inclination of modules relative to the horizon has been built measuring system (shown in Fig. 4) allowing the movement of cells and change of angle of their inclination. The research position consists of a rectangular frame with installed photovoltaic panels, the frame is placed on wheels and had two elements allowing to the change of the angle of inclination of the photovoltaic modules. On the frame has been installed three photovoltaic panels: polycrystalline with a power of 250 W_p, amorphous with a power of 120 W_p and monocrystalline with a power of 250 W_p. In order to determine the current-voltage characteristics and power-voltage characteristics of each panel to the position was attached a variable resistance resistor.

Fig. 4. Measuring system.

5 Determination of the efficiency of photovoltaic modules

In order to calculate the efficiency of photovoltaic cells should be determined the ratio of the maximum power generated by the photovoltaic cell on the conditions occurring at the time of measurement to the solar radiation power [11, 12], which can be expressed as the product of the intensity of solar radiation incident on the cell surface and the active surface area:

$$\eta_{max}=(V_{max}I_{max})/(q_rA) \quad (5)$$

Efficiency of the cell is determined for each series of measurements by performing calculations for a known value of solar radiation incident on the active surface of the panel and voltage and current value corresponding to the maximum power achieved in the measurement series.

6 Results

Tests using previously described photovoltaic modules were made in the spring. On the basis of the collected measurements were plotted current-voltage characteristics of examined photovoltaic panels. The characteristics of individual modules were collected in order to compare their charts depending on angle of inclination of panels relative to the horizon. Characteristics obtained for the amorphous photovoltaic cell are shown in Fig. 5, for the monocrystalline photovoltaic cell in Fig. 6 and for the polycrystalline photovoltaic cell in Fig. 7.

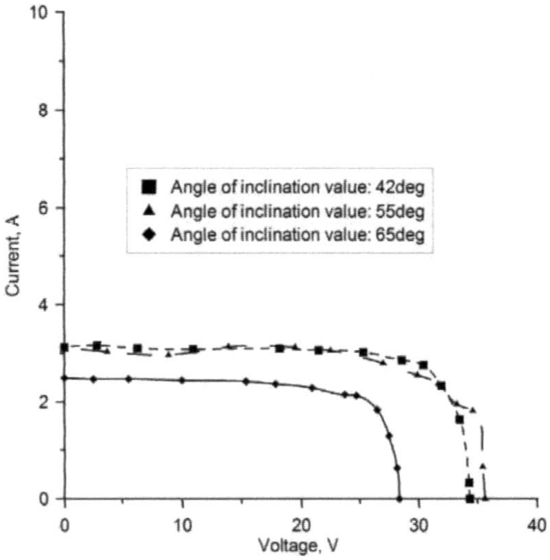

Fig. 5. Current-voltage characteristics obtained for the amorphous photovoltaic cell.

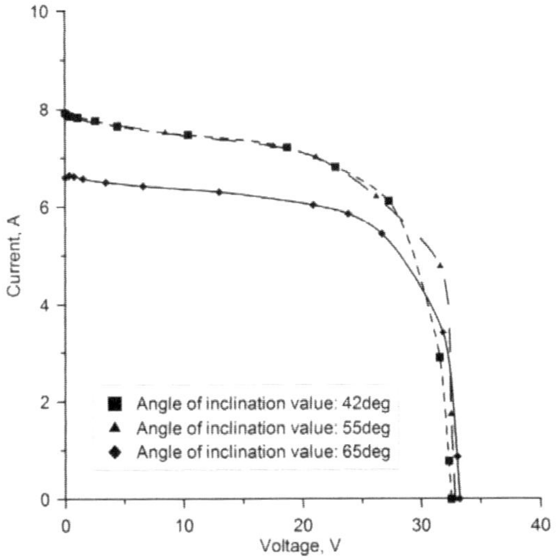

Fig. 6. Current-voltage characteristics obtained for the monocrystalline photovoltaic cell.

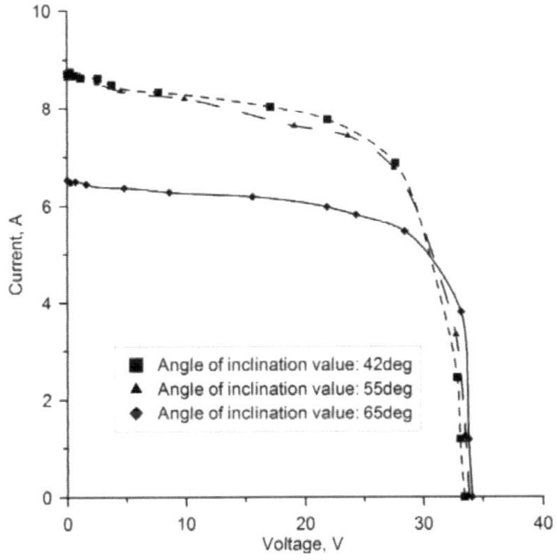

Fig. 7. Current voltage characteristics obtained for the polycrystalline photovoltaic cell.

In order to determine the efficiency of photovoltaic modules for each series of measurements were assigned maximum power points of characteristics, followed by a calculation based on the equation (5). During the calculation were used values of the active surface area of examined photovoltaic modules from their documentation and the average values of solar radiation measured by the pyranometer during each series of measurements. For each data set was also determined filling factor value. The results, with including absolute errors, which takes into account the uncertainty of used sensors, are presented in Table 1.

Table 1. Efficiency values for tested photovoltaic panels and values of fill factor for each measurement series.

	Mono-crystalline PV panel	Poly-crystalline PV panel	Amorphous PV panel
Efficiency of the panel for the angle of inclination of 65,0°	(16,145 ±0,161) %	(15,359 ±0,154) %	(6,287 ±0,063) %
Efficiency of the panel for the angle of inclination of 55,0°	(17,155 ±0,172) %	(16,423 ±0,164) %	(7,803 ±0,078) %
Efficiency of the panel for the angle of inclination of 42,0°	(16,011 ±0,160) %	(16,156 ±0,162) %	(7,488 ±0,075) %
Value of fill factor for the angle of inclination of 65,0°	(66,421 ±1,328) %	(67,563 ±1,351) %	(72,757 ±1,455) %
Value of fill factor for the angle of inclination of 55,0°	(63,428 ±1,268) %	(61,904 ±1,238) %	(67,496 ±1,350) %
Value of fill factor for the angle of inclination of 42,0°	(65,294 ±1,306) %	(62,771 ±1,255) %	(79,026 ±1,580) %

Conclusions

By comparing data collected in table Tab. 1 may be noted that, for studied angles of inclination, selected from the most commonly used in Poland inclination angles of photovoltaic panels, modules achieve maximum efficiency at an angle of inclination of 55°. The obtained value of the angle of inclination of panels relative to the horizon is consistent with the experiences of companies that use and assemble the photovoltaic panels on Polish territory - they prescribe use of inclination angles of modules of 35-45° in summer and 65° in winter. Measurements were made in the spring, so that the obtained value of an optimal angle of inclination of panels which fall within the scope between these values allows to be considered that the outcome of the research is realistic. It should also be noted that changes in the efficiency of each module are only about one order of magnitude greater than the differences resulting from the uncertainty of measurement for used sensors. This is mainly due to the participation of the scattered radiation in the total radiation at about 40-50% and also because the surface structure of the solar panels is adapted to absorb the maximum amount of the radiation regardless of the angle of incidence on the surface of the module. Due to such small dependence between the efficiency of solar panels and angle of inclination relative to the horizon can be bring up within the reasonableness of use, at least in the case of installations of low and medium power, the construction of the variable angle of inclination at the time of day and year, as well as the lag systems, which additionally reduce the net production of electricity through the energy to power the actuators that change the setting of panels and controllers directing the operation of the entire system.

References

1. Sarniak M., *Podstawy fotowoltaiki* (in polish) (2008), **ISBN: 978-83-7207-773-8**
2. Wacławek M., Rodziewicz T., *Ogniwa słoneczne* (in polish) (2014), **ISBN: 978-83-2043-700-3**
3. Kostas V., Platis A., Sapidis N., *Performability Assessment of Standalone Photovoltaic Residential Systems*, International Journal of Performability Engineering 12 (5), 485-500 (2016)
4. Sharma S., Tomar M., Kumar A., Puri N. K., Gupta V., *Enhanced ferroelectric photovoltaic response of $BiFeO_3/BaTiO_3$ multilayered structure*, Journal of Applied Physics **118**, 074103 (2015)
5. Mertens K., *Photovoltaics: fundamentals, technology and practice* (2014), **ISBN: 978-11-1863-416-5**
6. Łotocki H., *ABC systemów fotowoltaicznych sprzężonych z siecią energetyczną* (in polish) (2013), **ISBN: 978-83-62760-28-2**
7. Góralczyk I., Tytko R., *Fotowoltaika: urządzenia, instalacje fotowoltaiczne i elektryczne* (in polish) (2015), **ISBN: 978-83-7490-810-8**
8. Klugmann-Radziemska E., *Fotowoltaika: w teorii i praktyce* (in polish) (2010), **ISBN: 978-83-60233-58-0**
9. X. F. Wang, Y. W. Bao, Y. Qiu, X. G. Liu, *Durability of Photovoltaic Laminated Glass Modules Affected by Temperature-Humidity Cycles,* Key Engineering Materials 680, 323-326 (2016)
10. Scheer R., Schock H.-W., *Chalcogenide photovoltaics: physics, technologies, and thin film devices* (2011), **ISBN: 978-3-527-31459-1**
11. Plantić L.S., *Electrical energy generation with differently oriented photovoltaic modules as façade elements,* Thermal Science 20 (4), 1377-1386 (2016)
12. Mavromatakis F., Franghiadakis Y., Vignola F., *Modeling Photovoltaic Model,* Engineering, Technology & Applied Science Research **6** (5), 1115-1118 (2016)

E3S Web of Conferences 13, 03002 (2017) DOI: 10.1051/e3sconf/20171303002
WTiUE 2016

Application of ORC power station to increase electric power of gas compression ignition engine

Szymon Mocarski[1], Aleksandra Borsukiewicz-Gozdur[1,*]

[1] Zachodniopomorski Uniwersytet Technologiczny w Szczecinie, Katedra Techniki Cieplnej al. Piastów 17, 70-310 Szczecin, Polska

Abstract. The paper presents the calculation results of efficiency of the subcritical low temperature ORC power station powered by waste heat resulting from the process of cooling a stationary compression ignition engine. The source of heat to supply the ORC power station is the heat in a form of water jet cooling the engine at a temperature of 92°C, and the exhaust gas stream at a temperature of 420°C. The study considers three variants of systems with the ORC power stations with different ways of using heat source. The first variant assumes using just engine cooling water to power the ORC station. In the second variant the ORC system is powered solely by a heat flux from the combustion gases by means of an intermediary medium - thermal oil, while the third variant provides the simultaneous management of both heat fluxes to heat the water stream as a source of power supply to the ORC station. The calculations were made for the eight working media belonging both to groups of so-called dry media (R218, R1234yf, R227ea) and wet media (R32, R161, R152a, R134a, R22).

Symbols

cp – average heat capacity [kJ/(kg·K)]
h – specific enthalpy [kJ/kg]
$\dot{m}$ –mass flow of the substance [kg/s]
N – power [kW]
$\dot{Q}$ – heat flux [kW]
s – specific entropy [kJ/(kg·K)]
T, t – temperature [°C]
η – efficiency [%]
x – steam dryness degree [-]

subscripts:
C-R – refers to Clausius Rankine cycle,
c – refers to medium cooling the condenser,
d – refers to supplied stream,
g – refers to exhaust gases,
n – refers to working liquid,
o – refers to thermal oil,
p – refers to pump,
par – refers to turbine inlet steam,
s – refers to water as a heat carrier,
t – refers to turbine,
wyp – refers to output stream,
1, 2, 2s, 3, 4s, 5, 6 – characteristic points of the cycle.

abbreviations in figures:
G – generator,
P – pump,
PWC – intermediary heat exchanger,
S – condenser,
SW – compression ignition engine,
T – turbine,
WP – steam generator,
W1 – heater,
W2 – evaporator,
W3 – superheater,

1 Introduction

Waste energy, due to its low fuel prices, for many years was not considered as a potential source of driving energy for further processes. The reason for this was the lack of concept/technology enabling efficient management of waste energy, or lack of economic justification for the investment in energy management systems. At the moment, significant reversal of this trend can be observable, and the example for this may be the release of relevant legal acts on this issue [1].
Unfortunately, the implementation of low and medium temperature energy conversion processes is usually featured by not very high thermal efficiencies. Waste heat recovery is theoretically possible everywhere where such heat is produced. Effective use of waste energy, and thus, maximizing the overall efficiency of all the processes of energy exploitation, production and conversion, in the age of increasingly higher awareness of the environmental care, depletion of conventional energy resources and financial benefits out of efficient use of energy should be a priority for modern energy and modern industry. For this reason, reducing the consumption of fossil fuels in the production of useful energy is now a serious challenge and is not only a research but also geopolitical goal. This is important due to the overall increase in energy consumption in the world and the resulting increase in emissions to the environment. One of the ways enabling, partially, to achieve this goal is to use the ORC power stations, which in many cases allow to manage low and/or

*Corresponding author: aborsukiewicz@zut.edu.pl

medium temperature energy and effectively reduce or even eliminate a part of the energy loss to the environment. Using this type of energy in some other way could be difficult or impossible. The ORC stations allow for emission-free electricity production using geothermal, solar sources of energy, exhaust waste heat utilization or other medium at an increased temperature. An interesting idea might be the use of the ORC station as a secondary cycle, cooperating with a power station fed from a high temperature source, in order to increase the overall efficiency of the process. Another way to use the ORC station for waste heat management to increase the overall efficiency of the primary process is utilization of waste heat produced during combustion of gas fuel by a stationary compression ignition engine. Such an engine, working with an electric generator, creates a thermal-current set. The ways of using waste heat arising during the operation of the compression ignition engine to increase the efficiency of the electric generator set have been analyzed in papers [2] and [3]. This paper presents the calculation results for the compression ignition engine and the ORC station system while it was assumed that installation (adding a module) of the ORC does not interfere in the construction and operation of the compression ignition engine in any way, which means that it can be installed on already working engines. It was furthermore assumed that the ORC system is a system operating at relatively low temperatures. This assumption stems from the fact that the offer of low-temperature ORC stations is relatively rich at the moment because in systems of this type screw expanders (instead of turbines) may be used [4, 5].

2 Analysis of using waste heat generated by compression ignition engine to power the ORC station

The paper analyzes the possibility of using waste heat generated during the operation of the stationary compression ignition engine burning gaseous fuel. Mechanical power generated by such an engine is used to drive an electric generator in order to produce electricity. The efficiency of generating electrical power by such engines is relatively high (for modern engines up to 45% [1]) however, there is still thermal power in a form of a water jet cooling the engine and a stream of combustion gases emitted as waste heat. The paper analyzes the effectiveness of three variants to use this heat to power the ORC station in order to increase electric efficiency of the primary process carried out by the generator set. The analysis is based on the consideration of a possible use of both of these heat fluxes and determining to what extent the use of the ORC system improves the electric efficiency of the primary process. The analysis was carried out for eight different substances used as working medium in the ORC station, whose basic parameters are presented in Table 1.

The analysis examined the use of the ORC station operation according to the Clausius Rankine comparative subcritical cycle. A diagram of the ORC station, stating the basic elements of the system and the characteristic points of the cycle is shown in Figure 1a.

Table 1. Basic properties of working medium selected for analysis

No.	Medium	Type of medium	Critical temperature	Critical pressure	Saturation pressure at t=30°C
			[°C]	[MPa]	[MPa]
1	R218	Dry	71,87	2,6400	0,99032
2	R32	Wet	78,10	5,7820	1,92750
3	R1234yf	Dry	94,70	3,3822	0,78351
4	R22	Wet	96,15	4,9900	1,19190
5	R134a	Wet	101,06	4,0593	0,77020
6	R227ea	Dry	101,75	2,9250	0,52842
7	R161	Wet	102,15	5,0910	1,05860
8	R152a	Wet	113,26	4,5168	0,68982

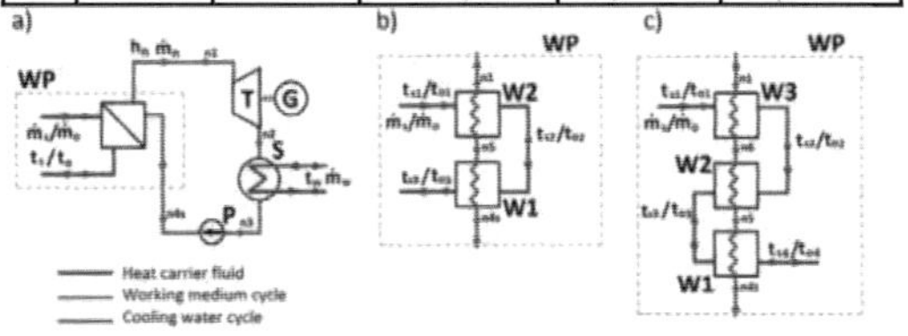

Fig. 1. Diagram of the ORC system a) and: the flow of fluid through steam generator exchangers using a dry medium b) and wet medium c).

A very important aspect of the discussion on the operation effectiveness of the ORC station is the shape of saturation curves of working media in the T-s system presented in Figure 2 because depending on what kind of working medium is examined, the operation of the station changes and results from the various solutions of the steam generator (WP), which entails differences in the way of determining the parameters of the working medium state and within the performed calculations.

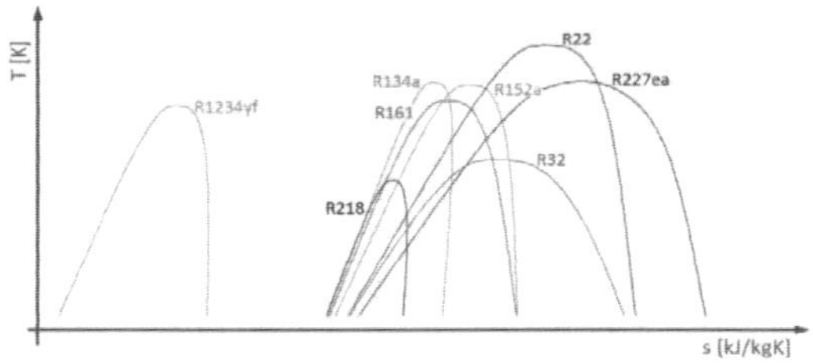

Fig. 2. Shapes of saturation curves of working media taken for analysis.

In case of using the dry medium the steam generator consists of two heat exchangers: the heater (W1) and the evaporator (W2) as shown in Figure 1b. While in the case of using the wet medium the steam generator needs to be supplemented with the third exchanger, the superheater (W3), in which the steam superheating takes place, as shown in Figure 1c. The heat source for the ORC station in this case is the DEUTZ POWERSYSTEMS TBG 620 V12K compression ignition engine adapted to burn gaseous fuel in the form of natural gas. The operating parameters necessary to treat the engine as a heat source for the ORC station are shown in Table 2.

Table 2. Quantities characterizing heat source

Quantity	Unit	Value
Heat flux absorbed with fuel	[kW]	2592
Mechanical power of the engine	[kW]	1050
Mechanical efficiency of the engine	[%]	40,5
Efficiency of the electric generator (by the engine manufacturer)	[%]	96,2
Electric power of the generator set (after taking into account the efficiency of the generator)	[kW]	1010
Electric efficiency of the generator set	[%]	39,0
Thermal efficiency of the engine	[%]	47,2
Temperature of the engine cooling water at the inlet	[°C]	78
Temperature of the engine cooling water at the outlet	[°C]	92
Heat flux carried by cooling water	[kW]	690
Flow rate of the engine cooling water	[m3/h]	36 - 56
Mean specific heat of water	[kJ/kg·K]	4,19
Exhaust gas temperature at the outlet of the collector	[°C]	420
Exhaust gas temperature after having used	[°C]	120
Heat flux carried by exhaust gas	[kW]	532
Flow rate of exhaust gas	[kg/h]	5726
Unusable power (energy loss)	[kW]	320
Mean specific heat of oil heat carrier	[kJ/kgK]	2,70

In case of this engine the thermal power, that is available flux of waste heat possible to manage, is 47,2% of the total power supplied to the engine with the fuel. This power is transferred in an amount of: 690 kW carried along with the engine cooling water and 532 kW carried with the stream of combustion gases. Mechanical power of the engine is 1050 kW, which represents 40,5% of the power supplied from the fuel while the electric efficiency of the primary process is 39,0%. The remaining part of the energy flow supplied with the fuel are the losses of 320 kW. The study contains analysis of three possible variants of using both heat flux to supply the ORC power station in order to increase the total electric power.

Variant I assumes that the source of power supply to the steam generator of the ORC system is only a heat flux of the compression ignition engine cooling water. A method of the heat input from the engine to the ORC station is shown in Figure 3.

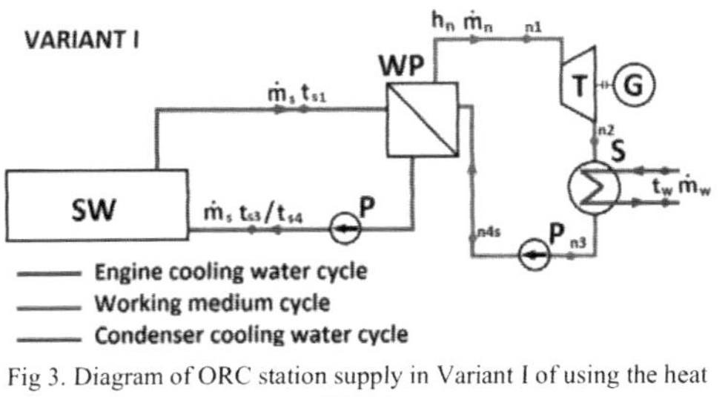

Fig 3. Diagram of ORC station supply in Variant I of using the heat source

This option implies the direct use of cooling water with the temperature of 92°C to power the ORC station without using intermediary heat exchanger and intermediary liquid, and such a design of the power station and heat exchanger WP that water after leaving the steam generator of the station had a temperature of t_{s3}/t_{s4}=78°C, that is the temperature at the inlet to the engine cooling system given be the engine manufacturer. Assuming that the minimum possible value of the temperature difference between the factors in the heat exchangers is at Δt=5K, the steam temperature of the working medium at the inlet of the turbine will amount maximum at t_{par}= 87°C.

All assumed and calculated temperature values of individual characteristic points of the heat input from the heat source to the ORC station for variant I are given in Table 3.

Table 3. Water temperature as heat carrier in individual characteristic points of the heat supply for variant I

Medium	t_{s1}	t_{s2}	t_{s3}	t_{s4}	$\dot{m}_s$
	[°C]	[°C]	[°C]	[°C]	[kg/s]
R218	92,00	84,58	78,00		11,763
R32	92,00	85,97	84,63	78,00	11,763
R1234yf	92,00	84,16	78,00		11,763
R22	92,00	90,17	82,62	78,00	11,763
R134a	92,00	91,32	83,41	78,00	11,763
R227ea	92,00	84,70	78,00		11,763
R161	92,00	90,56	82,19	78,00	11,763
R152a	92,00	91,07	82,16	78,00	11,763

Variant II assumes using exhaust gas energy to power the steam generator of the station through the use of intermediary heat exchanger in which the exhaust stream transfers heat to the intermediary oil circuit supplying the steam generator of the steam power station. A diagram of such a solution is shown in Figure 4.

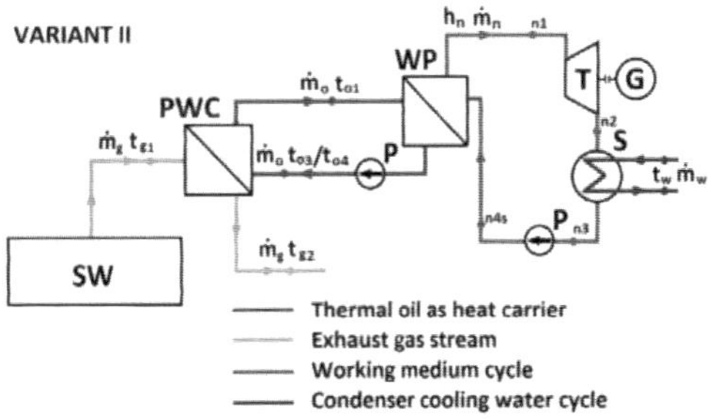

Fig. 4. Diagram of ORC station supply in Variant II with the use of exhaust gas energy as heat source

It was assumed that in order to obtain the heat flux of 532 kW from exhaust gases the exhaust gases must be cooled to a temperature of t_{g2}=120°C. It was also implied that the ORC cycle in variant II is a low-temperature circuit and the temperature of the oil heat carrier at the inlet to the steam generator will amount t_{o1}=115°C. That in turn determines the maximum possible steam temperature value of working medium at the inlet to the turbine at t_{par}=110°C.

All assumed and calculated temperature values of individual characteristic points of the heat input from the heat source to the ORC station for variant II are given in Table 4.

Table 4. Temperatures of oil heat carrier in individual characteristic points of the heat supply system for variant II

Medium	t_{o1}	t_{o2}	t_{o3}	t_{o4}	$\dot{m}_o$
	[°C]	[°C]	[°C]	[°C]	[kg/s]
R218	115,00	95,40	78,00		3,432
R32	115,00	99,06	95,52	78,00	3,432
R1234yf	115,00	95,41	80,00		3,628
R22	115,00	108,04	93,41	78,00	3,432
R134a	115,00	109,79	105,20	88,00	4,703
R227ea	115,00	99,78	78,00		3,432
R161	115,00	111,59	103,84	96,00	6,683
R152a	115,00	112,06	99,11	88,00	4,703

Variant III involves the simultaneous use of both heat fluxes by heating the cooling water jet before referring it to the steam generator of the steam power station using the exhaust stream. A diagram of such a solution is presented in Figure 5.

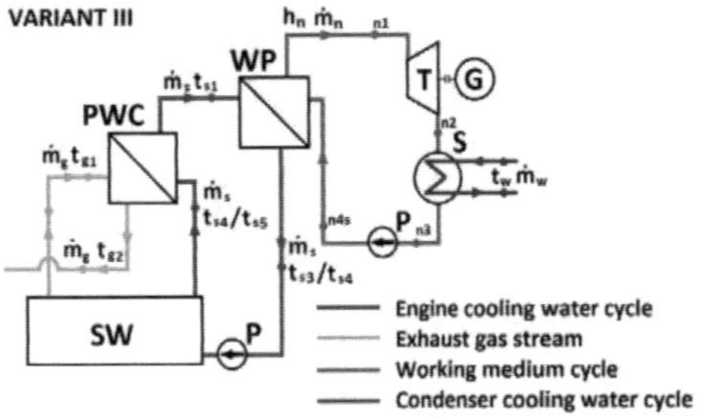

Fig. 5. Diagram of the ORC station supply in Variant III of using the heat source

In case of such a solution the maximum value of the evaporation temperature of working medium also depends on the temperature of the heat carrier at the inlet of the steam generator t_{s1}, however here the value of the temperature should be determined on the basis of the energy balance equations of components included in the circuit of engine cooling water, which is a heat carrier from the heat source to the steam generator of the steam power station, too. In case of this engine and the parameters given by the engine manufacturer, which are contained in Table 1, the water temperature at the inlet of the steam generator is t_{s1} = 102,8°C, thus, taking into account the condition that the minimum temperature difference between the factors in the heat exchangers is Δt=5K, the maximum obtainable steam temperature of working medium at the inlet to the turbine in this variant amounts t_{par} = 97,8°C.

All assumed and calculated temperature values of individual characteristic points of the heat input from the heat source to the ORC station for variant III are given in Table 5.

Tables 3 – 5 also give the values of the mass flow of the heat carrier in the heat supply circuit, depending on the variant of a heat source because in the case of variants I and III it is important to determine the mass flow of the engine cooling water which is at the same time the energy carrier in the circuit $\dot{m}_s$, as the size of this flux must be within the size range specified by the engine manufacturer, which is 36 - 56m^3/h .

Table 5. Temperatures of water as the heat carrier in individual characteristic points of the heat supply system for variant III

Medium	t_{s1}	t_{s2}	t_{s3}	t_{s4}	$\dot{m}_s$
	[°C]	[°C]	[°C]	[°C]	[kg/s]
R218	102,80	89,66	78,00		11,760
R32	102,80	92,12	89,74	78,00	11,760
R1234yf	102,80	88,92	78,00		11,760
R22	102,80	98,94	87,19	78,00	11,760
R134a	102,80	101,27	88,67	78,00	11,760
R227ea	102,80	92,01	78,00		11,760
R161	102,80	99,88	86,11	78,00	11,760
R152a	102,80	100,95	86,01	78,00	11,760

Calculation methodology is based on the energy balance equation of individual elements of the station scheme and the power system, on the assumption that the energy losses during heat transfer are negligible. The energy balance equations use the specific enthalpy values of working medium in the individual characteristic points of the system. The values of specific enthalpy were determined through the database of thermodynamic properties of the working media, REFPROP version 9.0 [6], taking into account the characteristics of the Clausius Rankine cycle and dependencies between the key parameters of the factor state such as temperature, pressure, specific enthalpy and specific entropy, resulting from the specificity of the Clausius Rankine cycle.

The course of thermodynamic processes of the Clausius Rankine cycle, according to which the parameters of working medium in the individual points of the station have been designated, is shown schematically in Figure 6a) in the case of using wet medium and 6b) in the case of dry medium.

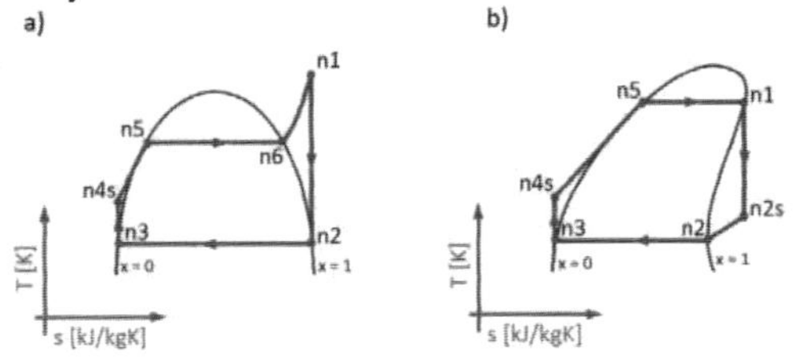

Fig. 6. Thermodynamic processes of Clausius Rankine subcritical cycle using wet medium a) and dry medium b)

Specific enthalpy values for individual variants of using heat source with corresponding temperature values are given in the table 6.

It should be noted that Table 6 shows only the specific enthalpy values at a steam temperature at the inlet to the turbine tn1, at which it was possible to obtain the most advantageous operating parameters of the ORC station for the particular working medium.

Table 6. Results of specific enthalpy values and temperatures in characteristic points of the ORC cycle

Medium	h_{n1} [kJ/kgK]	t_{n1} [°C]	h_{n2} [kJ/kgK]	t_{n2} [°C]	h_{n2s} [kJ/kgK]	t_{n2s} [°C]	h_{n3} [kJ/kgK]	t_{n3} [°C]	h_{n4s} [kJ/kgK]	t_{n4s} [°C]	h_{n5} [kJ/kgK]	t_{n5} [°C]	h_{n6} [kJ/kgK]	t_{n6} [°C]
VARIANT I														
R218	316,96	60,00	306,53		310,29	33,77	233,84		234,64	31,02	273,35	60,00		
R32	557,19	108,83	515,72				255,32		259,39	33,00	400,38	78,00	428,90	78,00
R1234yf	235,71	75,00	218,33		217,93	30,00	77,09		78,47	31,07	147,69	75,00		
R22	439,94	95,09	414,26	30,00			236,62	30,00	238,56	31,60	304,96	77,00	413,62	77,00
R134a	438,51	83,48	414,82				241,72		243,19	31,03	318,63	78,00	428,98	78,00
R227ea	367,97	79,00	343,35		352,71	40,25	234,64		235,57	30,76	298,94	79,00		
R161	638,80	89,89	591,91				263,23		266,12	31,10	377,72	77,00	600,49	77,00
R152a	563,95	88,13	525,96				252,80		254,50	30,84	346,40	77,00	543,32	77,00
VARIANT II														
R218	316,96	60,00	306,53		310,29	33,77	233,84		234,64	31,02	273,35	60,00		
R32	557,19	108,83	515,72				255,32		259,39	33,00	400,38	78,00	428,90	78,00
R1234yf	235,71	75,00	218,33		217,93	30,00	77,09		78,47	31,07	147,69	75,00		
R22	444,73	108,76	414,26	30,00			236,62	30,00	239,19	32,15	324,77	87,00	406,07	87,00
R134a	446,69	109,83	414,82				241,72		244,40	31,85	373,30	100,00	407,68	100,00
R227ea	369,75	89,00	343,35		352,36	39,87	234,64		235,88	31,01	314,70	89,00		
R161	651,64	109,15	591,91				263,23		267,62	31,64	426,03	92,00	582,75	92,00
R152a	575,23	108,93	525,96				252,80		255,53	31,34	387,07	94,00	540,47	94,00
VARIANT III														
R218	316,96	60,00	306,53		310,29	33,77	233,84		234,64	31,02	273,35	60,00		
R32	557,19	108,83	515,72				255,32		259,39	33,00	400,38	78,00	428,90	78,00
R1234yf	235,71	75,00	218,33		217,93	30,00	77,09		78,47	31,07	147,69	75,00		
R22	442,36	101,93	414,26	30,00			236,62	30,00	238,86	31,88	314,27	82,00	410,68	82,00
R134a	440,49	89,35	414,82				241,72		243,43	31,20	328,21	83,00	428,30	83,00
R227ea	369,62	87,00	343,35		352,61	40,14	234,64		235,82	30,96	311,38	87,00		
R161	642,30	94,99	591,91				263,23		266,48	31,23	389,35	81,00	598,00	81,00
R152a	566,71	93,01	525,96				252,80		254,72	30,95	355,43	81,00	543,42	81,00

3 Analysis of the calculation results

Table 7 below shows the calculation results of the basic operating parameters of the ORC station, depending on the working medium and a heat source variant. As can be seen best results, regardless of the variant of a heat source used, were obtained for R227ea factor.

Table 7. Calculation results of the ORC station parameters depending on working medium and the heat source variant

Medium	$\dot{m}_n$ [kg/s]	$\dot{Q}_{W1}$ [kW]	$\dot{Q}_{W2}$ [kW]	$\dot{Q}_{W3}$ [kW]	$\dot{Q}_{wyp}$ [kW]	N_t [kW]	N_p [kW]	N_{C-R} [kW]	η [%]
VARIANT I									
R218	8,38	324,46	365,54		609,28	87,42	6,71	80,72	11,70
R32	2,32	326,67	66,08	297,25	603,34	96,09	9,43	86,66	12,56
R1234yf	4,39	303,74	386,26		619,80	76,27	6,07	70,20	10,17
R22	3,43	227,51	372,31	90,18	608,66	87,99	6,65	81,34	11,79
R134a	3,53	266,50	389,83	33,67	611,50	83,69	5,19	78,50	11,38
R227ea	5,21	330,25	359,75		566,54	128,31	4,85	123,46	17,89
R161	1,85	206,62	412,45	70,93	608,54	86,81	5,35	81,46	11,81
R152a	2,23	204,92	439,08	46,00	609,08	84,71	3,79	80,92	11,73
VARIANT II									
R218	6,46	250,17	281,83		469,77	67,40	5,17	62,23	11,70
R32	1,79	251,87	50,95	229,18	465,19	74,08	7,27	66,81	12,56
R1234yf	3,38	234,19	297,81		477,88	58,80	4,68	54,12	10,17
R22	2,59	221,51	210,43	100,06	459,79	78,87	6,65	72,21	13,57
R134a	2,63	338,99	90,42	102,59	455,23	83,81	7,05	76,77	14,43
R227ea	3,97	313,23	218,77		432,01	104,91	4,93	99,99	18,79
R161	1,39	219,45	217,11	95,44	455,34	82,75	6,08	76,66	14,41
R152a	1,66	218,89	255,27	57,84	454,55	81,99	4,54	77,45	14,56
VARIANT III									
R218	14,84	574,63	647,37		1079,05	154,83	11,88	142,95	11,70
R32	4,10	578,54	117,03	526,43	1068,53	170,17	16,70	153,47	12,56
R1234yf	7,77	537,93	684,07		1097,68	135,07	10,75	124,32	10,17
R22	6,00	452,83	578,93	190,24	1066,71	168,74	13,45	155,29	12,71
R134a	6,20	525,73	620,67	75,59	1073,42	159,18	10,60	148,58	12,16
R227ea	9,13	690,09	531,91		992,85	239,92	10,78	229,15	18,75
R161	3,25	399,52	678,44	144,04	1068,72	163,85	10,57	153,28	12,54
R152a	3,92	394,46	736,32	91,22	1069,91	159,61	7,52	152,09	12,45

 DOI: 10.1051/e3sconf/20171303002

Figure 7 shows a comparison of the ORC capacity, depending on the working medium used and the variant of the station power supply. As can be seen, the best results were obtained for the R227ea medium and the best variant of using a heat source is variant III, that is as a result of supplying the steam generator of the ORC station with the engine cooling water further heated with exhaust gas energy.

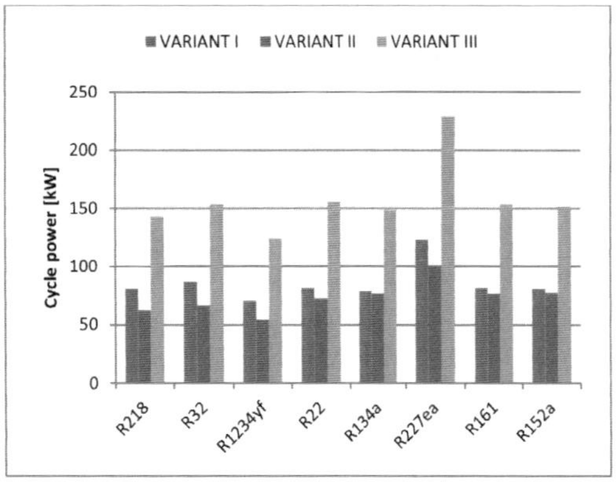

Fig. 7. Comparison of the ORC power system according to the variant of heat source and working medium used

Analyzing Figure 7 it can also be noted that the lowest power values of the station were obtained in variant II which is connected with the fact that the heat flux fed with the exhaust gas stream was much lower than in the case of variants I and III.

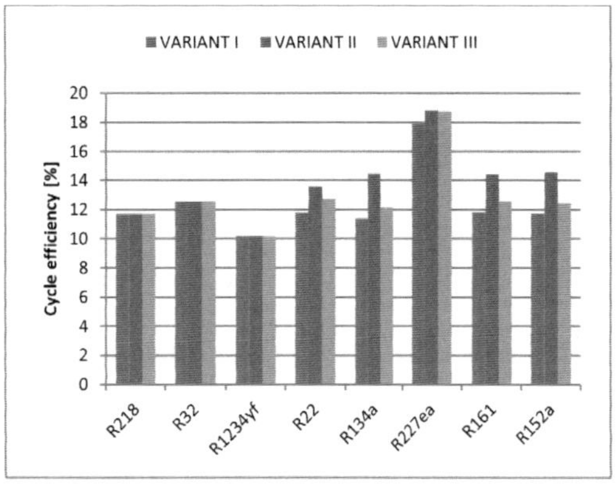

Fig. 8. Efficiency comparison of the ORC system according to the variant of heat source and working medium used

Figure 8 shows the comparison of efficiencies achieved by the ORC station, depending on the working medium and the variant of the station power supply.

In case of efficiency the highest values were obtained in variant II since in this variant it was possible to obtain the highest steam temperature of the organic medium at the inlet to the turbine.

Figure 9 shows a graph comparing the power of the ORC station in variant III and the total value of the station power of the other two variants. The graph shows that the common use of both waste heat streams produced in the engine to power the ORC station is in the case of three factors more advantageous or slightly lower for the other five working fluids than separable management of waste heat streams from the compression ignition engine in two separate ORC systems.

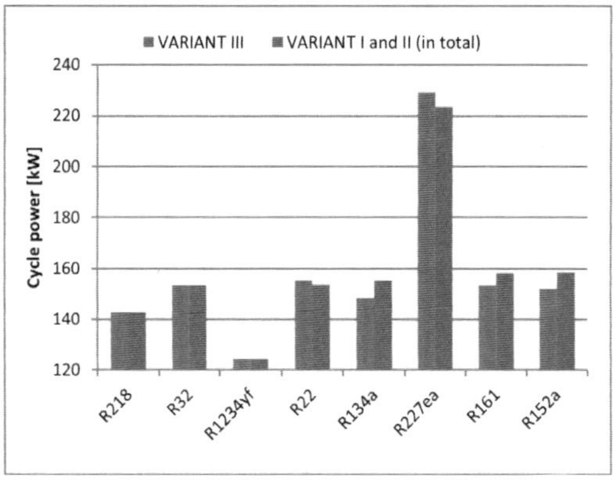

Fig. 9. Comparison of the ORC system power in variant III and power sums of the systems from variants I and II according to the working medium used. \

Table 8 presents a summary on how the use of the ORC station to utilize the waste heat, which arises in the stationary compression ignition engine, affects the overall efficiency increase of the electric energy generation process by the generator set.

The highest power, as shown in Table 7, was obtained for R227ea medium in variant III, that is why the biggest increase in the overall efficiency of the generator set was noted exactly in the case of the station working with this medium, in variant III, and amounts 8,81%, which reflects 22,58 % increase of efficiency.

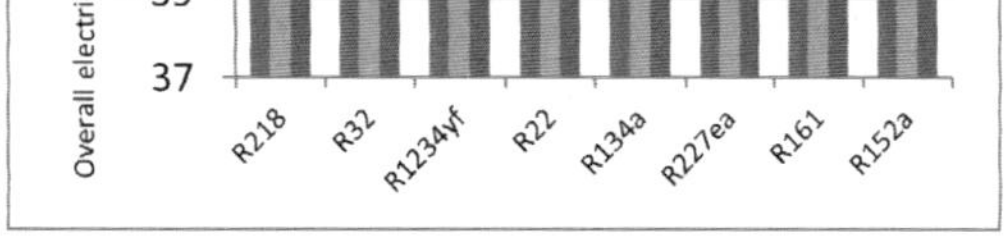

Fig. 10. Comparison of the generator set efficiency after using the ORC system with the primary efficiency

Figure 10 shows comparison of the overall electric efficiency of the generator set after using waste heat recovery through the ORC station, by anticipated variants of using this heat, and the primary efficiency (without using heat recovery) which in the case of this generator set amounts 39,0%.

Table 8. Efficiency increase of the generator set after the ORC system use for waste heat utilization.

electrical efficiency of the generator plant							
before		after					
power	efficiency	working fluid	ORC power	power in total	efficiency	efficiency increase	
[kW]	[%]		[kW]	[kW]	[%]	[point %]	[%]
VARIANT I							
1010	39	R218	80,72	1090,72	42,08	3,08	7,90
		R32	86,66	1096,66	42,31	3,31	8,49
		R1234yf	70,20	1080,20	41,67	2,67	6,86
		R22	81,34	1091,34	42,10	3,10	7,96
		R134a	78,50	1088,50	41,99	2,99	7,68
		R227ea	123,46	1133,46	43,73	4,73	12,13
		R161	81,46	1091,46	42,11	3,11	7,97
		R152a	80,92	1090,92	42,09	3,09	7,92
VARIANT II							
1010	39	R218	62,23	1072,23	41,37	2,37	6,07
		R32	66,81	1076,81	41,54	2,54	6,52
		R1234yf	54,12	1064,12	41,05	2,05	5,27
		R22	72,21	1082,21	41,75	2,75	7,06
		R134a	76,77	1086,77	41,93	2,93	7,51
		R227ea	99,99	1109,99	42,82	3,82	9,80
		R161	76,66	1086,66	41,92	2,92	7,50
		R152a	77,45	1087,45	41,95	2,95	7,57
VARIANT III							
1010	39	R218	142,95	1152,95	44,48	5,48	14,05
		R32	153,47	1163,47	44,89	5,89	15,09
		R1234yf	124,32	1134,32	43,76	4,76	12,21
		R22	155,29	1165,29	44,96	5,96	15,27
		R134a	148,58	1158,58	44,70	5,70	14,61
		R227ea	229,15	1239,15	47,81	8,81	22,58
		R161	153,28	1163,28	44,88	5,88	15,08
		R152a	152,09	1162,09	44,83	5,83	14,96

Conclusion

After the analysis presented in the work the following conclusions have been formulated:
recovery of waste heat through the ORC station can improve the electric efficiency of the generator set based on a compression ignition engine,
the best energy results are obtained by simultaneous use of both available waste heat streams which arise in a compression ignition engine, in a form of the engine cooling water jet and exhaust gas stream,
an important aspect of the effective use of waste heat through the ORC station is the proper selection of working medium. The most advantageous results were obtained for the ORC system with the use of R227ea working medium.

References

1. Decree of the Ministry of Economy dated 10 August 2012 on the specific scope and method of preparation of the energy efficiency audit, a card design of the energy efficiency audit and the methods for energy savings calculations.
2. A. Borsukiewicz-Gozdur, W. Nowak, Increasing of electricity generation capacity of biogas power generator by application of sub- and supercritical modules of Organic Rankine Cycle. Archives of Thermodynamics, Vol. 30(2009), No. 4, pp. 175-188.
3. R. Kaczmarek, Wykorzystanie ciepła powstającego w kogeneracji w silniku gazowym do zwiększenia produkcji energii elektrycznej, Energetyka 2013, nr 9, s. 664-668.
4. F. Ayachi, E. B. Ksayer, P. Neveu, A. Zoughaib, Experimental investigation and modeling of a hermetic scroll expander, Applied Energy, Vol. 181, 2016, pp. 256–267.
5. E. Galloni, G. Fontana, S. Staccone, Design and experimental analysis of a mini ORC (organic Rankine cycle) power plant based on R245fa working fluid, Energy 90, pp. 768-775.
6. E. W. Lemmon, M. L. Huber, and M. O. McLinden, Refprop, NIST Standard Reference Database 23, ver. 9.1, National Institute of Standards and Technology, USA.

Analysis and selection criteria of photovoltaic panels for DHW

Artur Nemś[1,*] and *Magdalena* Nemś[1]

[1] Wroclaw University of Science and Technology, Faculty of Mechanical and Power Engineering, 50-370 Wroclaw, Poland

Abstract. The article discusses the issue of selecting power of photovoltaic panels for heating domestic hot water (DHW). It is indicated that the price of installation is several times lower than a couple of years ago. The authors present a method of selecting system power depending on DHW installation's heating requirements and the building's energy consumption. This algorithm allows for central heating requirements that are crucial in case of heating a building with solid fuel appliances. The described selection method uses heating power details obtained while drawing up Energy Performance Certificates, or the so-called heating degree-days, which enable to precisely determine the heating requirements in individual days of the year and select appropriate power of PV installation for the building's energy performance. In addition, the authors have calculated the heating losses occurring in different elements of installation and showed their influence on the system's power selection. The last part discusses the possibility of integrating PV installation used for heating DHW with the building's central heating system, in order to use the energy surplus.

1 Nomenclature

A – area of a photovoltaic panel, m^2
c_w – specific heat of water, c_w = 4.19 kJ/ kg·K
D_{ext} – external diameter of the tank, m
d_{int} – internal diameter of the tank, m
E_p – amount of primary energy, kWh/ m^2·year
Gr – Grashof number
I – current, A
$I_{S\_45^\circ}$ – solar radiation at the angle of 45° from the southern direction, W/m^2
k_t – corrective multiplier for hot domestic water temperature, -
L – electrical cable length, m
L_i – number of reference units
n – number of photovoltaic panels
Nu – Nusselt number
Pr – Prandtl number
Q_{cab} – heat lost in electrical cables, MJ
$\overline{Q_{ch}}$ – average annual heat demand for central heating, MJ/day
Q_{ch_i} – average daily heat demand for central heating, MJ/day
Q_{dhw} – heat demand, MJ/day
Q_{loss} – heat losses of the tank's surface, MJ/day
Q_{PV} – amount of heat obtained from photovoltaic panels, MJ/day
$\dot{Q}_{bot}$ – heat losses of the tank's bottom surface, W
$\dot{Q}_{cyl}$ – heat losses of the tank's side surface, W
$\dot{q}_l$ – stream of electrical cable heat loss, W/m
$\dot{Q}_{loss}$ – heat losses of the tank's surface, W
$\dot{Q}_{top}$ – heat losses of the tank's top surface, W
R_{el} – cable resistance, Ω
s – cable cross-section, mm^2
T' – temperature of the tank's internal wall, K
T'' – temperature of the tank's external wall, K
T_1 – temperature of water in the tank, K
T_2 – temperature outside the tank (ambient), K
t_{cw} – temperature of hot water in the faucet, ^{0}C
t_d – daily average ambient temperature, ^{0}C
t_{ow} – temperature of cold water, ^{0}C
t_r – annual average ambient temperature, ^{0}C
t_w – average temperature inside a building, ^{0}C
V_{dhw} – unit usage of domestic hot water, dm^3/day
V_w – unit usage of water, dm^3/day
α_1 – heat transfer coefficient on the inside of the tank, W/ m^2·K
α_2 – heat transfer coefficient on the outside of the tank, W/ m^2·K
δ_{ins} – thickness of the tank's insulation, m
η_o – proportion of hot water in total water demand per person
η_{PV} – efficiency of photovoltaic panels
λ_{ins} – heat transfer coefficient of the tank's insulation, W/ m·K
ν_1 – kinematic viscosity of water, m^2/s
ν_2 – kinematic viscosity of air, m^2/s
ρ – cable resistivity, Ω·mm^2/m
ρ_w – water density, g/m^3
τ – time, s
τ_{us} – usage time, day

* Corresponding author: artur.nems@pwr.edu.pl

DOI: 10.1051/e3sconf/20171303003

2 Introduction

There is a conviction in Poland that the best solar system allowing to prepare domestic hot water DHW are solar collectors. However, the constantly changing technology of photovoltaic panel production makes it necessary to consider, whether such a system could be a good alternative for collectors. As given in [1] the price of 3 photovoltaic cells of total power of 690W in 2010 was PLN 9,000. Now, a 250W panel costs approximately PLN 500-800. What is more, it should be noted that their efficiency has increased, albeit slightly.

Both the installation of solar collectors and photovoltaic panels have advantages and disadvantages. However, one of the differences between them is that the set of photovoltaic panels heating water is less complicated. As shown in Figure 1, this system consists only of a set of panels, electrical cables, a control with temperature sensor and an electrical heater. Heater regulator, determining the maximum power point (MPP) can be included, yet, it is not indispensable.

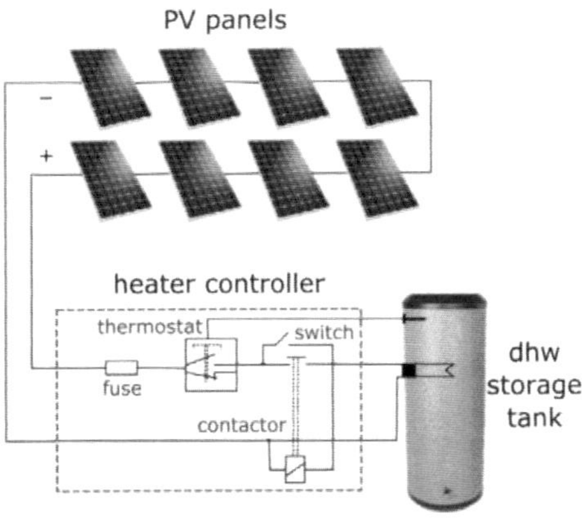

Figure 1. A simplified scheme of PV system used for heating DHW

3 Domestic hot water demand

It is of fundamental importance to determine the amount of energy required to prepare hot water when selecting the power of the heating installation. Currently, there are a few methods of determining the demand for DHW. One of them is described in [2] or in the Regulation on drawing up Energy Performance Certificates [3]. The formula determining the value is described with equation (1).

$$Q_{dhw} = V_{dhw} \cdot L_i \cdot c_w \cdot \rho_w \cdot (t_{cw} - t_{ow}) \cdot k_t \cdot \cdot \frac{\tau_{us}}{1000 \cdot 3600}, MJ/day \quad (1)$$

According to the regulation, it is assumed that the demand for hot water per one inhabitant of a detached house is 35dm^3/day.

A slightly different methodology of calculating the demand is presented in EN 15316-3 standard [4] described by relationship (2).

$$Q_{dhw} = \frac{V_w \cdot L_i \cdot c_w \cdot \rho_w \cdot (t_{cw} - t_{ow}) \cdot \eta_o}{10^6}, MJ/day \quad (2)$$

It does not take account of the usage time nor the coefficient resulting from the temperature of domestic hot water. However, it uses the information about total water demand. Such a solution allows for individual determination of needs, depending on the water used by a single person. In case of selecting installation for a building that is being built, it is not possible to determine the demand on the basis of water consumption data read from the counter, hence - average values can be adopted, being approximately 100 dm^3/person·day for single family houses. For such values, it is assumed that the proportion of hot water in total demand for water is 60÷70%. Assuming that there are 4 people living in the house, each using 70 dm^3/day, we get a daily demand on the level of 41.06 MJ. For simplification reasons it is assumed that water in the waterworks has a fixed annual temperature $10°C$, and it is heated up to 45°C , so the demand is fixed throughout the whole year.

As it results from the above discussion, a minimal volume of accumulative tank should be 280 dm^3, so that is possible to accumulate the whole necessary energy and distribute it at any time during the day. Such a DHW tank is characterised with heat losses that need to be added to the required demand for water used to prepare DHW. When selecting the tank, one can use the product's manufacturer data sheet and read the value of heat losses. However, the authors' experience shows that these values differ from those calculated while determining heat transfer. Thus, the authors suggest determining the value of heat losses related to accumulation in the tank, which is not perfect, with the use of heat transfer equations resulting from Fourier's and Newton's laws. They assume, that the transfer of heat from the tank to the surroundings is done by way of convection in space that is unlimited inside and outside, as well as by heat conduction through insulation. Calculations are done for the determined one-dimension heat transfer, i.e.:

- temperature field does not change in time, it only depends on the location,
- temperature changes in one direction in space.

The following are not taken into account:

- heat transfer by way of radiation due to relatively small difference of temperatures,
- heat conduction through the tank's construction due to the fact that heat resistance of steel is more than 1,000 times smaller than that of polyurethane foam, which constitutes insulation.

Due to existing type series, it is assumed that the tank's volume is 300 dm^3, other assumptions are presented in Table 1.

Table 1. Data adopted for heat calculations

T_1, K	T_2, K	Insulating material	δ, m	λ, W/mK
328	2933	polyurethane foam	0.1	0.023

Water and air characteristics depending on temperature are shown in Table 2.

DOI: 10.1051/e3sconf/20171303003

Table 2. Values of water and air characteristics required for calculations of heat losses

Pr_1, -	3.3
Pr_2, -	0.7
ν_1, m²/s	$0.52 \cdot 10^{-6}$
ν_2, m²/s	$15.06 \cdot 10^{-6}$
λ_1, W/m·K	0.654
λ_2, W/m·K	0.026
β, 1/K	$4.8 \cdot 10^{-4}$

The tanks' covers are treated like two flat walls, whereas, if a horizontal surface transfers heat upwards, then the calculated value of heat transfer coefficient α has been increased by 30%, and in case of a cover transferring heat downwards, decreased by 30%; whereas side wall was calculated on the basis of equations dedicated for cylindrical surfaces.

In order to obtain the final value of heat losses, we should sum up heat streams related to flat surfaces and cylindrical surface, indirectly using equations and relationships convergent with heat conduction and convective heat transfer, being the process of simultaneous transfer and penetration [5]. On the basis of the following equations, a system with eleven unknowns was written and solved in an iterative way. Conduction and transfer of heat through a single-layer flat partition is shown in Figure 2 and heat streams are described with equations (3÷5).

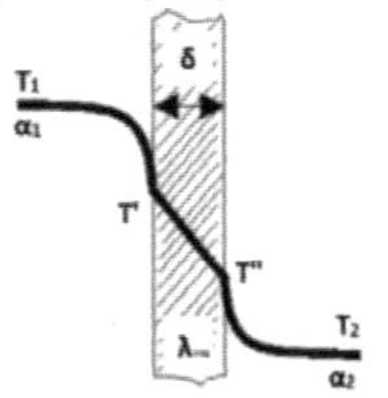

Figure 2. Conduction and transfer of heat through the tank's wall

$$\dot{Q}_{top} = (T_1 - T') \cdot \alpha_1 \cdot \left(\pi \cdot \frac{d_{int}^2}{4}\right) \quad (3)$$

$$\dot{Q}_{top} = \frac{(T' - T'')}{\frac{\delta_{ins}}{\lambda_{ins}}} \cdot \pi \cdot \frac{\left(\frac{d_{int}+D_{ext}}{2}\right)^2}{4} \quad (4)$$

$$\dot{Q}_{top} = (T'' - T_2) \cdot \alpha_2 \cdot \left(\pi \cdot \frac{D_{ext}^2}{4}\right) \quad (5)$$

Heat transfer coefficients α have been determined according to equations (6) and (7), whereas for convective heat transfer α depends on Nusselt number described with equations (9) and (10).

$$\alpha_{1kp} = 1.3 \cdot \frac{Nu_1 \cdot \lambda_1}{d_{int}} \quad (6)$$

$$\alpha_{2kp} = 1.3 \cdot \frac{Nu_2 \cdot \lambda_2}{D_{ext}} \quad (7)$$

$$Nu_1 = C \cdot (Gr_1 \cdot Pr_1)^n \quad (9)$$

$$Nu_2 = C \cdot (Gr_2 \cdot Pr_2)^n \quad (10)$$

Constant C and n in equations (9) and (10) depend on the scope of Gr·Pr product in the following way [5]:

for $(Gr \cdot Pr) \leq 10^{-3}$	C = 0.5	n = 0
$10^{-3} \leq (Gr \cdot Pr) \leq 500$	C = 1.18	n = 0.125
$500 \leq (Gr \cdot Pr) \leq 2 \cdot 10^7$	C = 0.54	n = 0.250
$2 \cdot 10^7 \leq (Gr \cdot Pr) \leq 1 \cdot 10^{13}$	C = 0.135	n = 0.333

In Gr·Pr product the first component - Prandtl number was read from the tables, whereas Grashof number was calculated on the basis of equation (11) and (12).

$$Gr_1 = \frac{g \cdot \beta_1 \cdot (T_1 - T') \cdot d_{int}^3}{\nu_1^2} \quad (11)$$

$$Gr_2 = \frac{g \cdot \frac{1}{\frac{T''+T_2}{2}} \cdot (T'' - T_2) \cdot D_{ext}^3}{\nu_2^2} \quad (12)$$

An almost identical system as the one above was solved for the bottom cover, the difference being only in the multiplier of equations (6) and (7), with number 0.7 instead of 1.3. Whereas for the side cylindrical wall, heat transfer is described with different equations (13÷15).

$$\dot{Q}_{cyl} = (T_1 - T') \cdot (\pi \cdot d_{int} \cdot \alpha_{1kp}) \cdot (L - 2\delta_{ins}) \quad (13)$$

$$\dot{Q}_{cyl} = (T' - T'') \cdot \frac{\pi \cdot 2 \cdot \lambda_{iz}}{ln\left(\frac{D_{ext}}{d_{int}}\right)} \cdot (L - \delta_{ins}) \quad (14)$$

$$\dot{Q}_{cyl} = (T'' - T_2) \cdot (\pi \cdot D_{ext} \cdot \alpha_{2kp}) \cdot L \quad (15)$$

The remaining equations are correlative with equations for the flat wall, however the difference is the characteristic dimension - diameter in case of the cover and height L in case of the cylindrical wall. Aggregate results for the systems of equations are shown in Table 3.

Table 3. Table of results for losses in DHW tank

Side wall - cylindrical						
T'	T''	Q_{cyl}	α_1	α_2	Nu_1	Nu_2
K	K	W	$\frac{W}{m^2 \cdot K}$	$\frac{W}{m^2 \cdot K}$	-	-
327.9	295.6	17.5	139.4	2.3	226.5	109.9
Top cover						
T'	T''	Q_{top}	α_1	α_2	Nu_1	Nu_2
K	K	W	$\frac{W}{m^2 \cdot K}$	$\frac{W}{m^2 \cdot K}$	-	-
327.9	294.8	2.5	174.4	3.1	112.8	69.5
Bottom cover						
T'	T''	Q_{bot}	α_1	α_2	Nu_1	Nu_2
K	K	W	$\frac{W}{m^2 \cdot K}$	$\frac{W}{m^2 \cdot K}$	-	-
327.9	295.8	2.4	108.7	2.0	130.6	80.4

According to equation (16) after summing up the values of three heat streams for a 300 l hot domestic water tank, the obtained value is $\dot{Q} = 22.5$ W.

$$\dot{Q}_{loss} = \dot{Q}_{top} + \dot{Q}_{bot} + \dot{Q}_{cyl} \quad (16)$$

According to equation (17) it corresponds to 1.9 MJ of energy lost during one day.

$$Q_{loss} = \dot{Q}_{loss} \cdot \tau \cdot 3600 \cdot 10^{-6} = 22.5 \cdot 24 \cdot 3600 \cdot 10^{-6} = 1.9\,MJ \qquad (17)$$

After including the losses, the value of an aggregate amount of energy required for DHW tank during one day is obtained.

Energy lost during a day is less than 5% of energy required to heat domestic hot water. As shown in Figure 3 it increases as the tank's volume increases. For a 600 dm^3 tank, appropriate for an 8-person family, the losses are 3.4%, whereas in case of 900 dm^3 volume, they are only 3%. In the given case it is possible to omit the complicated calculation process, yet, the tank's insulation is of great importance, because these losses may prove significant for a tank with less insulation.

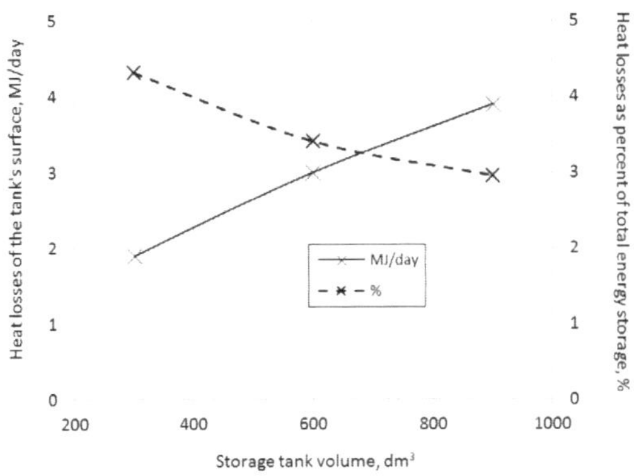

Figure 3. Daily heat losses from the tank

4 Solar installation gains

As the next step of the project, it is necessary to select data related to insolation in the given area. In Poland this information (selected from among at least 30 summer measurements) is shared on the Ministry of Infrastructure and Construction's website [6]. It informs about the amount of energy obtained in individual months of the year, as has been shown in Figure 4a. These data are rather demonstrative and do not provide information about insolation on individual days of the year. The Ministry's website also shows hourly data of radiation reaching the surface. They can be used to prepare a table of daily amount of energy in individual weeks (Figure 4b) or days (Figure 4c).

Average weekly data can be used for further project calculations and are especially valuable when one wants to make quick, but estimative calculations. Daily, not averaged data related to insolation allow to make proper calculations (only hourly data are more detailed than those, yet their amount can cause calculative problems). This data will be included in further calculations.

Assuming that the discussed system will consist of panels of 250W power each, the amount of energy obtained can be calculated with equation (18), The remaining parameters of a model panel have been put together in Table 4.

$$Q_{PV} = \eta_{PV} \cdot I_{S\_45^o} \cdot A \cdot n \cdot \frac{3600}{10^6}, MJ/day \qquad (18)$$

Table 4. Parameters of a selected PV module

N_{PV}	η	U_{mp}	I_{mp}	U_{oc}	I_{sc}	$a \cdot b \cdot c$
W	-	V	A	V	A	mm
250	0.153	30.4	8.23	37.6	8.81	1650·992·45

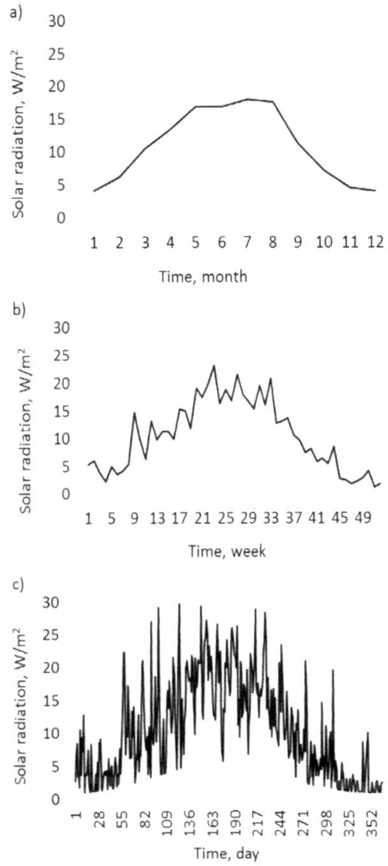

Figure 4. The amount of energy striking the surface at the angle of 45° from the south in Warsaw during a day: a) average monthly, b) average weekly, c) daily, not average

The amount of energy obtained during one day from one panel is small, so panels should be joined in order to obtain the demanded power.

It is worth noticing that the first solution (Fig. 5) is beneficial in terms of savings, as the whole energy gained is used to cover DHW needs. However, such a system needs to cooperate with an additional source of energy in order to cover the remaining needs.

The other version (Figure 6) assumes that the needs of DHW will be totally covered per year. It turns out to be possible to achieve after using over 100 panels. It causes the occurrence of huge surpluses of energy during summer, which need to be used (sold or used for other purposes) so that this system can be profitable at all.

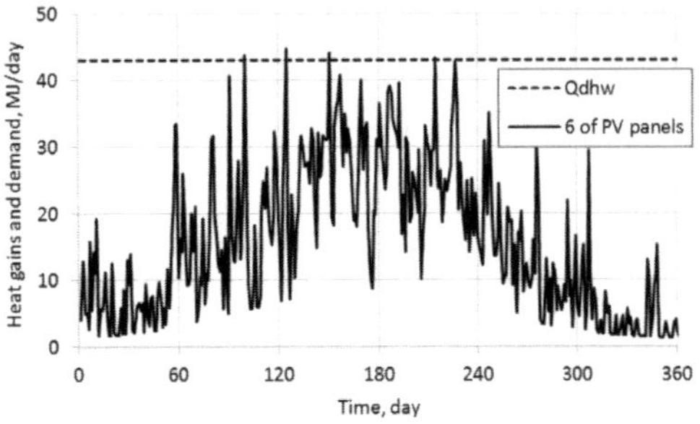

Figure 5. Annual summary of daily gains of energy obtained from panels for a minimum number of pieces (6) allowing to cover DHW requirements

If the PV are to cooperate with a heating device that is not automated (coal boiler, biomass boiler, fireplace, etc.) then their number should be selected in such a way that the additional device's work is necessary only in the heating season, when it has to fulfil the needs of central heating. In such a case, one needs to know the building's energy consumption to select a proper number of PVs. For this purpose, one should make heat calculations or use the data related to energy requirements, found, for instance, in energy performance certificates.

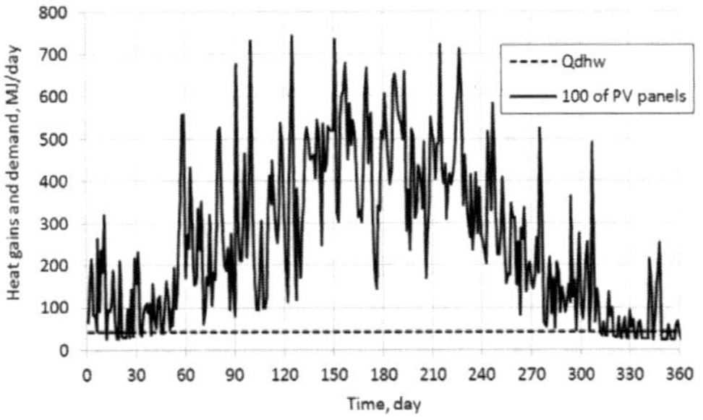

Figure 6. Annual summary of daily gains of energy obtained from panels for a maximum number of pieces (100) allowing to cover DHW requirements throughout the whole year

For the model calculations it is assumed that E_p building's energy consumption is 100 kWh/m^2·year, and its surface is 150m^2. It means that average daily requirements of energy necessary to provide comfort are 148 MJ/day. However, heating demands are not constant throughout the year, so, in order to determine them in individual days of the year, the method of average daily air temperature has to be used. It assumes that heating is turned off when the average daily ambient air temperature is higher than the base temperature (temperature inside the building). This method allows to determine the number of heating degree-days [7]. In order for the calculations of thermal energy requirements to be more reliable, the parameter of limit temperature is additionally used. Its use decreases the number of degree-days by months outside the heating season as well as first (September) and last (June) days of the heating season [8].

First, average heat demand for the purpose of central heating has to be determined, knowing the building's energy consumption and heat necessary to prepare domestic hot water.

$$\overline{Q_{ch}} = E_p - (Q_{dhw} + Q_{loss}), \mathrm{MJ/day} \quad (19)$$

Then, average daily heat demand for the purpose of central heating is calculated with the use of average daily temperatures (in the given example they are data for Warsaw, available at the Ministry of Infrastructure and Construction's website [6]).

$$Q_{ch_i} = \begin{cases} \frac{\overline{Q_{ch}} \cdot (t_w - t_d)}{t_w - t_r} \; for \, t_d < t_w \\ 0 \qquad\qquad\quad for \; t_d > t_w \end{cases}, \mathrm{MJ/day} \quad (20)$$

Calculated daily requirements for Q_{ch}, Q_{dhw} and the building's energy consumption E_p per year are shown in Figure 7.

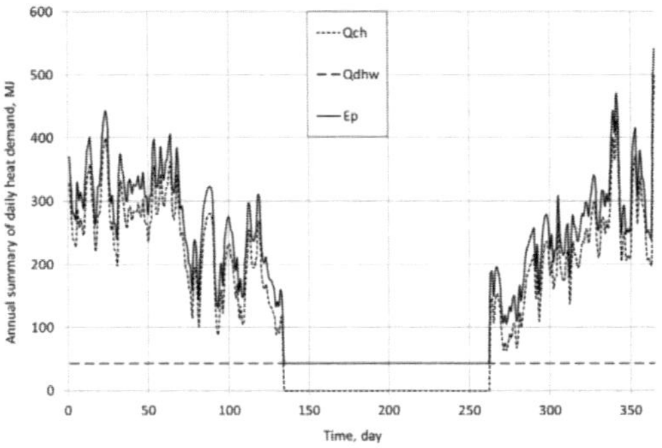

Figure 7. Daily heat demand per year

The data presented show the time when the heating season begins. That is why it is possible to select such a number of panels that allows to completely cover the DHW requirements during summer period, which is shown in Figure 8.

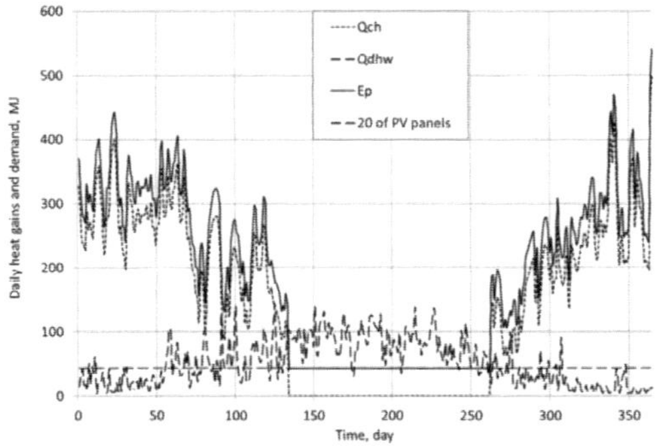

Figure 8. Annual summary of daily gains obtained from panels for the number of pieces (20) allowing to completely cover DHW requirements during summer period

It is worth noticing that PV efficiency depends on irradiance (Fig. 9) and on temperature to some extent. This should be kept in mind at the stage of calculations.

DOI: 10.1051/e3sconf/20171303003

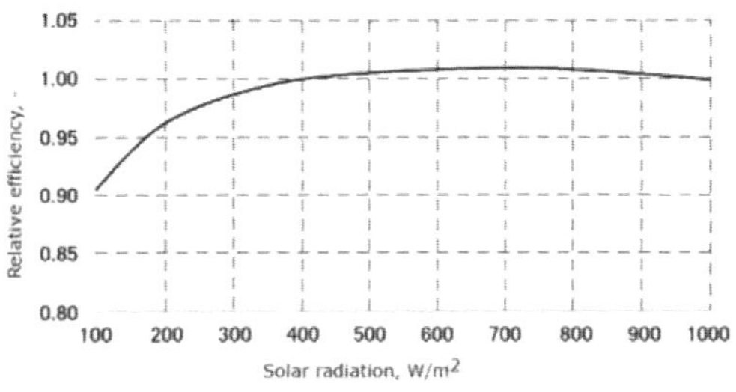

Figure 9. The typical change in module efficiency at an irradiance of 200 W/m^2 in relation to 1000 W/m^2 (at 20°C)

However, it turns out that it causes only slight changes in the amount of energy received, as a much bigger part of energy is obtained at an irradiance of over 400 W/m^2 than of approximately 200 W/m^2. That is why deviations from nominal efficiency can be excluded.

The last element used to establish the relation of necessary energy to obtained energy is the determination of losses on cables connecting panels with the electrical heater placed inside the DHW tank. The stream of lost energy due to heat emitted from the cable through which current flows is described by Joule heat (21):

$$\dot{q}_l = I^2 \cdot R_{el}, \quad (21)$$

where electrical resistance depends on cross section s and resistivity ρ and is calculated on the basis of equation (22).

$$R_{el} = \frac{\rho}{s}, \Omega \quad (22)$$

Assuming that the cable's diameter is 6 mm and resistivity is 0.017 Ω·mm^2/m it is necessary to determine current I flowing through the cables, in order to determine the total amount of heat emitted by cables (23) of assumed length of 20 m.

$$Q_{cab} = \frac{\dot{q}_l \cdot L \cdot \tau}{10^6}, MJ \quad (23)$$

In order to determine the current flowing through the cables it is necessary to determine the configuration of panels. Knowing that maximum voltage is 30.4 V, and current is 8.23 A one can notice that it is possible to make a serial connection of all panels, as the maximum voltage will be only slightly bigger than 600V (750 V in open system), and these installations are designed to work with maximum voltage of 1 kV. Serial connection will ensure lowest current possible, which, according to equation (21) will cause the smallest losses. Assuming that, if panels transferred energy to DHW tank with maximum current during the whole day, it turns out that daily losses in cables would be 0.33 MJ. Due to the fact that this value is much bigger than the factual value, these losses can be ignored.

Conclusion

A few years ago, photovoltaic panels were considered a relatively expensive system of obtaining electricity. Now, when their price is much lower, they are also used as systems for preparing DHW. Furthermore, PVs working for the purpose of DHW do not require such devices (still expensive) as an inverter, charging regulator or battery. Apart from a relatively simple construction, they have multiple advantages which surpass solar systems based on solar collectors. They include: the possibility of being connected to electrical installation, obtaining energy even during winter, no need to worry about overheating water in the installation or lack of operational costs of circulation pump or of covering energy losses in cables connecting solar system with DHW tank. The only disadvantage is a relatively big surface necessary to install them.

This paper shows that the process of selecting panels for DHW installation is not complicated. Having precise insolation data, it is possible to obtain very reliable results. A significant issue in the calculation process is the volume of accumulation tank that allows to accumulate energy surpluses from at least one day. A bigger DHW tank allows to accumulate energy during sunny days and store it for cloudy days. However, large over dimensioning generates unnecessarily big heat losses from the tank's surface. One of the most important aspects in the process of selecting panels is determining the criterion for choosing the installation's power. 3 variants have been suggested in this paper. The first one, covering the smallest part of DHW demand will be an installation with the shortest payback time, but it requires a full year of work of another heating device. The other proposition allows to cover DHW requirements for the whole year. However, it is the least profitable. Furthermore, in the discussed case, the area occupied by panels would exceed 160m^2. However, it does not require an interference of another heating device in the process of preparing DHW and it allows to use a significant surplus of energy to cover electricity needs of a house. The last proposition is used in houses with a non-automated heating installation. In such a case, the PV installation enables a self-operating preparation of DHW outside the heating season (during summer).

References

1. K. Stanisz, District Heating, Heating, Ventilation, **41**, 9, (2010, in Polish)
2. The Ordinance of the Minister of Transport, Construction and Maritime Economy of 5 July 2013 amending the ordinance on technical conditions which must be fulfilled by buildings and their locations, Journal of Laws of 13/8/2013, No. 926 (in Polish)
3. The Ordinance of the Minister of Infrastructure of 3 June 2013 on the methodology for calculating the energy performance of buildings and flats or parts of building constituting an independent technical and functional entity and the method of drawing up and samples of their energy performance certificates, Journal of Laws No. 2014, item 888 (in Polish)
4. J. Zimny, *Renewable sources of energy in energy-efficient constructions*, (Polish Geothermal Association, Krakow, 2010, in Polish)

DOI: 10.1051/e3sconf/20171303003

5. E. Kostowski, *Exercises on heat transfer*, (Silesian University of Technology Publishing House, Gliwice, 2006, in Polish)
6. Emission factors and calorific values of fuel as well as typical meteorological years and statistical climate data for energy performance calculations. http://mib.gov.pl/2-wskazniki_emisji_wartosci_opalowe_paliwa.htm#, (in Polish)
7. J. Dopke, Calculating the number of heating degree-days according to the method of Eurostat, (ogrzewnictwo.pl, 2010, in Polish)
8. D. G. Erbs, S.A. Klein, W. A. Beckman, Estimation of degree-days and ambient temperature bin data from monthly average temperature, ASHRAE J, June, (1983)

Heat transfer model for PV module with cooling channel

Anna Opiela, and *Zbigniew* Zapałowicz*

Department of Heat Engineering, Faculty of Mechanical Engineering and Mechatronics, West Pomeranian University of Technology, Szczecin, al. Piastów 17, 70-310 Szczecin, Poland

Abstract. The paper presents heat transfer model for a PV module with the cooling channel. The model has been applied to elaborate a calculation algorithm in order to estimate the power of PV module in relation to its temperature. The proposed calculation methodology allows to evaluate the amount of energy produced by PV installation in the fixed time scale.

Nomenclature

a	– thermal diffusivity, m^2/s
A	– surface, m^2
b	– distance, m
B	– temperature coefficient for PV module, 1/K
c	– specific heat, J/(kgK)
g	– gravity, m/s^2
G	– irradiance, W/m^2
h	– height, m
k	– overall heat transfer coefficient, $W/(m^2K)$
L	– characteristic dimension, m
$\dot{m}$	– mass rate, kg/s
N	– power, W
$\dot{Q}$	– heat flow rate, W
r	– reflectivity,
R	– thermal resistance, $(m^2K)/W$
s	– width, m
t	– temperature, °C
w	– velocity, m/s
α	– convective heat transfer coefficient, $W/(m^2K)$
β	– tilt angle for module,
β^*	– volumetric thermal expansion coefficient, 1/K
δ	– thickness, m
η	– efficiency,
λ	– thermal conductivity, W/(mK)
μ	– dynamic viscosity, Pa·s
ν	– kinematic viscosity, m^2/s
ξ	– coefficient,
ρ	– density, m^3/kg

Subscripts

A	– ambient
B	– lateral wall
c	– contact (p-n junction)
el	– electrical
f	– fluid
F	– front
K	– channel
M	– module
N	– horizon
R	– room
S	– sun
STC	– Standard Test Condition
T	– rear
w	– wind
W	– wall

Indices

k	– convection
p	– conduction
r	– radiation

Criterial number

Gr	– Grashof number, $Gr = \left(\beta^* g L^3 \Delta T\right)/\nu^2$
Nu	– Nusselt number, $Nu = \alpha L/\lambda$
Pr	– Prandtl number, $\mathrm{Pr} = \nu/a$
Re	– Reynolds number, $\mathrm{R}e = wL/\nu$

1. Introduction

One of development tendency in modern solar power engineering is increase of efficiency of PV modules. Technical data of PV modules given by manufacturers in their catalogues are defined in STC conditions (Standard Test Conditions), and they can only by chance correspond with real conditions. Therefore, nominal power parameters of PV modules assumed by the choice of elements for the installation differ from power values obtained in practical applications. Temperature of

*Corresponding author: zbigniew.zapalowicz@zut.edu.pl

p-n junction is one of relevant factors that decide about the instantaneous power of PV module. The higher the value of this temperature, the lower the power of PV module, and so the lower the energy production in the PV installation [1]. That is the reason, why it is aimed at lowering the temperature of p-n junction by cooling the rear wall of PV module. There are a few ways to obtain the aim: convection cooling by a fluid stream (air, water, low-boiling fluids), cooling with application of PCM materials, thermoelectric cooling, or spraying the surface with a droplets stream [2-4]. Authors of the present paper propose the heat transfer model with cooling channel, which allows to determine in an iterative way the approximate temperature of the p-n junction for the fixed time scale. Which, in turn, enables to determine the instantaneous electric power of PV module. Further, if time discretization is made, determination of approximate energy production in fixed time scales is possible. Of course, amount of produced energy depends on the local value of irradiance. The proposed model is a modified version of the model discussed in paper [2].

2. Model

In order to estimate the instantaneous electric power of PV module with cooling channel, energy balance was made for PV module and for the fluid in cooling channel, separately.

In case of PV module, it is assumed that, as a result of instantaneous balance of fluxes of input and output energies, module's temperature is the same in all its volume, with exception for the p-n junction only. Energy balance equation for PV module treated as one body has the form (Fig.1):

$$\dot{Q}^r_{MF} = \dot{Q}^r_{MF-N} + \dot{Q}^k_{MF-A} + \dot{Q}^k_{MB-A} + \dot{Q}^k_{MT-f} + \\ + \dot{Q}^r_{MT-f} + N_{el} + \dot{Q}^p_{el} + \Delta\dot{Q}_M \quad (1)$$

The following simplifying assumptions were applied in further analysis:

- steady state of PV module; the state occurs when $\Delta\dot{Q}_M = 0$ in PV module, if the heat capacities of PV layers are neglected,
- heat flow rates: carried away by conductors – $\dot{Q}^p_{el}$, transferred by radiation between the rear wall of PV module and the air in the channel – $\dot{Q}^r_{MT-f}$, transferred by radiation between the front wall of PV module and horizon – $\dot{Q}^r_{MF-N}$, transferred through the lateral walls of the module – $\dot{Q}^k_{MB-A}$, and can be neglected,
- solar radiation – $\dot{Q}^r_{MF}$, on the p-n junction.

Then, after considering the above assumptions, energy conservation equation (1) has the form:

$$\dot{Q}^r_{MF} = \dot{Q}^k_{MF-A} + \dot{Q}^k_{MT-f} + N_{el} \quad (2)$$

or

$$\frac{\dot{Q}^k_{MF-A}}{\dot{Q}^r_{MF}} + \frac{\dot{Q}^k_{MT-f}}{\dot{Q}^r_{MF}} + \frac{N_{el}}{\dot{Q}^r_{MF}} = 1 \quad (3)$$

If using respective coefficients equation (3) has the form:

$$\xi^k_{MF-A} + \xi^k_{MT-f} + \xi_{el} = 1 \quad (4)$$

Solar radiation reaching the front side of PV module is partially reflected from the anti-reflective coating therefore:

$$\dot{Q}^r_{MF} = (1 - r_{MF}) \cdot G^r_{S-MF} \cdot A_{MF} \quad (5)$$

Heat flow rates transferred between PV module, and environment and fluid on the other hand are given by equations:

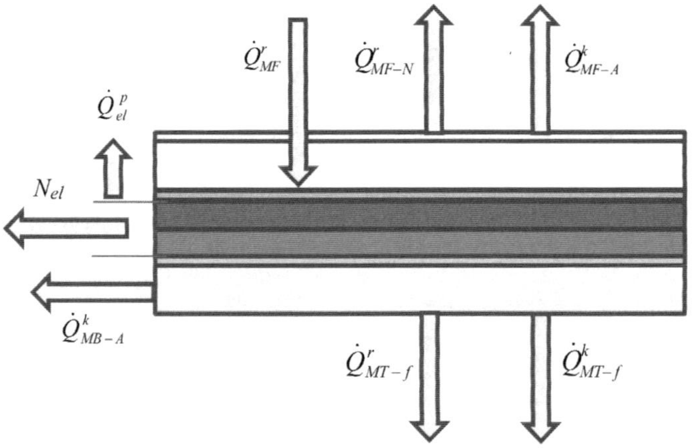

Fig.1. Energy streams in PV module.

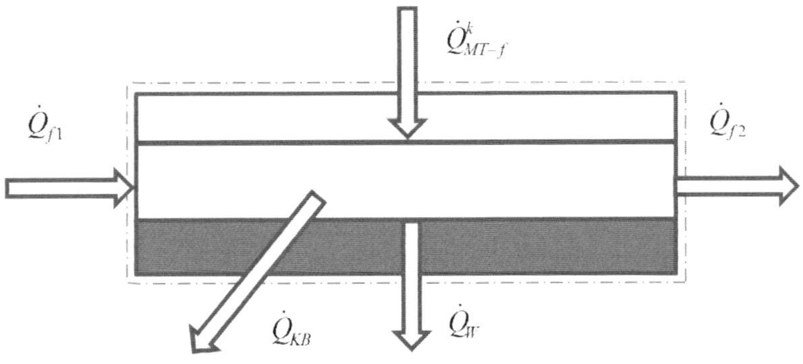

Fig.2. Heat flow rates for cooling fluid flowing in the channel.

$$\dot{Q}^k_{MF-A} = \alpha_{MF-A} \cdot A_{MF-A} \cdot (t_{MF} - t_A) \quad (6)$$

$$\dot{Q}^k_{MT-f} = \alpha_{MT-f} \cdot A_{MT-f} \cdot (t_{MT} - t_f) \quad (7)$$

In turn, energy balance for the cooling fluid flowing in the channel (Fig.2) has the form:

$$\dot{Q}^k_{MT-f} + \dot{Q}_{f1} = \dot{Q}_{f2} + \dot{Q}_{KB} + \dot{Q}_W \quad (8)$$

or

$$\xi^k_{MT-f} \cdot \dot{Q}^r_{MF} + \dot{Q}_{f1} = \dot{Q}_{f2} + \dot{Q}_{KB} + \dot{Q}_W \quad (9)$$

Energy flow rates of the cooling fluid at the inlet and outlet from the channel can be determined from relations:

$$\dot{Q}_{f1} = \dot{m}_{f1} \cdot c_{f1} \cdot t_{f1} = w_{f1} \cdot A_{f1} \cdot \rho_{f1} \cdot c_{f1} \cdot t_{f1} \quad (10)$$

$$\dot{Q}_{f2} = \dot{m}_{f2} \cdot c_{f2} \cdot t_{f2} = w_{f2} \cdot A_{f2} \cdot \rho_{f2} \cdot c_{f2} \cdot t_{f2} \quad (11)$$

Heat flow rate carried away from the fluid in the channel as a result of heat transfer through the wall is equal to:

$$\dot{Q}_W = k_{f-W} \cdot A_W \cdot (t_f - t_R) \quad (12)$$

Heat flow rate conducted through lateral protective walls of the channel can be estimated as:

$$\dot{Q}_{KB} = 2 \cdot k_{f-B} \cdot A_B \cdot (t_f - t_A) \quad (13)$$

Besides, it is assumed that temperature of the fluid in the channel is a mean of its temperatures at the inlet and at outlet of the channel:

$$t_f = \frac{t_{f1} + t_{f2}}{2} \quad (14)$$

After substitution of relation (10) - (14) in energy conservation equation for the fluid (9) the following equation is obtained:

$$\xi^k_{MT-f} \cdot \dot{Q}^r_{MF} + w_{f1} \cdot A_{f1} \cdot \rho_{f1} \cdot c_{f1} \cdot t_{f1} =$$
$$= w_{f2} \cdot A_{f2} \cdot \rho_{f2} \cdot c_{f2} \cdot t_{f2} +$$
$$+ 2 \cdot k_{f-B} \cdot A_B \cdot \left(\frac{t_{f1} + t_{f2}}{2} - t_A \right) +$$
$$+ k_{f-W} \cdot A_W \cdot \left(\frac{t_{f1} + t_{f2}}{2} - t_R \right) \quad (15)$$

From the above equation, temperature of fluid at the outlet from the channel can be determined as:

$$t_{f2} = \frac{\xi^k_{MT-f} \cdot \dot{Q}^r_{MF}}{w_{f2} \cdot A_{f2} \cdot \rho_{f2} \cdot c_{f2} \cdot + k_{f-B} \cdot A_B + 0{,}5 \cdot k_{f-W} \cdot A_W} +$$
$$+ \frac{(w_{f1} \cdot A_{f1} \cdot \rho_{f1} \cdot c_{f1} - k_{f-B} \cdot A_B - 0{,}5 \cdot k_{f-W} \cdot A_W) \cdot t_{f1}}{w_{f2} \cdot A_{f2} \cdot \rho_{f2} \cdot c_{f2} \cdot + k_{f-B} \cdot A_B + 0{,}5 \cdot k_{f-W} \cdot A_W} +$$
$$+ \frac{2 \cdot k_{f-B} \cdot A_B \cdot t_A - k_{f-W} \cdot A_W \cdot t_R}{w_{f2} \cdot A_{f2} \cdot \rho_{f2} \cdot c_{f2} \cdot + k_{f-B} \cdot A_B + 0{,}5 \cdot k_{f-W} \cdot A_W}$$
(16)

Then, temperature t_f can be determined from equation (14), and next, temperature of rear wall of PV module can be determined from equation (7):

$$t_{MT} = t_f + \frac{\xi^k_{MT-f} \cdot \dot{Q}^r_{MF}}{\alpha_{MT-f} \cdot A_{MT-f}} \quad (17)$$

Temperature of p-n junction can be calculated from equation [5]:

$$t_c = t_{MT} + \frac{G^r_{S-MF}}{G_0} \cdot \Delta t \quad (18)$$

where: G_0 – referential irradiance ($G_0 = 1000$ W/m^2), Δt – temperature difference of PV module n-p junction and of its rear wall at referential irradiance.

Thus, electrical efficiency of PV module resulting from current temperature of the p-n junction is described by equation:

$$\eta_{el} = \eta_{el\_STC} \cdot \left[1 - \frac{B_N}{100}(t_c - t_{STC}) \right] = = \eta_{el\_STC} \cdot \left[1 - \frac{B_N}{100}(t_c - 25) \right] \quad (19)$$

where B_N is coefficient of power change with temperature (given by the producer).

Power of PV module can be calculated from equation:

$$N_{el} = N_{el\_STC} \, B_{PV} \left(\frac{G^r_{S-MF}}{G_{STC}} \right) \cdot \left(\frac{\eta_{el}}{\eta_{el\_STC}} \right) \quad (20)$$

Coefficient B_{PV} in equation (20) describes the degree of drop of nominal parameters of the module depending on operation conditions. The coefficient relates to such factors like smudges on the module, losses on conductors, shading, snow covering, ageing etc. $B_{PV} = 1$ is assumed in the paper.

It is also necessary for calculations to determine convective heat transfer coefficients, overall heat-transfer coefficients among particular media, or respective thermal resistance values.

Overall heat-transfer coefficient through the lateral wall of the channel depends on its thermal resistance and can be calculated from equation:

$$k_{f-B} = \frac{1}{R_{f-B}} = \frac{1}{\frac{1}{\alpha_{f-B}} + \frac{\delta_B}{\lambda_B} + \frac{1}{\alpha_{B-A}}} \quad (21)$$

In turn, heat transfer coefficient through the building wall can be determined from the relation:

$$k_{f-W} = \frac{1}{R_{f-W}} = \frac{1}{\frac{1}{\alpha_{f-W}} + \frac{\delta_W}{\lambda_W} + \frac{1}{\alpha_{W-R}}} \quad (22)$$

Convective heat transfer coefficient from the front wall of PV module to the environment is given by the equation:

$$\alpha_{MF-A} = \frac{\lambda_A}{h_A} \cdot Nu_A \quad (23)$$

In order to determine this coefficient, knowledge of Nusselt empirical correlations describing overflowing the module's front wall by air is required. The relations are given among other in paper [6] (e.g. relations of Mitchel, Mc Adams, or Shewen). The choice of Nusselt empirical correlation results from site and kind of mounting of the PV module on the wall of the building.

Also, a simplified relation considering additional effect of wind can be applied for calculations, e.g.:

$$\alpha_{MF-A} = 2{,}8 + 3{,}0 \cdot w_w \quad (24)$$

Besides, it has been assumed that the process of overflowing the lateral walls of the channel by air is the same as the process of overflowing the front wall of PV module, hence:

$$\alpha_{B-A} = \alpha_{MF-A} \quad (25)$$

Convective heat-transfer coefficient from rear wall of PV module to the fluid in the channel can be determined from relations valid for the flow of the fluid in the narrow channel [7]. Heat transfer in a confined channel by natural convection is treated as heat conduction. The equivalent convective heat-transfer coefficient λ^*_f relates to both: the process of heat conduction and heat convection in the channel.

General empirical correlation describing heat transfer in the narrow channel has the form:

$$\frac{\lambda^*_f}{\lambda_f} = C \cdot (Gr \cdot \mathrm{Pr})^a \quad (26)$$

The choice of empirical correlation for calculations depends on the width and length of the channel; in case of PV module, it depends on b / h_M ratio.

Then, convective heat-transfer coefficient can be determined from the equation:

$$\alpha_{MT-f} = \frac{\lambda^*_f}{b} \quad (27)$$

In case of forced convection in the channel, the choice of empirical relation depends on the kind of flow. Then, general empirical equation has the form:

$$Nu_f = C \cdot \mathrm{Re}^b \cdot \mathrm{Pr}^c \quad (28)$$

It is assumed for calculations that values of convective heat-transfer coefficients from the liquid in the channel to the lateral wall of the channel, and from the liquid to the building wall meet the following relation:

$$\alpha_{MT-f} = \alpha_{f-B} = \alpha_{f-W} \quad (29)$$

Convective heat transfer coefficient from the building wall to the air indoors can be determined from the equation:

$$\alpha_{W-R} = \frac{\lambda_R}{h_R} \cdot Nu_R \quad (30)$$

Paper [6] gives empirical correlations describing the heat transfer between environment and indoors, relating to direction of heat flow and inclination of building wall.

3. CALCULATION METHODOLOGY

In order to determine the instantaneous working parameters of PV module, data for calculations and initial conditions should be given.

The first group of data requires the technical data of the module (power N_{el_STC}, efficiency η_{el_STC}, coefficient B_N, temperature difference Δt at referential irradiance, reflexivity coefficient r_{MF} and module's width s_M and length h_M). It is easy to determine the gross surface area of PV module on the basis of its geometrical dimensions, from the equation:

$$A_M = s_M \cdot h_M \tag{31}$$

It can be assumed for further calculations that:

$$A_M = A_{MF} = A_{MF-A} = A_{MT-f} = A_W .$$

Cooling channel can be an integral part of PV module, but it can also be a separate heat exchanger mounted on its rear wall, as well. It is assumed in the present paper that the channel is created between the rear wall of the module and the building wall. Then, PV module is placed on a frame fastened to the building wall in distance *b*. Lateral walls of the channel are additional constructional elements. Parameters of the lateral walls of the channel are known: δ_B, λ_B. Thus, it is possible to determine cross section areas on inlet and outlet from the channel, as well as surface area of lateral walls of the channel, from the following equations, respectively:

$$A_{f1} = A_{f2} = b \cdot s_M \tag{32}$$

$$A_B = 2 \cdot b \cdot h_M \tag{33}$$

The latter data group concerns parameters of solar radiation and parameters of environment. The data depends on the site of PV installation location (latitude and longitude). Database collecting parameters of solar radiation and of environment supplies among other: global irradiance reaching the horizontal or inclined surface directed respectively to the sun, for the fixed day of the year and for the fixed hour. The above data allows to determine the irradiance G^r_{S-MF}, ambient temperature t_A, and wind velocity w_w.

It is also required for calculations to know the kind of fluid and its initial temperature t_{f1}. Then, it is possible to determine physical parameters of the fluid.

Flow of the fluid in the channel can result from natural convection, or it can be the motion constrained by a transporting device (pump, fan). In case of natural convection, mean velocity of transportation of fluid in the channel can be determined from equation [8]:

$$w_f = \frac{b^2 g \beta^* \cdot \sin\beta}{12\mu_f} \tag{34}$$

As it has been mentioned, the channel is limited by building wall made of materials with known physical parameters, that is, thickness δ_W and thermal conductivity λ_W. It can be assumed in calculations that air temperature indoors is constant and relates to conditions of thermal comfort, that is $t_R = 20 \div 22$ °C.

In the first calculation step, it can be assumed that initial value of p-n junction temperature equals $t_c = t_{STC} = 25°C$, then, initial electrical efficiency of PV module calculated from equation (19) is η_{el_STC}. Thus, initial power of PV module N_{el} depends only on the current value of parameter G^r_{S-MF}. Next, coefficient ξ_{el} can be determined from the equation:

$$\xi_{el} = \frac{N_{el}}{\dot{Q}^r_{MF}} \tag{35}$$

Besides, it is initially assumed that heat flow rates transferred from PV module to environment and to the air in the channel have the same values, that is $\dot{Q}^k_{MF-A} = \dot{Q}^k_{MT-f}$, thus it results from Eq.4 that:

$$\xi^k_{MT-f} = \xi^k_{MF-A} = 0{,}5 \cdot (1 - \xi_{el}) \tag{36}$$

In the first calculation step, determination of physical properties of fluid on the outlet from the channel is impossible while temperature t_{f2} is unknown. That is why it can be assumed that physical properties of fluid are equal on inlet and outlet. In further calculations steps, properties of fluid on output are determined for temperature t_{f2} calculated in the previous calculations step.

Output temperature t_{f2} is calculated in succeeding iterative steps up to the moment when its value reaches the demanded accuracy. If verifying condition is met, it is due to calculate all parameters relevant for operation of PV installation, and move on to the next cycle of calculations.

4. SUMMARY

The paper presents the model and calculations methodology that allow to determine working parameters for PV module with cooling channel. Proposed model has a general character and allows to choose calculations relations depending on required technical situation. The model, as well as given calculations methodology will be verified experimentally.

REFERENCES

1. S. Dubey, J.N. Sarvaiya, B. Seshadri, Energy Procedia, **33**, 311-322, (2013)
2. A. Opiela, Z. Zapałowicz, Polska Energetyka Słoneczna. No.1-4, 13-17, (2013)
3. J. Park, T. Kim, S-B. Leigh, Solar Energy, **105**, 561-574, (2014)
4. K.A. Moharram, M.S. Abd-Elhady, H.A. Kandil, H. El-Sherif. Ain Shams Engineering Journal, No.4, 869-877, (2013)
5. https://pvpmc.sandia.gov/modeling-steps/2-dc-module-iv/module-temperature. (25-02-2016)
6. D. Chwieduk, *Energetyka słoneczna budynku* (Wyd. Arkady, Warszawa, 2011)
7. B. Staniszewski: *Wymiana ciepła. Podstawy teoretyczne* (PWN, Warszawa, 1980)
8. E. Klugmann-Radziemska, E. Klugmann, *Systemy słonecznego ogrzewania i zasilania elektrycznego budynków* (Wyd. Ekonomia i Środowisko. Białystok, 2002).

Small Scale Gasification of Biomass and Municipal Wastes for Heat and Electricity Production using HTAG Technology

Jan Stasiek[1,*], *Marcin* Jewartowski[1], and *Weihong* Yang[2]

[1]Gdansk University of Technology, Faculty of Mechanical Engineering, Dep. of Energy and Industrial Apparatus, ul. G. Narutowicza 11/12, 80-233 Gdansk, Poland
[2]Royal Institute of Technology, Department of Materials Science and Engineering, Div. of Energy and Furnace Technology, Brinellvägen 23, S-100 44 Stockholm, Sweden

Abstract. Combustion and gasification technology utilizing high-cycle regenerative air/steam preheater has drawn increased attention in many application areas. The process is to be realized at temperature level above ash melting point using highly preheated agent. The use of highly preheated media above 900°C provides additional energy to conversion processes and results in considerable changes to the design of combustion and gasification equipment and its performance. This paper presents an advanced gasification system that utilizes high-temperature air and steam to convert biomass and municipal wastes into syngas production as well as selected results from experimental studies of high temperature air/steam gasification.

1 Introduction

Biomass is an organic material coming from living organisms and does not add carbon dioxide to the environment. It can be used as a source of energy and is one of the most commonly used fuels around the world and can be good supplement or even alternative for fossil fuels due to increasing demand for energy. It can be directly burned or converted into fuel trough pyrolysis or gasification and used to generate heat or electricity. Most of biomass can be converted by gasification and the process is generally more efficient than direct combustion. The value of a biomass depends on the chemical and physical properties of the large molecules from which it is made [1]. The product of gasification has lower heating value compared to fossil fuels but can be used in similar power plants for heat and electricity production.

The automotive industry generates yearly millions tones of wastes. Many components are recovered, but remaining parts are mechanically processed automotive shredder residue (ASR). It is a complex mixture of materials, which may be toxic. Combustion is used commonly for its utilization, however it has disadvantages, and pollutant emission is the one, which has to be considered. Therefore, pyrolysis and/or gasification are more promising methods of ASR utilization and generation of energy [2]. Municipal solid wastes (MSW), associated with nonindustrial human activity, are continually generated in large amounts around the world, creating problems with their disposal. Also here, gasification is the technique, which provides better solution to a problem than incineration technology [3].

In biomass gasification, the fuel gas can contain significant amounts of tar, formation of which can be suppressed by high temperature or additionally decomposition can be expected by thermal cracking and steam reforming. Tars can be also eliminated by the use of a catalyst. However tar behavior must be further investigated in order to meet requirements of acceptable tar content, less than 100 mg/Nm^3 for satisfactory internal combustion engine operation.

2 High Temperature Air/Steam Gasification

The use of biomass and wastes for energy production is a subject of many ongoing research and development projects. There are at least four main methods of biomass and waste utilization: direct combustion, pyrolysis, biodegradations and gasification. Direct combustion of biomass or wastes is not acceptable due to different pyrolysis and oxidative pyrolysis time for the various materials present in the stream of the fuel. The gasification process converts raw biomass into a combustible gas, with energy content within the region of 60-70% of the original biomass feedstock. This gas can then be burnt to produce thermal energy, or to run engines for mechanical or electrical power. Gasification is the incomplete combustion of solid biomass forming a mixture of gases referred to as Fuel Gas. Carbon monoxide is the predominant component of useful heat value. Other gaseous products depend on whether the conversion is achieved with air or pure oxygen, whether water is introduced into the feedstock and the temperatures reached in the process. Hydrogen will be produced if water is present and temperature in excess of 900°C is reached. Excess water will permit the creation of carbon dioxide which will weaken the heating value of the mix. Air facilitates a low temperature conversion,

* Corresponding author: jstasiek@pg.gda.pl

DOI: 10.1051/e3sconf/20171303005

however a low heat value gas will be created as nitrogen will be present in the final mix. Fuel gas has a relatively low heat value (4-11 MJ/Nm^3) and is therefore best used in situ for heating or power generation.

Gasification can be made with air or steam as a gasification media. Air gasification for heat and power production is the most common technology developed and applied. Much less developed is technology for the synthetic fuels (syngas) production from biomass and wastes. Users expect technology, which can transfer whole solid waste fuel to clean fuel gas. Therefore production of tar, soot and char should be minimal or ideally prevented. In general, gasification could be divided into two processes according to the temperature of the feed gas (gasification agent):

- low temperature gasification - where cold or slightly preheated air is used as gasifying agent,
- high temperature gasification, which air preheated or mixture of air and steam preheated above 1000°C is used as gasifying agent (Fig. 1).

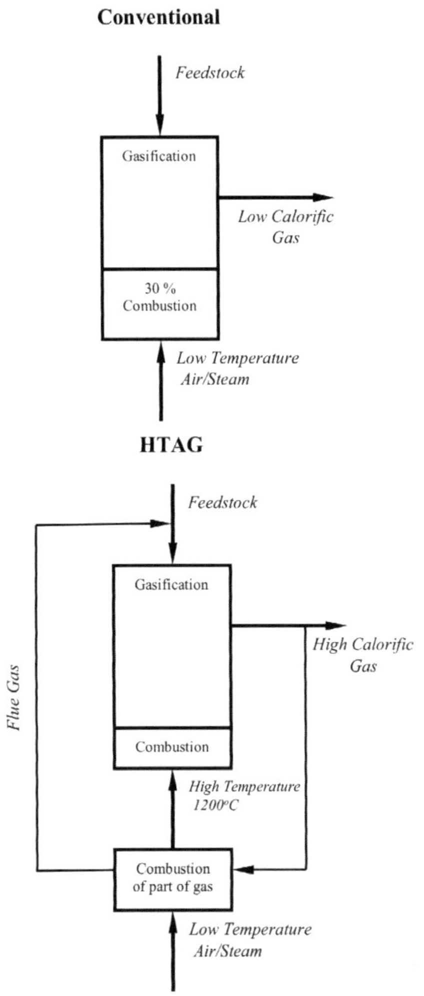

Fig. 1. Comparison between conventional and HTAG gasification

Low temperature combustion processes (grate combustion, fluidized bed combustion or gasification) will have difficulties to fulfill new European Community regulations, particularly regarding ash residues, which cannot be landfilled in the nearest future. Also it is important to stress that using the low-temperature processes is the most common to produce only heat in a form of hot water or steam. What is needed is the product, which can be used freely. This is fuel gas from biomass or wastes (syngas), which can then be used in boilers (for example as reburning gas to reduce NOx), in industrial furnaces as a substitute of natural gas or mixed with natural gas, in gas turbines, chemical plants and last but not least for small scale production of electricity using the internal gas combustion engines, gas turbines or fuel cells.

Research and Development on High Temperature Air Combustion (HiTAC) of gas and liquid fuels is followed by development of a new gasification processes with high temperature air and steam called High Temperature Air/Steam Gasification (HTAG or HiTAG). This new technology has been applied in performing extensive studies on thermal conversion of biomass and wastes into fuel gas, since year 2001 at Royal Institute of Technology (KTH) [4-6]. It uses high-cycle regenerative air/steam preheater developed by Nippon Furnace Kogyo Kaisha Ltd. to convert biomass and municipal wastes into hydrogen-rich fuel gas (Fig. 2). The generator allows preheating of air or mixture of air and steam up to 1300°C. High temperature gasification is carried out with air or steam as a gasification agent (feed gas).

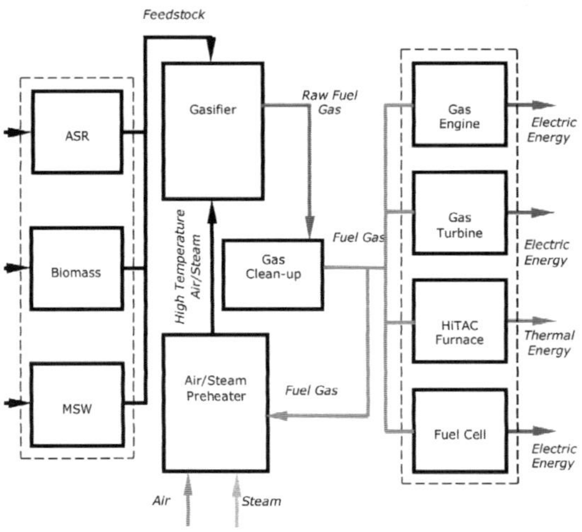

Fig. 2. General scheme of HTAG plant for different utilization alternatives of the syngas (fuel gas) from biomass and wastes

Using highly preheated air or mixture of air and steam changes energy balance of the gasifier and phenomena of gasification process. Less feedstock is burned thus less flue gas is produced what results in less diluted fuel gas. Calorific value of the fuel gas from high temperature air/steam gasification is much higher comparing to any low temperature air gasification process. In case of dry biomass it is normally above

8 MJ/Nm³ in case of high temperature air/steam gasification. Since pyrolysis and char residue gasification is carried out at much higher temperature the tar is already decomposed in much higher degree. It results thus in much lower contamination of the fuel gas with tar and dust. Gasification of high calorific wastes (as plastics, rubber or automotive shredder residues) with high temperature air and steam results in much higher calorific value of the fuel gas compared to gasification of low calorific wastes. Produced fuel gas can be used in many applications for example to produce electricity in gas engine, gas turbine or in near future, even in solid fuel cell.

The main features of the HTAG process in converting biomass and wastes to fuel gas are:

- conversion of solid wastes and low grade fuels to clean gaseous fuels,
- high calorific value and high amounts of hydrogen and hydrocarbons in the composition of fuel gas,
- high efficiency of conversion from solid to gas fuel,
- no landfilling of ash residues is required,
- minimum amount of pollutions generated,
- high intensity of process (conversion rate).

2.1 Gasifier system

Investigations on HTAG in KTH were performed initially using batch type downdraft fixed-bed gasifier [5] following by continuous downdraft one, which was later transformed into updraft-type gasifier. Fig. 3 presents concept of an advanced HTAG installation in view of fuel gas (syngas) production. The fuel was fed from the top of the reactor using continuous feeding system.

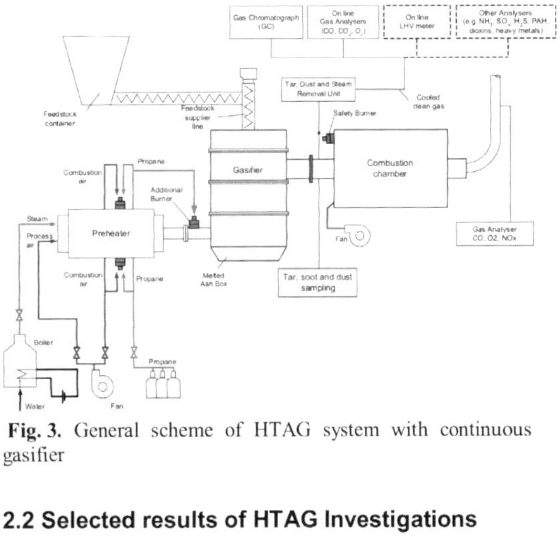

Fig. 3. General scheme of HTAG system with continuous gasifier

2.2 Selected results of HTAG Investigations

Results of wood pellets gasification with HTAG show, that highest concentration of desired gaseous products (hydrogen, carbon monoxide and methane) in produced syngas can be achieved only with steam as gasifying agent. The steam controls the gasification temperature, if the preheated air and steam mixture is used [5,7]. The gasification temperature response to the change of the feed gas composition and temperature of the feed gas is also discussed. The results showed that the use of steam in the feed gas decreases the process temperature sharply. Increase of the feed gas temperature reduces production of tars, soot and char residue as well as increases the heating value of the dry fuel gas produced, (Fig. 4 and 5) [8]. Results of mathematical modelling of gasification using high-temperature air show, that increase of the feed gas temperature results in a higher gasification rate, higher molar fractions of H_2, CO and C_mH_n, hence higher LHV and a faster ignition rate [4].

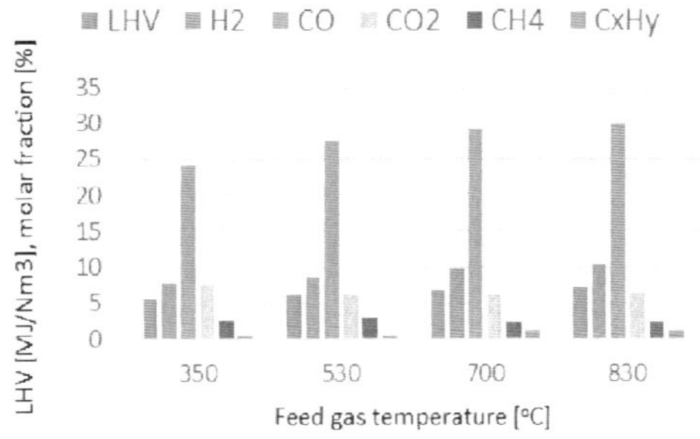

Fig. 4. Influence of the temperature of the feed gas on the average values of LHV and composition of dry fuel gas for HTAG of wood pellets: feedstock - wood pellets 12 mm, feed gas flow - 50 Nm³/h, fuel mass - 20 kg, X_{H2O}: 0%.

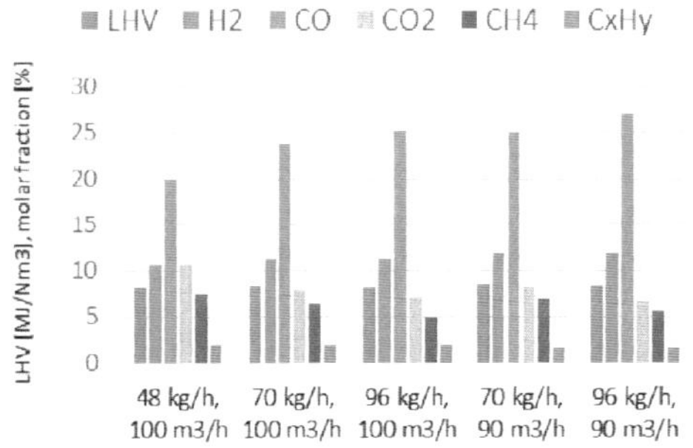

Fig. 5. Composition of produced fuel gas, its temperature and LHV for different continuous gasification conditions of wood pellets: different pellets flow [kg/h], oxidizer flow [m3/h] and process temperature (from left to right: 810°C, 586°C, 431°C, 654°C, 415°C).

The high temperature air/steam gasification technique has been tested for a fuel in pellet form made from waste material of woody and plastic origin. The feedstock was gasified in an updraft fixed bed reactor by mixtures of air and steam (4% to 82% steam) preheated to 1400°C, a temperature well above the fluid temperature of the feedstock. HTAG was reported as a good way of utilizing plastic containing fuel [9]. This technology was also used in experimental studies that were conducted using tropical coffee husks under high temperature conditions [10]. It was found the positive influence of high temperature on increasing the gasification process. Increased gasification temperature led to a linear increment of CO concentration in the syngas.

The use of HTAG has been proposed for two-staged method of ASR treatment [11]. ASR was primarily treated by microwave pyrolysis and later the liquid and solid products become the feedstock for the high temperature gasification process. Concept was proposed to maximize the conversion of organic contents of ASR into valuable products, while preserving metals during each step of the process. Results of experiments conducted in a batch-type, lab-scale gasifier using high temperature steam show that HTAG can be used as the complimentary method to pyrolysis and in particular microwave pyrolysis. Further investigation using two different types of gasifying agent: pure steam and 3% of oxygen at temperature between 700°C and 950°C show that two undesirable phenomena occurred: with increasing temperature of the process the reaction rate decreases and there was no significant differences between steam and oxygen gasification rates. That indicates, that increasing the temperature of reaction is not always the best option to improve the gasification process and increase its rate, since this is not the only effect influencing the process performance [12].

3 Conclusions

High Temperature Air/Steam Gasification proved to be very useful technology for gas fuel generation, which can be used later to produce heat and electricity. HTAG is applicable to biomass gasification as well as to treatment of municipal wastes and automotive shredder residue and successfully converting its energy into heat and/or electricity.

Acknowledgement

The work reported in this paper includes scientific cooperation between Gdansk University of Technology (GUT), Poland and Royal Institute of Technology (KTH), Stockholm, Sweden. The personal assistance of Professor Wlodzimierz Blasiak and his help in experimental performance is kindly acknowledged.

References

1. P. McKendry, Bioresource Technol. 83, 37-46 (2002)
2. P. Donaj, W. Blasiak, W. Yang, C. Forsgren, Journal of Hazardous Materials 185, 472-481 (2011)
3. C. Zhou, T. Stuermer, R. Gunarathne, W. Yang, W. Blasiak, Fuel 122, 36-46 (2014)
4. W. Yang, A. Ponzio, C. Lucas, W. Blasiak, Fuel Processing Technology 87, 235-245 (2006)
5. W. Blasiak, S. Kalisz, D. Szewczyk, C. Lucas, R. Abeyweera, High Temperature Air/steam Gasification of Biomass and Wastes – Stage 2, Technical Report No.2, ISRN KTH/MSE-03/61-SE+ENERGY/TR, Royal Institute of Technology, Stockholm, Sweden, (2003)
6. A. Ponzio, W. Yang, W. Blasiak, J. Stasiek, Hydrogen generation from biomass using high temperature agent gasification from biomass using high temperature agent gasification, 16th European Biomass Conference and Exibition, Feria Valencia, Spain, 2-6 June (2008)
7. D. Mikielewicz, J. Stasiek, W. Blasiak, A. Ponzio, W. Yang, Advanced Technology for Thermo-chemical Conversion of Biomass and Solid Muncipial Wastes for Hydrogen-Rich Gas Production, 8th European Converence on Industrial Furnaces and Boilers, Vilamura, Portugal, 25-28 March, (2008)
8. C. Lucas, D. Szewczyk, W. Blasiak, S. Mochida, Biomass and Bioenergy 27, 563-575 (2004)
9. A. Ponzio, S. Kalisz, W. Blasiak, Fuel Processing Technology 87, 223-233 (2006)
10. L. Wilson, G.R. John, C.F. Mhilu, W. Yang, W. Blasiak, Fuel Processing Technology 91, 1330-1337 (2010)
11. P. Donaj, W. Yang, W. Blasiak, C. Forsgren, Journal of Hazardous Materials 182, 80-89 (2010)
12. P. Donaj, W. Yang, W. Blasiak, C. Forsgren, Journal of Hazardous Materials 185, 472-481 (2011)

Reference diameter in calculations of creep strain for steam pipelines

Stanisław Łopata[1,*]

[1]Cracow University of Technology, al. Jana Pawła II 37, 31864 Kraków, Poland

Abstract. Recommended methods of the operational safety assessment of high-pressure steam pipelines include periodic testing that enables determination of creep strain and the creep rate. The accuracy with which the two quantities are determined depends, among others, on the control of the testing conditions and on the assumed reference diameter. The analysis conducted herein concerns the impact of the reference diameter on the results characterizing the creep phenomenon in the pipeline elements (straight sections and bends). In this respect, the initial ovality of the pipeline cross-section is an important parameter. The calculations are made using own results obtained from many years of the steam pipeline creep testing, including tests performed after the expiry of the computational service life.

1 Introduction

Thermal cycles realized in commercial and industrial power plants incorporate not only their main components, i.e. boilers, turbines and generators, but also auxiliary equipment including steam pipelines. These may be combined into unit- or collector-type systems which may sometimes be very sophisticated. Their failure usually involves the power unit shutdown even though the boiler and the turbine are in good working order. Due to high values of temperature and pressure of superheated steam transmitted along them, such pipelines are often referred to as high-temperature or high-pressure ones. They are often affected by creep and thermo-mechanical fatigue phenomena which, as their operation time gets longer, change the properties of materials they were made of. The creep process is the dominant one in pipelines, and it manifests itself by an increment in their diameter.

According to the standard [1], the steam pipeline creep can be tested by means of periodic measurements of the diameter on what is referred to as measuring pins. The pins are installed on the external surface of pipes, typically in two directions which are perpendicular to each other and to the pipe axis [2-5]. Although rare in practice, more directions can occasionally be used [6]. The principle that governs the measurement of the steam pipeline creep-related plastic strain is illustrated in Fig. 1 [4, 5]. It may be added that the presented method belongs to non-destructive testing, which certainly is its advantage.

In the power industry in Poland, periodic testing of creep for steam pipelines is an element of their diagnostics. It is required by regulations of the Technical Inspection Office (UDT) [7] and diagnostic tests have to be performed throughout the pipeline operating lifetime. In practice, apart from due care, the testing sufficient accuracy is ensured by appropriate identification of conditions in which tests are carried out. This comes down to measuring the temperatures of the pipeline wall and of the micrometer frame, which – as results from many years of in-house experience – usually differ from each other and their values obtained in subsequent tests are different, too [4].

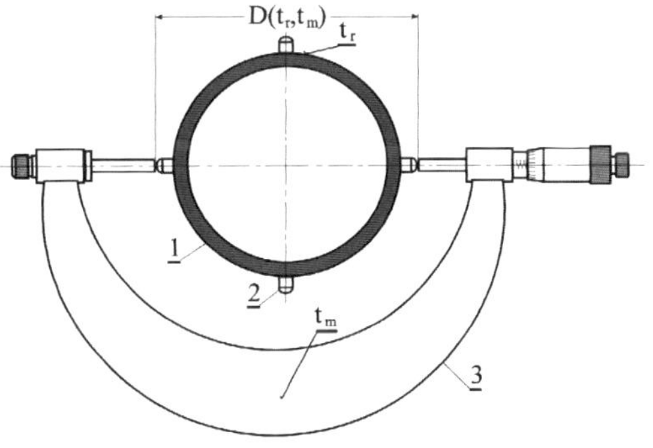

Fig. 1. Location of measuring pins on the steam pipeline and the principle of testing the increment in its outer diameter: 1 - pipeline, 2 - measuring pin, 3 - micrometer, t_r, t_m - temperatures of the pipe wall and of the micrometer frame, respectively.

Knowing temperatures t_r and t_m, it is possible to recalculate the diameter obtained from measurements $D(t_r, t_m)$ – Fig. 1 – to its value corresponding to the assumed reference condition, i.e. the temperature of 20°C. This enables an assessment by comparing the results with the values determined earlier and with those adopted (recognized) as permissible. The recalculation is most often performed using the following relation [2-5]:

* Corresponding author: lopata@mech.pk.edu.pl

$$D(t_o) = \frac{D(t_r, t_m) + b}{1 + \beta_r (t_r - 20) - \beta_m (t_m - 20)} \qquad (1)$$

where:

- b – micrometer correction (average value of corrections determined prior to and after the measurement), mm,
- $D(t_o)$ – diameter measured on measuring pins, after recalculation to the reference condition, i.e. temperature $t_o = 20$°C, mm,
- $D(t_r, t_m)$ – diameter measured on measuring pins in real conditions of the measurement, i.e. at temperatures t_r and t_m, mm,
- t_m, t_r – temperature of the micrometer frame and of the pipeline, respectively, °C,
- t_o – reference temperature, °C,
- β_r, β_m – linear thermal expansion coefficient of the pipeline and the micrometer material, respectively, mm/(mm·K).

The results obtained after different times of the steam pipeline operation make it possible to find quantities that characterize the pipeline creep, i.e. the creep-related plastic strain (or in short: creep strain, as it is often referred to in practice and even in reference literature), and/or the creep rate.

2 Reference diameter in calculations of the steam pipeline creep strain

One of the above-mentioned quantities that characterize the pipeline material, taking account of its real operating time, is the creep strain understood as circumferential strain and defined by the following relation (for the outer generating line):

$$\varepsilon = \frac{\Delta O_z}{O_z} \cdot 100\% = \frac{\Delta D_z}{D_z} \cdot 100\% \qquad (2)$$

where:

- D_z – pipeline outer diameter (according to the measurement made prior to the pipeline commissioning), mm,
- O_z – pipeline outer perimeter (according to the measurement made prior to the pipeline commissioning), mm,
- ΔD_z – increment in the pipeline outer diameter (after τ of operating hours), mm,
- ΔO_z – increment in the pipeline outer perimeter (after of operating hours), mm.

The results obtained during the steam pipeline operation from systematic measurements of diameters on the measuring pins make it possible to calculate strain values $\varepsilon = \varepsilon_i$, which arise in the time between subsequent measurements (where i denotes the measurement number), using the following equation:

$$\varepsilon = \varepsilon_i = \frac{D_i - D_1}{D_z} \cdot 100\% \qquad (3)$$

where:

- D_1, D_i – diameter on measuring pins, recalculated to the reference condition, for the first and the i-th measurement (measurements performed before the pipeline commissioning and after time τ of operating hours, respectively), mm,
- D_z – pipeline outer diameter (according to the measurement made prior to the pipeline commissioning), mm,
- ε – pipeline creep strain, %,
- ε_i – steam pipeline creep strain obtained after time τ of operating hours, %.

As it was already mentioned, the pipeline strain measurements are performed in two directions perpendicular to each other. In this situation relation (2) is correct assuming that the pipe has the shape of a perfect circle. The increment in the diameter is then independent of the measurement direction. In reality, however, this is not the case because pipelines are characterized by their cross-section ovality [4, 5, 8, 9], which means that the value of diameter D_z is different in each direction. Consequently, using relation (3) and depending on the assumed definition of this quantity, i.e. D_z (to which the increase of diameter measured on the measuring pins is referred), the obtained values of ε_i will differ from each other. This will naturally concern both the values calculated based on the data for each measurement direction and for the average ones. The latter can be achieved by averaging the final results obtained earlier for each of the two "directions" independently or by prior averaging of quantities D_i, D_1 and D_z appearing in relation (3). Therefore, the computational relations for the two possible methods will be as follows:

- in the first method, which is more common in the power engineering practice in Poland, it is assumed that D_z is the average of the values obtained for the two directions [2-5, 10], and the increments determined in both directions are related thereto (by the following equations):

$$D_{z(sr)} = 0.5\left(D_{z(1)} + D_{z(2)}\right) \qquad (4)$$

$$\varepsilon_1 = \varepsilon_{i(1)} = \frac{D_{i(1)} - D_{1(1)}}{D_{z(sr)}} \cdot 100\% \qquad (5a)$$

$$\varepsilon_2 = \varepsilon_{i(2)} = \frac{D_{i(2)} - D_{1(2)}}{D_{z(sr)}} \cdot 100\% \qquad (5b)$$

$$\varepsilon = \varepsilon_{i(sr)} = 0.5\left(\varepsilon_{i(1)} + \varepsilon_{i(2)}\right) \qquad (6)$$

or (if for each of the measurement directions the quantities measured on the measuring pins are also averaged first):

$$D_{i(sr)} = 0.5\left(D_{i(1)} + D_{i(2)}\right) \qquad (7)$$

$$D_{1(sr)} = 0.5\left(D_{1(1)} + D_{1(2)}\right) \qquad (8)$$

$$\varepsilon = \varepsilon_{i(sr)} = \frac{D_{i(sr)} - D_{1(sr)}}{D_{z(sr)}} \cdot 100\% \quad (9)$$

- in the second method, in every case D_z is the value obtained for a given direction of the measurement [11, 12]:

$$\varepsilon'_1 = \varepsilon'_{i(1)} = \frac{D_{i(1)} - D_{1(1)}}{D_{z(1)}} \cdot 100\% \quad (10a)$$

$$\varepsilon'_2 = \varepsilon'_{i(2)} = \frac{D_{i(2)} - D_{1(2)}}{D_{z(2)}} \cdot 100\% \quad (10b)$$

$$\varepsilon' = \varepsilon'_{i(sr)} = 0.5\left(\varepsilon'_{i(1)} + \varepsilon'_{i(2)}\right) \quad (11)$$

where (in equations (4) - (11)):

D_1, D_i – diameter on measuring pins, recalculated to the reference condition, for the first and the i-th measurement (measurements performed before the pipeline commissioning and after time τ of operating hours, respectively), mm,

D_z – pipeline outer diameter (according to the measurement made prior to the pipeline commissioning), mm,

$D_{z(1)}$, $D_{z(2)}$ – pipeline outer diameter in direction (1) and (2), respectively (the directions are perpendicular to each other), mm

ε, ε' – pipeline creep strain, %,

ε_i – steam pipeline creep strain obtained after time τ of operating hours, %.

In equations (4) - (11), in nomenclatures with the meaning as defined in the marking list above, subscripts sr, 1 and 2 (in parentheses) correspond to the following values, respectively: the average one and the ones in the two assumed directions – 1 and 2 – which are perpendicular to each other (for horizontal pipelines it may be assumed that the directions will be in the vertical and in the horizon planes, and for vertical ones – they will be assumed in relation to e.g. the building longitudinal axis, i.e. the vertical or the horizontal direction thereto).

It should be noted that formulae (6) and (9) are equivalent. Consequently, they will give identical values. It may also be added that:

- sometimes it is not the pipeline initial diameter (i.e. the one measured prior to the pipeline commissioning) that is assumed as the reference value, but the average determined based on the two diameters measured on the measuring pins during the first measurement (recalculated for the reference temperature) and adopted as the reference measurement [3]. In such cases, the general structure of the formulae corresponds to relations (5a), (5b) and (9), respectively, where $D_{z(sr)}$ should be replaced by $D_{1(sr)}$. However, this approach is inappropriate because the dimension additionally takes account of the height of both measuring pins;
- in some cases, the pipeline nominal diameter was assumed as the reference parameter;
- as a basis for the calculation of creep strain ε_i, measurement results obtained after a certain time of the pipeline operation, e.g. after 5000÷8000 h, were also used [3] instead of those obtained prior to the pipeline commissioning (in this method the assumption is that the pipeline material is characterized by the second stage of the creep process). Such an approach is also inappropriate because in this situation the values calculated based on the data obtained from subsequent measurements are in fact strains that exceed a certain strain value obtained prior to such a reference measurement. While assessing creep, it is not right to omit this „initial" value. It should also be noted that such a „delay" is not equivalent to the steam pipeline material entering the second stage of the creep process [4, 5, 13-15]. Strain values obtained in this way are underrated.

The issue of different possible definitions of the diameter to be adopted as the reference quantity in calculations of the steam pipeline creep strain has not been analysed yet (and therefore there is no reference literature on the problem). It should be supposed that its impact on determined values may be negligible at low values of the pipeline cross-section ovality. At higher values, however, this is no longer so obvious. This is due to the fact that the quantity, i.e. the initial ovality (and the pipe wall non-uniform thickness on the perimeter), has a great effect on creep strain values, as proved by earlier analyses of the problem [9, 16, 17]. The initial ovality characterizing the profile of the pipeline cross-section in the place where the measuring pins are installed (Fig. 1) can be determined using the following relation:

$$a_p = \frac{2(D_{zw} - D_{zm})}{D_{zw} + D_{zm}} \cdot 100\% \quad (12)$$

where:

a_p – initial ovality of the pipeline cross-section profile, %,

D_{zm}, D_{zw} – pipeline outer diameter (in a given cross-section, smaller and bigger one, respectively, in directions perpendicular to each other; it may therefore be assumed that e.g. $D_{zm} = D_{z(1)}$ and $D_{zw} = D_{z(2)}$, mm.

3 Calculation results and analysis

Using own results of creep testing of high-pressure steam pipelines obtained during many years of measurements performed in one of the Polish thermal power plants, creep strain was calculated by means of the two methods presented above, i.e. using formulae (4)-(11). Two facilities were selected for the analysis. They are made of the same (Cr-Mo-V) steel grade, operated at similar superheated steam parameters (temperature about 530-535°C, pressure about 9.2-9.4

MPa) and tested systematically from the time of commissioning. They are straight pipeline sections which differ from each other in initial ovality calculated from relation (12).

The creep strain values (both average ones and those obtained in two different directions) are presented in Figs. 2-5. The figures also include data concerning the tested pipelines, i.e. the pipe nominal dimensions, the operating parameters (for transported superheated steam), the initial diameters (in both measuring directions) and the initial ovality value.

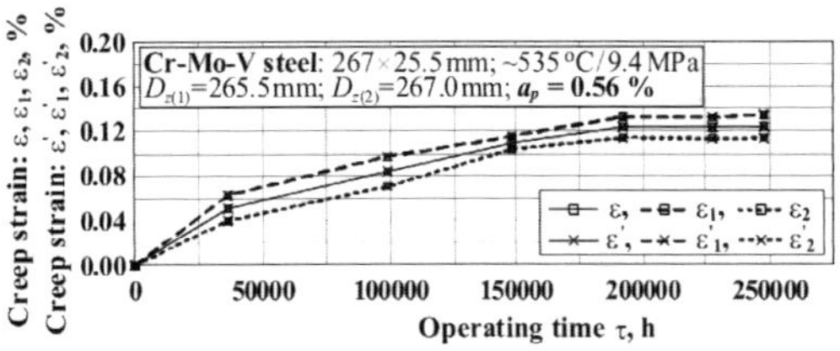

Fig. 2. Pipeline creep curves (for a_p = 0.56 %) plotted according to formulae (4)-(6) and (10a)-(11).

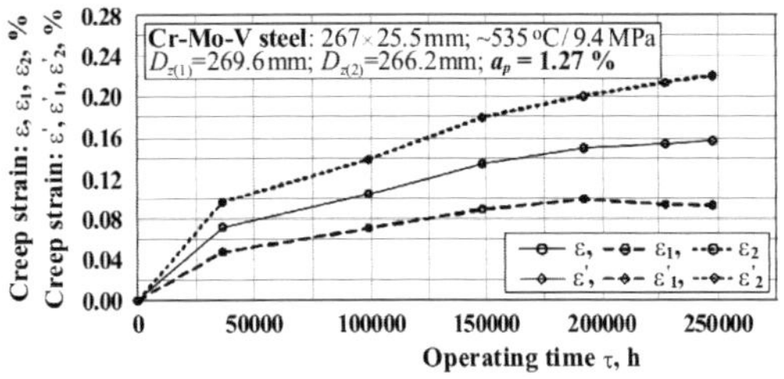

Fig. 3. Pipeline creep curves (for a_p = 1.27 %) plotted according to formulae (4)-(6) and (10a)-(11).

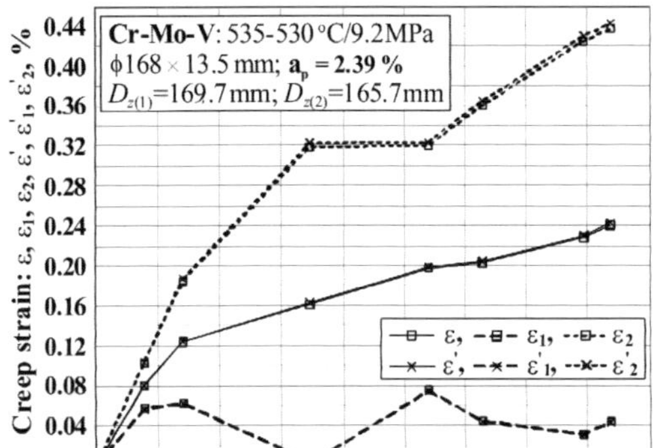

Fig. 4. Pipeline creep curves (for a_p = 2.39 %) plotted according to formulae (4)-(6) and (10a)-(11).

An analysis of Figs. 2-5 indicates that the history of changes in average values of creep strain (ε, ε'), i.e. the creep curves, are characteristic of the creep phenomenon [13-15]. One of its features is the slow, time-dependent plastic deformation of the pipeline material (here represented by the above-mentioned increment in the pipe diameter). Two of the phenomenon three stages are visible: the entire first stage (the unsteady-state process taking place at a creep rate continuously decreasing with time until no more changes occur in it) and the second one (the steady-state process with a constant creep rate). It may be added that the last stage of the phenomenon is the accelerated increase in the material deformation (the process occurs with a rise in the creep rate leading to destruction, which makes it undesirable in the power equipment operation – the elements operating in creep conditions should be replaced early enough). All the creep process stages mentioned above are also referred to as first-order (primary) creep, second-order (secondary or steady-state) creep and third-order (tertiary or pre-destruction) creep.

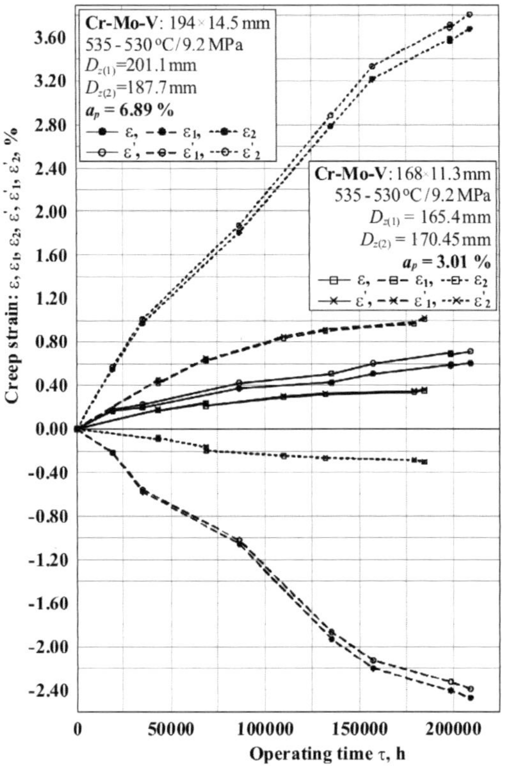

Fig. 5. Pipeline creep curves (for a_p = 3.01 % and a_p = 6.89 %) plotted according to formulae (4)-(6) and (10a)-(11)

The above remark has a general nature, but more specific characteristics can be formulated based on Figs. 2-5, such as:

- average values of creep strain in steam pipelines (made of similar materials and with similar operating parameters) are reached after comparable times of operation and they depend on the initial ovality value (the higher it is, the bigger the deformation). For example for the facilities under consideration: for a_p of 0.59%, 1.27%, 2.39%, 3.01% and 6.89% after 100,000 hours of operation the values of ε are: 0.084%, 0.104%, 0.170%, 0.280% and 0.395%, respectively;

- the curves plotted independently for the two measuring directions are substantially dependent on the initial ovality value. As it rises, the differences between the creep curves, e.g. ε_1 and ε_2 or ε'_1 and ε'_2, get bigger; it can also be seen that for the direction in which the pipeline diameter ($D_{z(1)}$ or $D_{z(2)}$) is smaller, the deformations are bigger and vice versa. In the cases under analysis where $a_p > 3\%$ (cf. Fig. 5) even „negative" strain values can be observed for the direction in which the pipeline diameter is bigger (and, respectively, high positive values in the other direction). Therefore, it may be stated that with time the pipeline profile gets closer to the round one (its ovality decreases);
- the determined creep strain values (both the average ones and those determined independently for the two measuring directions) are affected by the reference diameter, and the impact gets bigger with a rise in the initial ovality value. Relations (10a)-(11) give higher values compared to relations (4)-(6). However, analysing the average creep strain values (which are reliable in terms of the assessment of the pipeline suitability and operational safety), it should be noted that from the practical point of view the differences between the values obtained by means of relation (6) for ε and relation (11) for ε' are negligible for low values of the initial ovality. For the facilities under consideration, the relative values of those differences (denoted as δ) are shown in Fig. 6; the value is calculated from the following formula:

$$\delta = \frac{\varepsilon' - \varepsilon}{\varepsilon} \cdot 100\% \tag{13}$$

where:

δ – relative difference in creep-related strain determined using relations (11) and (6), %

ε, ε' – pipeline creep strain, %

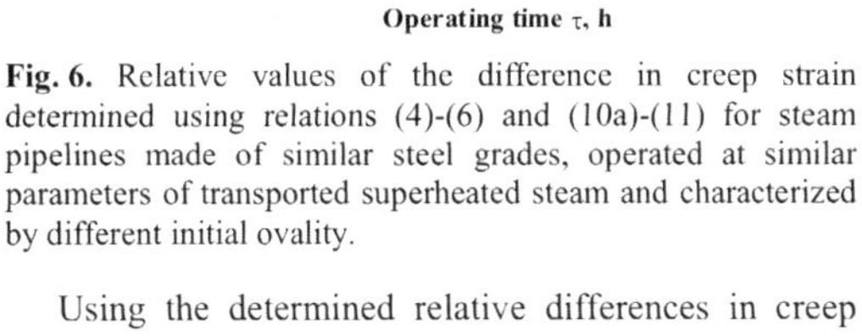

Fig. 6. Relative values of the difference in creep strain determined using relations (4)-(6) and (10a)-(11) for steam pipelines made of similar steel grades, operated at similar parameters of transported superheated steam and characterized by different initial ovality.

Using the determined relative differences in creep strain δ, their average values were calculated in the second stage of the creep process of the analysed pipelines. For certainty it is assumed that in each case the second stage occurred after 75,000 hours of operation. The obtained results are presented as a function of initial ovality in Fig. 7.

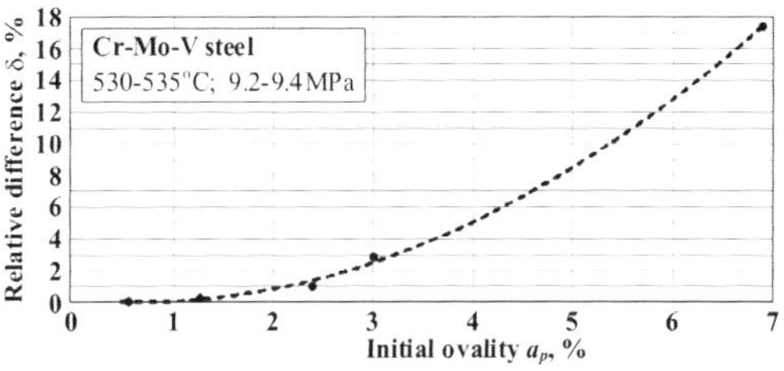

Fig. 7. Relative value of the difference in creep strain determined using relations (4)-(6) and (10a)-(11) depending on initial ovality of the steam pipelines under analysis.

Based on Fig. 7 it can be noticed that the impact of the adopted reference diameter in the calculations of the steam pipeline creep strain is more essential if the pipeline initial ovality is bigger than 2.5÷3.0%. Below this limit the differences in the obtained values are relatively small and do not exceed 1.6÷2.9%. A rise in initial ovality over 3.0% involves a „fast" increase in the difference – for $a_p \approx 4.5\div6.0\%$ the difference is: $\delta \approx 7\div13\%$, respectively, and in the case of the tested pipeline with ovality $a_p = 6.89\%$ it gets as high as $\delta \approx 17.4\%$.

4 Concluding remarks

Determination of parameters characterizing the creep process of materials which are used to make pressure elements of power engineering installations and equipment is an essential task of technical diagnostics. One of the parameters is creep-related plastic strain (usually referred to as creep strain). The strain value, which rises with the time of the steam pipeline operation, is a reliable indicator in the assessment of the pipeline suitability for further operation and of operational safety. For this reason, the accuracy with which the quantity is determined is of great importance, too.

In the calculations of the steam pipeline creep strain, performed independently for each of the two mutually perpendicular directions in which the measuring pins are installed (identical to the directions in which periodic tests are carried out), the reference parameter is the diameter defined prior to the pipeline commissioning. The diameter can be determined as an average value of diameters measured in the two directions (which means that for both directions it is the same) or it can be the diameter corresponding to one specific direction only. Due to that, the creep strain calculated in one specific direction has different values. Consequently, the average values obtained based on data from the two testing directions are also different. The differences are strongly affected by the tested pipeline initial ovality (profile) and they get bigger with a rise in it.

Practically, the steam pipeline creep strain calculation results obtained by means of the methods presented above differ only slightly, i.e. they do not exceed 3% provided that the pipe initial ovality is smaller than 3.0%, which means that it does not matter which method is actually applied. In the case of profiles with initial ovality higher than 3.0%, the differences are substantially bigger (e.g. for ovality of 4.5, 6.0 and 6.9% they total about 7, 13 and 17%, respectively, and it can be seen that the rate of the rise is characteristically „fast"). Therefore, it is recommended that the method described by formulae (10a)-(11) should be used because it gives higher results, thus ensuring higher safety in terms of the performed assessment of the pipeline suitability for further operation (especially in periods extended beyond the computational operating lifetime). The results of periodic creep strain testing are compared with permissible values and the steam pipeline diagnostics based on this particular method offers a wider safety margin.

References

1. EN 45510-7-1:1999. Guide for procurement of power station plant, equipment and systems – Part 7-1: Pipework and valves – High pressure piping systems
2. F. Młynarski, *Changes in steel properties of the boiler equipment with high steam parameters* (in polish), Cracow University of Technology, Zeszyt Naukowy No 12, series: Mechanika No 11 (1963)
3. M. Drop, H. Kruczalak, E. Włoczyk, *Guidelines for plastic strain testing - creep measurements* (in polish), ZP-BE „Energopomiar", Gliwice (1977)
4. S. Łopata, *Lifetime of high-pressure steam pipelines* (in polish), Zeszyty Naukowe Politechniki Krakowskiej, series: Mechanika No 81, Kraków (2001)
5. S. Łopata, Proceedings of the Ninth International Conference on Pressure Vessel Technology, **2**, p. 621, Sydney, Australia (2000)
6. G. Thoraval, Nuclear Engineering and Design **116**, North-Holland, Amsterdam (1989)
7. Technical Inspection Office, *Technical conditions of technical inspection. Pressure equipment. Steam pipelines connecting the boiler to the turbine* (in polish) TD-UC-90/RC, Warszawa (1991)
8. S. Łopata, Zeszyty Problemowe. Badania nieniszczące, No 2, pp. 77-82, Warszawa (1997)
9. S. Łopata, M. Augustyniak, K. Kwintowski, Proceedings of the Sixth International Congress on Thermal Stresses, **1**, pp. 409-412, Austria (2005)
10. W. Bęczkowski, *Power pipelines* (in polish), WNT, Warszawa (1963)
11. *Steel for boilers and organization of control over their behavior in the high steam parameters. Installation and operation of high-pressure boilers* (in russian), Ed. by S.C. Faerman, S.M. Schuher, GEI, Moscow, Leningrad (1949)
12. N.F. Osipowski: *High-pressure boilers* (in polish), PWT, Warszawa (1956)
13. F. Garofalo, *Fundamentals of Creep and Creep Rupture in Metals*. The Macmillan Company, New York, Collier-Macmillan Limited, London (1961)
14. M.F. Ashby, D.R.H. Jones, *Engineering Materials. An Introduction to their Properties and Applications*. Butterworth-Heinemann Ltd. (1993)
15. A. Jakowluk, *Creep and fatigue processes in materials* (in polish), WNT, Warszawa (1993)
16. S. Łopata, *Creep of the steam pipeline with wall thickness non-uniformity*, Advanced Materials Research, **875-877**, pp. 500-506 (2014)
17. S. Łopata, K. Kwintowski, *Creep of steam pipelines with ovality cross-section*, Czasopismo Techniczne. Mechanika (Technical Transactions, Mechanics), Cracow University of Technology, z. 2-M, pp. 181-190 (2008)

Numerical simulations of fluidization dynamics in a hot model of a CLC process

Anna Żyłka[1,*], *Jarosław* Krzywański[2], *Tomasz* Czakiert[1], *Kamil* Idziak[1], *Krzysztof* Kulicki[1], *Sylwia* Jankowska[1], and *Wojciech* Nowak[3]

[1]Czestochowa University of Technology, Faculty of Infrastructure and Environment, Institute of Advanced Energy Technologies, J.H. Dabrowskiego 73, 42-200 Czestochowa, Poland
[2]Jan Dlugosz University in Czestochowa, Institute of Technology and Safety Systems, Armii Krajowej 13/15, 42-200 Czestochowa, Poland
[3]AGH University of Science and Technology, Department of Thermal Machinery and Flow, Faculty of Energy and Fuels, Czarnowiejska 50B, I, 30-054 Cracow, Poland.

Abstract. Chemical Looping Combustion (CLC) is one of the most promising alternatives for solid fuel combustion. CO_2 concentration in the exhaust gas is high in CLC technology which enables high efficiency of CO_2 capture from flue gas. The use of solid oxygen carriers is a characteristic feature of a CLC process. Oxygen carriers are mainly metal oxides which are characterized by high oxygen transfer capacity and high mechanical resistance.
Since the CLC technology is not sufficiently recognized due to its complexity the development of models with real conditions of the CLC equipment is of practical significance.
The paper presents numerical simulations of the dynamic fluidized bed for Chemical Looping Combustion using CeSFaMB software. The model was validated on the basis of the results obtained from experiments, which were carried out on the Fluidized-Bed Chemical-Looping-Combustion of Solid-Fuels (FB-CLC-SF) unit. The studies were conducted in air atmosphere at temperature of 850°C. The validation of the 1.5D model showed that the maximum relative error between experiment and simulations results does not exceed 12%.

1 Introduction

Burning of fossil fuels is one of the main sources of CO_2 emissions into the atmosphere. It is therefore essential to develop new combustion technologies, which allows the use of fossil fuels while reducing CO_2 emissions [1]. One of the interesting alternative to this problem is Chemical Looping Combustion (CLC). This technology is characterized by using solid oxygen carriers for fuel combustion [2]. The absence of air in the combustion chamber allows to obtain the exhaust gas containing mostly CO_2 and H_2O (Fig. 1). The almost pure CO_2 is produced of the flue gas cleaning and condensation of steam. Therefore, the chemical looping combustion process allows to avoid expensive CO_2 absorption installations [3, 4].

This paper presents the results of numerical simulations of the fluidization dynamics in a hot model of a CLC process which are a continuation of the previous works carried out on the cold model [3, 5].

The test units for cold and hot model (FB- CLC-SF) are located in Institute of Advanced Energy Technologies at the Czestochowa University of Technology and they were built in the framework of Polish - Norwegian research project NewLoop "Innovative Idea for Combustion of Solid Fuels via Chemical Looping Technology" [6].

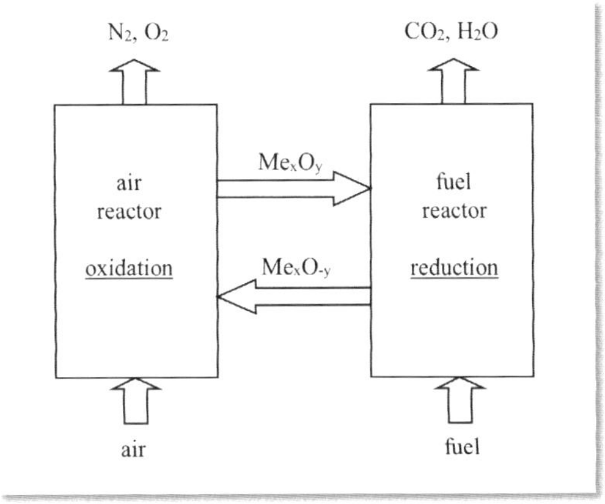

Fig. 1. The scheme of chemical looping combustion process [4]

* Corresponding author: azylka@is.pcz.pl

2 Materials and Methods

2.1 An object of investigations

The innovative ideology of NewLoop project combines two complementary technologies: CLOU *(Chemical looping with Oxygen Uncoupling)* and iG-CLC *(In-situ Gasification Chemical Looping),* which create a so-called *HCLC (Hybrid Chemical Looping Combustion)*. The CLOU process is characterized by the use of solid oxygen carrier, which does not require direct contact with the fuel to release oxygen in the fuel reactor. By contrast, the iG-CLC process requires that the oxygen carrier is in contact with the fuel [5, 7].

The experimental studies, as well as validation of the model were carried out on a Fluidized-Bed Chemical-Looping-Combustion of Solid-Fuels (FB-CLC-SF) unit (Fig. 2a). The system consists of two main circular contours. The first circular contour is composed of an air reactor with circulating fluidized bed, riser, cyclon and return system. The second circular contour consists of a fuel reactor with bubbling fluidized bed and particulate collector. (Fig. 2b). The main dimensions of the FB-CLC-SF unit are presented in Tables 1 and 2.

Table 1. The main dimensions of the fuel reactor

	Chamber I	Chamber II
The inner equivalent diameter of combustion chamber, [m]	0.074	0.074
The height of combustion chamber, [m]	0.500	0.500

Table 2. The main dimensions of the air reactor and cyclon

The diameter of the air reactor, [m]	0.098
The total height of the riser and air reactor, [m]	2.55
The diameter of the riser, [m]	0.04
The efficiency of the cyclone, [%]	99
The inner diameter of the cyclone, [m]	0.066
The height of the cylindrical portion of the cyclone, [m]	0.190
The height of the conical part of the cyclone, [m]	0.130
The diameter of the downcomer, [m]	0.028

a)

b)

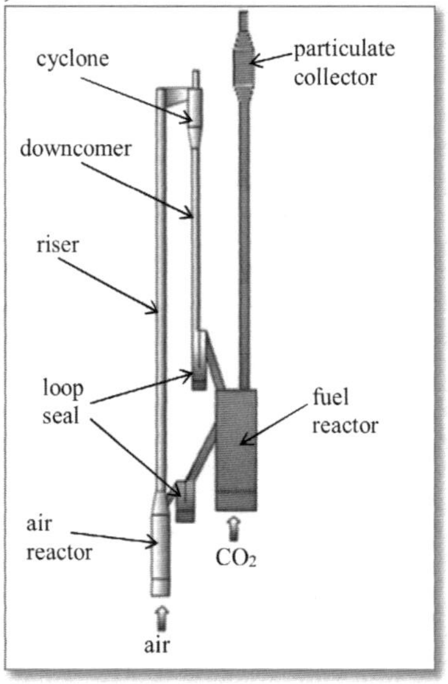

Fig. 2. The hot model ofFB-CLC-SF unit: (a) Photography of the FB-CLC-SF unit in the Institute of Advanced Energy Technology at the Czestochowa University of Technology; (b) The scheme of FB-CLC-SF unit [5]

2.2 Research material

The sand with the Sauter mean diameter of particles of 166 μm, true density of 3200 kg/m^3 and the sphericity of 0.7 are used during the investigations (Fig. 3).

Fig. 3. Microscopic photograph of sand - research material used in the study

2.3 Characteristics of CeSFaMB software

The numerical simulations of the CLC process were carried out using the CeSFaMB software. (Comprehensive Simulator of Fluidized and Moving Bed equipment), which is especially dedicated to the modeling of fuel conversion implemented in bubbling and circulating fluidized bed (Fig. 4). CeSFaMB simulator requires the formulation of the so-called initial and boundary conditions. In order to confirm the reliability of the emerging model, both introduced and designated values should be verified by experimental studies. Otherwise, the results of the model calculations are purely hypothetical [5].

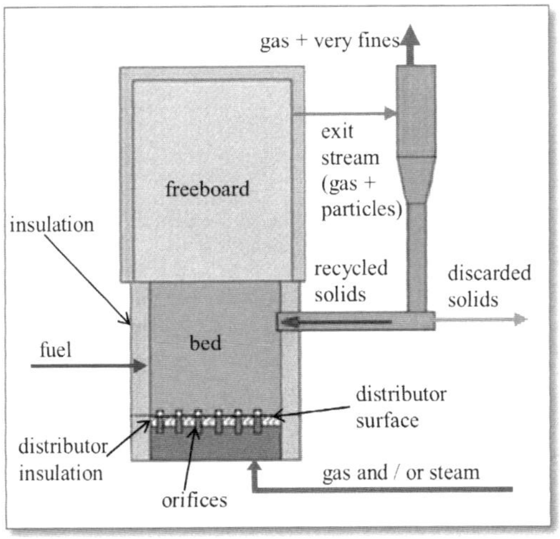

Fig. 4. The fluidized bed system considered by the CeSFaMB simulator [8]

CeSFaMB allows to determine the following operating parameters of the fluidized systems: [8-10]:

- temperature profile,
- particle-size distribution profiles,
- concentration and mass flow profiles of 20 gas components (Ar, CO_2, CO, O_2, N_2, H_2O, H_2, CH_4, SO_2, NO, N_2O, NO_2, HCN, C_2H_6, H_2S, NH_3, C_2H_4, C_3H_6, C_3H_8, C_6H_6) and tar throughout the bed in the emulsion and bubble phases
- velocity of the bubbles in the bed,
- diameter of bubbles depending on the height of the fluidized bed,
- the minimum fluidization velocity,
- superficial velocities in the bed and freeboard,
- mass flow of gases in the bed,
- mass flow of particles,
- void fractions in the bed and freeboard.

2.4 Research methodology

The experimental studies were carried out in three tests in which the CLC unit is operated stably. In each test, the velocity of the gas was changed in the air reactor. The study aimed to determine the effect of change of gas velocity in the air reactor on the basic parameters of the CLC system.

Table 3. Main operational conditions of the FB-CLC-SF unit

	Test No. 1	Test No. 2	Test No. 3
Gas flux in air reactor, [$\times 3600^{-1}$ m^3 s^{-1}]	9.00	21.84	25.67
Gas flux in fuel reactor, [$\times 3600^{-1}$ m^3 s^{-1}]	8.6	8.6	8.6
Temperature, [K]	1123	1123	1123
Absolute pressure below the gas distributor in air reactor, [Pa]	104 727	105 391	105 739
Absolute pressure below the gas distributor in fuel reactor, [Pa]	104 863	105 350	105 768
Total mass of solids in the air reactor [kg]	2.69	2.44	2.38
Total mass of solids in the fuel reactor [kg]	2.00	2.09	2.14

Modeling the dynamics of fluidized bed for CLC process performed using a CeSFaMB software. Since the geometry is rather complex: Air Reactor (with the Circulation Fluidizing Bed approach) and Fuel Reactor (with the Bubbling Fluidizing Bed approach) were considered separately and boundary conditions between these two main parts of the system were used to link them.

The method of modeling CLC process using CeSFaMB simulator was presented in the papers [3, 5]. These articles show the results for the CLC cold model where the process is carried out at a temperature of 25 °C. In contrast, this work presents the results for CLC hot model - at a temperature of 850 °C.

The main operating parameters for these tests, including the information on the gas flux, pressure and

DOI: 10.1051/e3sconf/20171304002

temperature as well as the total mass of circulating solids are given in Table 3.

3 Results and discussion

3.1 Model validation

Numerical simulations were carried out taking into account the process conditions during experimental tests. The comparison and verification of the calculation and experiment results involves the following parameters:
- main pressure losses,
- solids recirculation flux,
- void fractions,
- superficial gas velocity,
- gas mass flow rate,
- bubble diameter,
- rising velocity of gas bubbles.

3.1.1 Main pressure losses and solids recirculation flux

The average pressure losses across the beds for tests No. 1 - 3, predicted by the model as well as the errors of the predictions are given in Table 4. The relative errors between experimental and calculated results are lower than 10 %.

Table 4. Pressure loss across the bed, [Pa]

	Air Reactor	Fuel Reactor - Chamber I	Fuel Reactor - Chamber II
Test No. 1	2 819.54	1 971.31	1 971.31
Err [%]	1.05	0.95	0.95
Test No. 2	2 699.34	2 245.90	2 245.90
Err [%]	10.03	0.18	0.18
Test No. 3	2 711.59	2 315.16	2 315.16
Err [%]	2.52	0.19	0.19

Total mass of solids recycled to the bed from the cyclone system in the main cycle equal 0.0026 kg s^{-1}, 0.0033 kg s^{-1} and 0.0034 kg s^{-1}, for tests No. 1, 2 and 3, respectively. The relative errors between experimental and calculated results are lower than 1 %.

3.1.2 Void fractions

The dense fluidized bed is formed in the lower part of the reaction chamber. By contrast, the lean region is located above the dense region and covers a predominant part of the chamber [11-14].

The comparison between experimentally determined and predicted by the model void fractions in dense region in the air reactor and fuel reactor is given in Figures 5 and 6. A similar comparison in lean region in the air reactor and fuel reactor is shown in Figures 7 and 8.

The relative errors between experimental and calculated results are lower than 12% for dense zone and 1% for lean region, respectively.

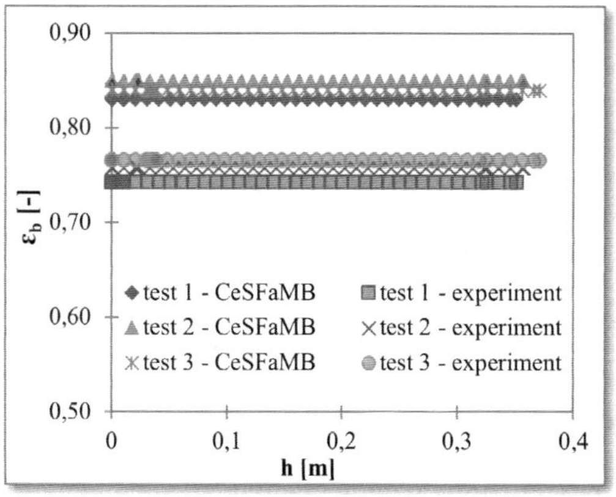

Fig. 5. Comparison between desired and predicted void fractions in dense region of the air reactor

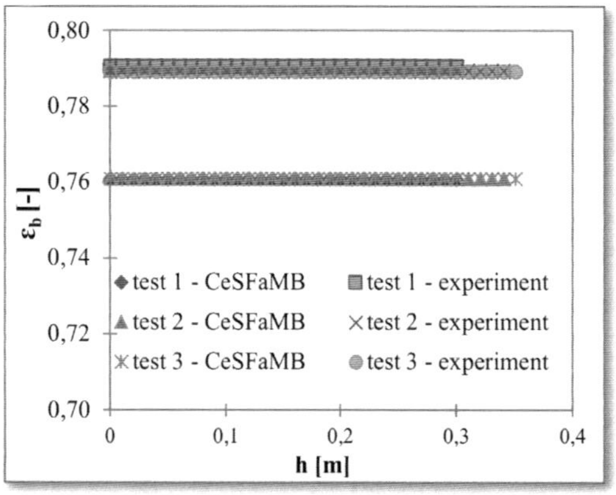

Fig. 6. Comparison between desired and predicted void fractions in dense region of the fuel reactor

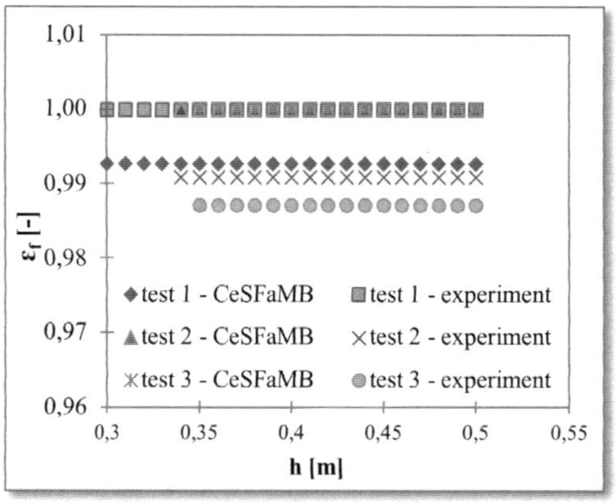

Fig. 7. Comparison between desired and predicted void fractions in lean region of the air reactor

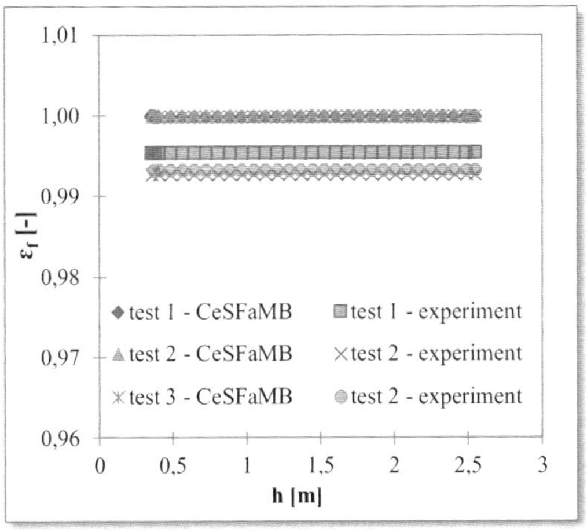

Fig. 8. Comparison between desired and predicted void fractions in lean region of the fuel reactor

The height of the bed is for each test in the air reactor are as follows: 0.35 m (test 1), 0.36 m (test 2), 0.37 m (test 3) and fuel reactor: 0.30 m (test 1), 0.34 m (test 2), 0.35 m (test 3), respectively.

3.1.3 Superficial gas velocity and gas mass flow rate

The superficial gas velocity depends on the geometry of the reaction chamber whereas the gas mass flow rate is highly dependent on the process temperature [5].

The calculation results are shown in Figures 9 and 10. The maximum relative errors between experimental and calculated superficial gas velocities for the dense bed region of AR and FR do not exceed 2% and 5%, respectively.

The maximum relative errors of gas mass flow rate estimation are located within the range of ± 5% compared to the experimental data. This comparison is shown in the Figures 11 and 12.

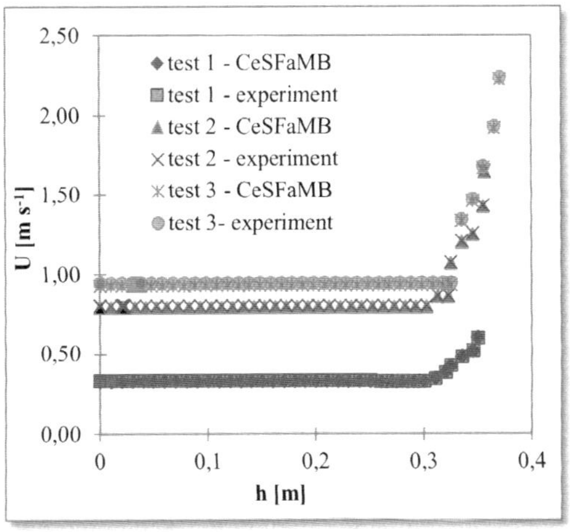

Fig. 9. Comparison between desired and predicted superficial gas velocity in dense bed region of air reactor

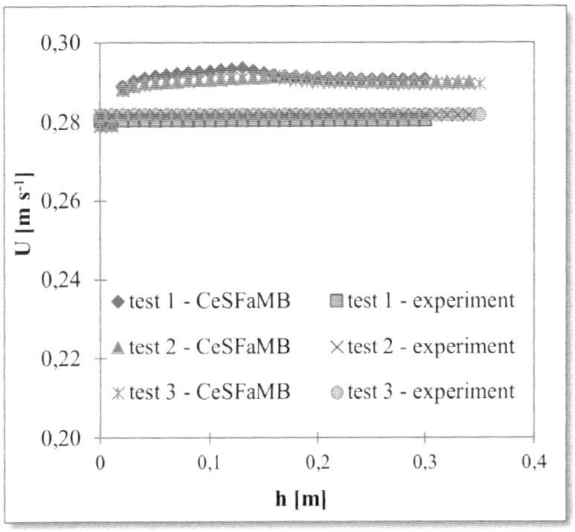

Fig. 10. Comparison between desired and predicted superficial gas velocity in dense bed region of fuel reactor

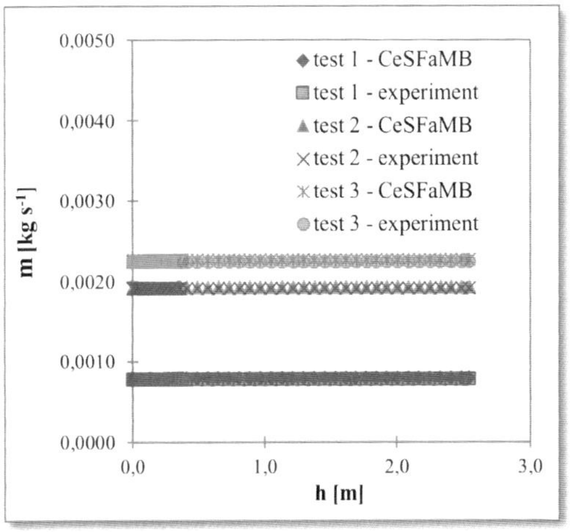

Fig. 11. Comparison between desired and predicted gas mass flow rate in the air reactor

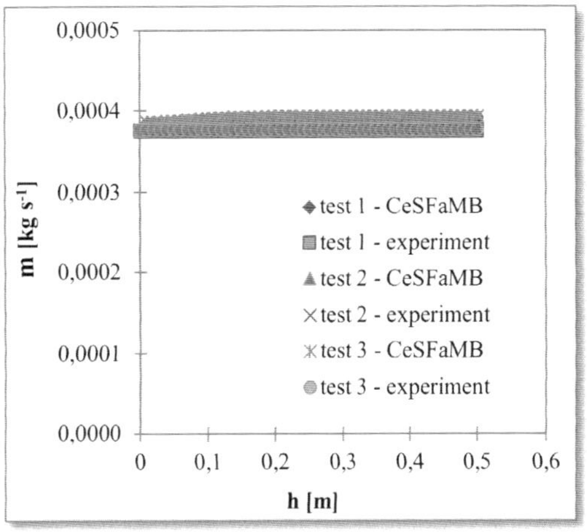

Fig. 12. Comparison between desired and predicted gas mass flow rate in the fuel reactor

3.1.4 Bubble diameter and rising velocity of gas bubbles

In the bubbling and circulating fluidized bed the increase of bubbles diameter occurs with the height of the reaction chamber. Similarly, the rising velocity of gas bubbles increases with the height of the reaction chamber. Furthermore, the rising velocity of gas bubbles is dependent on the bubble diameter. These correlations are described by the following formulas [3, 15]:

$$d_B = 0.430(U - U_{mf})0.4(h + 0.1272)0.8g^{-0.2} \quad (1)$$

$$U_B = U - U_{mf} + 0.711(gd_B)^{0.5} \quad (2)$$

The comparison between experimentally determined and predicted by the model for bubble diameter and rising velocity of gas bubbles are shown in Figures 13-16.

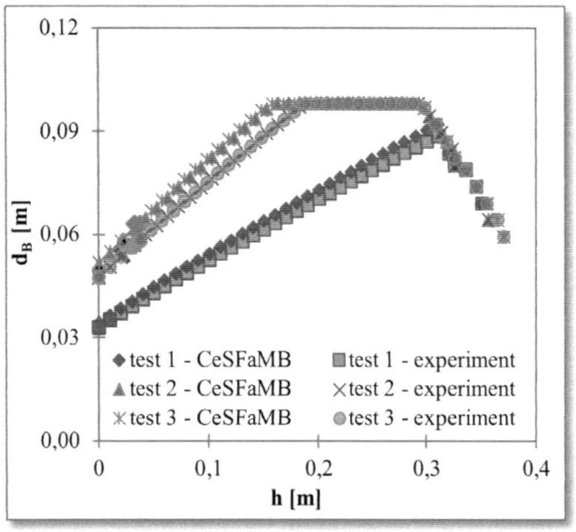

Fig. 13. Comparison between desired and predicted bubble diameter in dense bed of the air reactor

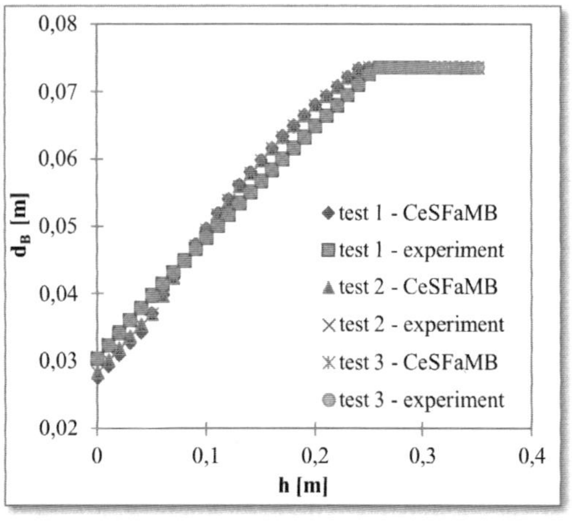

Fig. 14. Comparison between desired and predicted bubble diameter in the fuel reactor.

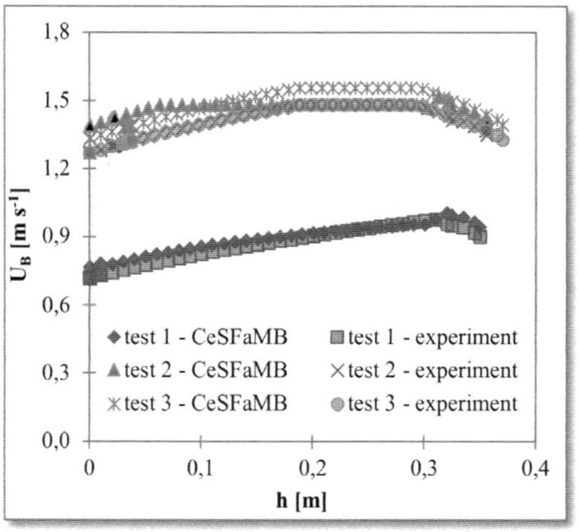

Fig. 15. Comparison between desired and predicted rising velocity of bubble in the air reactor.

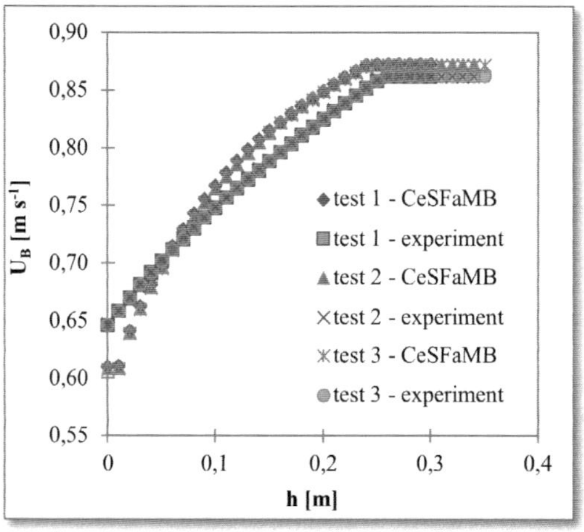

Fig. 16. Comparison between desired and predicted rising velocity of bubble in the fuel reactor

The calculated results do not exceed 10 % compared to the experimental data, for bubble diameter and velocity of a bubble rise.

3.2 Empirical models for basic dynamic parameters in CLC process

The successfully validated mathematical model allows to determine the influence of the superficial gas velocity on the following basic dynamic parameters of a CLC process [15]:

- solid circulating rate (G_S) in the air reactor,
- number of a particle cycles in air and fuel reactor,
- particles residence time in air and fuel reactor,
- mass of solids in air and fuel reactor.

The influence of the superficial gas velocity on solid circulating rate is given in Figure 17. An increase in superficial gas velocity in air reactor causes an increase in the solids circulating rate. This dependence can be described by the following formula (3):

$$G_S = 1.91\, e^{0.36\, v^{AR}} \tag{3}$$

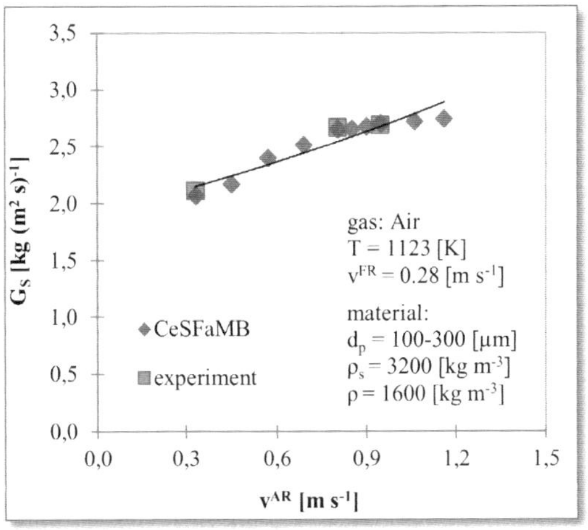

Fig. 17. Solid circulating rate versus superficial gas velocity in the air reactor

The relations of the superficial gas velocity and number of cycles in air and fuel reactors are given in Figures 18 and 19. The analyzed relations is described in equations (4 and 5):

$$n_c{}^{AR} = 3.02\, e^{-0.12\, v^{AR}} \tag{4}$$

$$n_c{}^{FR} = 4.62\, e^{-0.12\, v^{AR}} \tag{5}$$

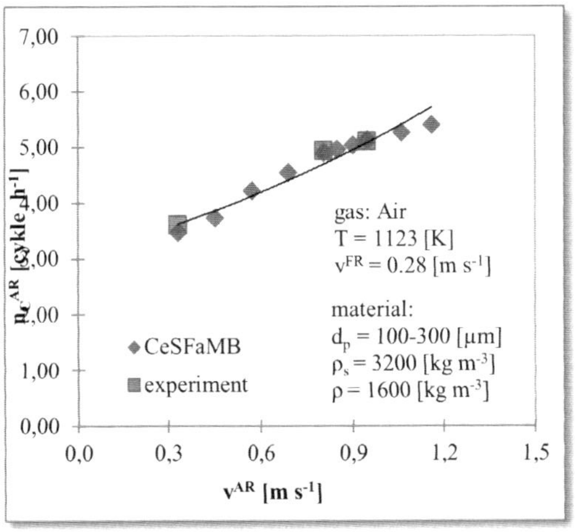

Fig. 18. The number of cycles in air reactor versus superficial gas velocity in the air reactor

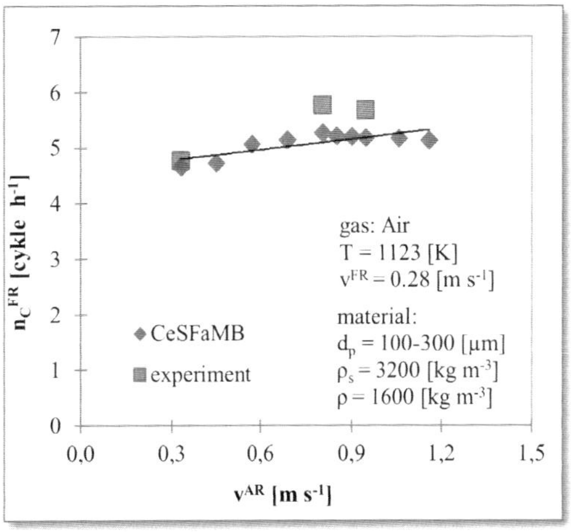

Fig. 19. The number of cycles in fuel reactor versus superficial gas velocity in the air reactor

The results of the reduction in residence time of particles both in air and fuel reactors are shown in Figures 20 and 21. This can be described by the following correlations:

$$t_R{}^{AR} = 1193\, e^{-0.6\, v^{AR}} \tag{6}$$

$$t_R{}^{FR} = 780\, e^{-0.12\, v^{AR}} \tag{7}$$

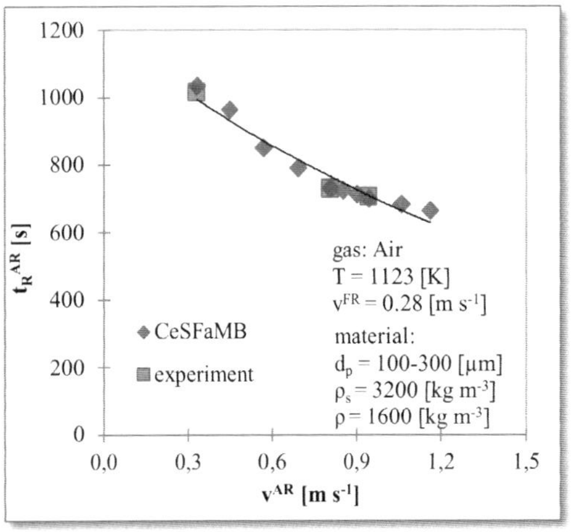

Fig. 20. The particles' residence time in air reactor versus superficial gas velocity in the air reactor

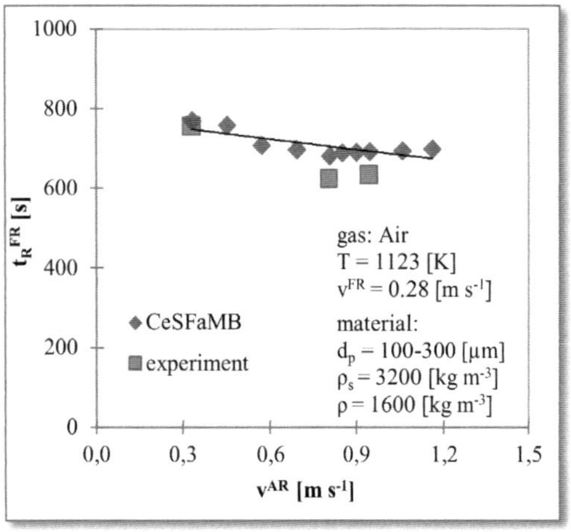

Fig. 21. The particles' residence time in fuel reactor versus superficial gas velocity in the air reactor

The dependences between superficial gas velocity and mass of solids contained in reactors are given in Figures 22 and 23.

The increase of the superficial gas velocity in air reactor causes the decrease in the total mass of solids contained in air reactor (Fig. 22) and an increase in the total mass in fuel reactor (Fig. 23). These behaviour are given in the forrmulas 8 and 9:

$$m^{AR} = 0.48\, v^{AR} + 2.84 \quad (8)$$

$$m^{FR} = 0.51\, v^{AR} + 1.85 \quad (9)$$

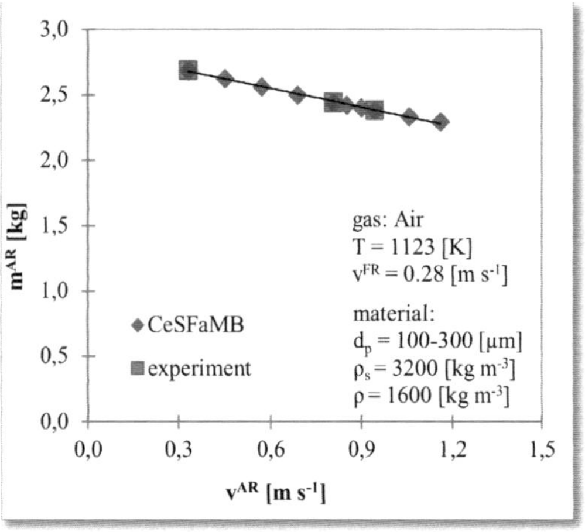

Fig. 22. The total mass in air reactor versus the superficial gas velocity in the air reactor

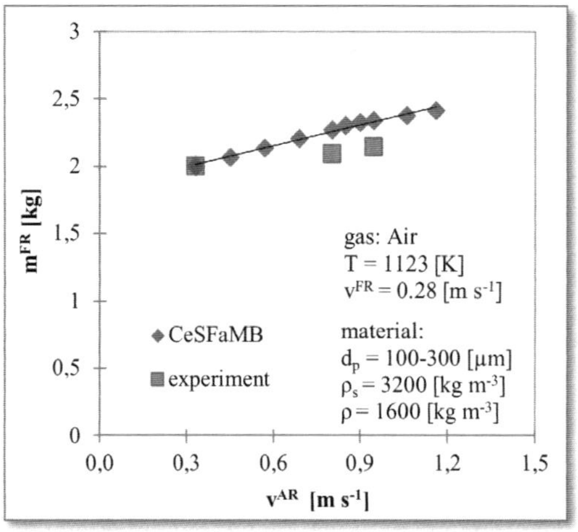

Fig. 23. The total mass in fuel reactor versus superficial gas velocity in the air reactor.

4 Summary and Conclusions

The first results of numerical simulations of the dynamic fluidized bed for the hot model of FB-CLC-SF unit are discussed in the paper. The simulations are carried out by the use of CeSFaMB Simulator. The standard operating conditions for the bubbling and circulating fluidized bed are used during the study.

A successfull validated model allows to carry out the simulation of system operation and determine the influence of the superficial gas velocity on the solids circulating rate, number of a particle cycles, particles residence time, as well as mass of solids in air and fuel reactor.

The increase of the superficial gas velocity in air reactor causes the increase in the solid circulating rate, number of cycles in both reactors, total mass in fuel reactor and decrease in the particles' residence time in both reactors, as well as total mass in air reactor.

The comparison of the experiment and calculations results involves the following parameters: main pressure losses, solids recirculation flux, void fractions, superficial gas velocity, gas mass flow rate, bubble diameter and rising velocity of gas bubbles. The validation of the model showed that the maximum relative error between the simulations and experiment results does not exceed 12%, which confirms the correctness of the CLC model. Therefore, the CeSFaMB simulator is a useful tool for the analysis of fluidization dynamics in CLC units.

List of symbols

d – diameter, m
g – acceleration of gravity, m s^{-2}
h – vertical coordinate, m
Gs – solid circulating rate, kg m^{-2} s^{-1}
m – mass of solids, kg
t_R – particles' residence time, s
U – velocity, m s^{-1}

ε – voidage, -
n_C – number of cycles in the reactors, cycles h^{-1}

Subscripts

b – bed,
B – bubbles,
f – freeboard,
mf – minimum fluidization conditions
p – particles

Acronyms

AR – Air Reactor
FB-CLC-SF – Fluidized-Bed Chemical-Looping - Combustion of Solid-Fuels,
CeSFaMB – Comprehensive Simulator of Fluidized and Moving Bed equipment,
CLC – Chemical Looping Combustion,
CLOU – Chemical looping with Oxygen Uncoupling,
FR – Fuel Reactor,
iG-CLC – In-situ Gasification Chemical Looping.

The Project "Innovative Idea for Combustion of Solid Fuels via Chemical Looping Technology" (Agreement No. POL-NOR/235083/104/2014) is funded from Norway Grants in the Polish-Norwegian Research Programme operated by the National Centre for Research and Development. CCS 2013 Call "New innovative solutions for CO_2 capture". The support is gratefully acknowledged.

References

1. E. Ksepko, M. Sciazko, P. Babinski, *Studies on the redox reaction kinetics of Fe_2O_3-CuO/Al_2O_3 and Fe_2O_3/TiO_2 oxygen carriers*, Applied Energy **115**, 374–383 (2014)
2. K. Kulicki, T. Czakiert, A. Żyłka, S. Jankowska, J. Krzywański, W. Nowak, *Characteristics of operating the fluidized bed CLC cold model*, Energy Market (Rynek Energii) 6 (**121**), 35-40 (2015)
3. A. Żyłka, J. Krzywański, T. Czakiert, K. Kulicki, S. Jankowska, W. Nowak, *Modeling for Chemical Looping Combustion of Solid Fuels*, Energy Market (Rynek Energii) 1 (**122**), 75-79 (2016)
4. E. Ksepko, *Perovskite-type $Sr(Mn_{1-x}Ni_x)O_3$ materials and their chemical-looping oxygen transfer properties*, International Journal of Hydrogen Energy **39**, 8126-8137 (2014)
5. A. Żyłka A., J. Krzywański, T. Czakiert, K. Kulicki, S. Jankowska, W. Nowak, Reducing CO_2 emissions - Combating climate change, *Numerical modeling of fluidized bed combustion of solid fuels in chemical looping using CeSFaMB software*, Czestochowa University of Technology, Monographs Series No. **308**, 220-234 (2016)
6. T. Czakiert, K. Kulicki, J. Krzywanski, A. Zylka, S. Jankowska, W. Nowak, *First Experience in Operation of Cold Model of FB-CLC-SF (Fluidized-Bed Chemical-Looping-Combustion Solid-Fuels) Facility*, [in:] Chaouki J., Berruti F., Bi X., Cocco R. (Eds.) Fluidization XV, ECI Symposium Series (2016)
7. J. Adanez, A. Abad, G. Garcia-Labiano, P. Gayan, F. Luis F., *Progress in chemical-looping combustion and reforming technologies*, Progress in Energy Combustion Science, **38(2)**, 215-282 (2012)
8. Marcio L.de Souza-Santos, Solid Fuels Combustion and Gasification. Modeling, Simulation, and Equipment Operation (2005)
9. Marcio L. de Souza-Santos, *Comprehensive Simulator (CSFMB) Applied to Circulating Fluidized Bed Boilers and Gasifiers*, The Open Chemical Engineering Journal **2**, 106-118, (2008)
10. Marcio L. de Souza-Santos, *CSFB applied to fluidized-bed gasification of special fuels*, Fuel **88**, 826-833 (2009)
11. J. Krzywański, R. Rajczyk, W. Nowak, *Model research of gas emissions from a lignite and biomass co-combustion in a large scale CFB boiler*, Chemical and Process Engineering, **35**, (2), 217-231 (2014)
12. W. Muskała, J. Krzywański, T. Czakiert, W. Nowak, *The Research of CFB boiler operation for oxygen–enhanced dried lignite combustion,* Energy Market (Rynek Energii), 1 (92), 172–176 (2011).
13. KK. Win, W. Nowak, H. Matsuda, M. Hasatani, Z. Bis, J. Krzywański, W. Gajewski, *Transport Velocity of Coarse Particles in Multi – Solid Fluidized Bed*, Journal of Chemical Engineering of Japan, 28, 5, 535–540 (1995).
14. W. Muskała, J. Krzywański, R. Rajczyk, M. Cecerko, B. Kierzkowski, W. Nowak, W. Gajewski, I*nvestigation of Erosion in CFB Boilers,* Energy Market (Rynek Energii), 2 (87), 97 – 102, (2010)
15. J. Krzywanski, A. Żyłka, T. Czakiert, K. Kulicki, S. Jankowska, W. Nowak, *A 1.5D model of a complex geometry laboratory scale fuidized bed CLC equipment*, Powder Technology, DOI: 10.1016/j.powtec.2016.09.041 (in press)

Issues to improve the safety of 18K370 steam turbine operation

Grzegorz Bzymek[1], *Janusz* Badur[2], and *Paweł* Ziółkowski[2,*]

[1]Opole Department, PGE GiEK S.A., Brzezie near Opole, 46-021 Opole, Poland
[2]Energy Conversion Department, The Szewalski Institute Fluid-Flow Machinery PAS-ci, Fiszera 14, 80-231 Gdańsk, Poland

Abstract. The paper presents the process of improving the safety and reliability of operation the 18K370 steam turbines Opole Power Plant since the first failure in 2010 [1], up to install the on-line monitoring system [2]. It shows how the units work and how to analyse the contol stage as a critical node in designing the turbine. Selected results of the analysis of the strength of CSD (Computational Solid Dynamic) and the nature of the flow in different operating regimes - thanks to CFD (Computational Fluid Dynamic) analysis have been included. We have also briefly discussed the way of lifecycle management of individual elements [2,3]. The presented actions could be considered satisfactory, and improve the safety of operating steam turbines of type 18K370.

1 Introductions

Both due to the emerging failures as well as variable movement conditions, it is necessary to conduct research in improving the safety of power units. After analysing the load status of individual devices of the cycle, it should be noted that particularly vulnerable to variable parameters of steam are boilers [4-9], then valves [10] and finally the first stages of the turbine [11-15]. To face these threats, it is necessary to increase the safety of operation of energy cycles including steam turbine type 18K370 [1-3]. The work regimes for load of control stage are different, so we have to rely upon more accurate calculation tools [3,16] including predicting further safe operation of steam turbine rotors after the calculated working time [14] as well as the increased expenditure and even entry into resonance resulting from the non-stationary operating conditions [17,18]. Another aspect is the interaction of fatigue and creep [12,13].
The new computational tools such FSI (Fluid Solid Interaction) allow determining the fluid - solid interactions [3,6,16,18]. Of particular importance is the transfer of information on the interface (with working medium - construction) which represents the connection, and at the same time the boundary condition, between the CSD (Computational Solid Dynamic) and CFD (Computational Fluid Dynamic). Thus, the analysis of the strength of CSD derive data from CFD and could, as a result of movements, at the same time affecting on nature of the flow.

The aim of this article is to present results of analysis of failures of two casing internal parts of the HP turbines, type of 18K370 and the presentation of the activities (including numerical calculation) taken to improve the safety of operating parameters of turbines in power plant PGE GiEK SA Branch in Opole.

1.1 Main motives

The main motivations for seeking areas of safe operation of the turbine 18K370 were two failures that took place respectively: 13 May 2010 and 26 October 2010. The scope of the devastation are shown in Fig.1 and Fig.2.

a)

b)
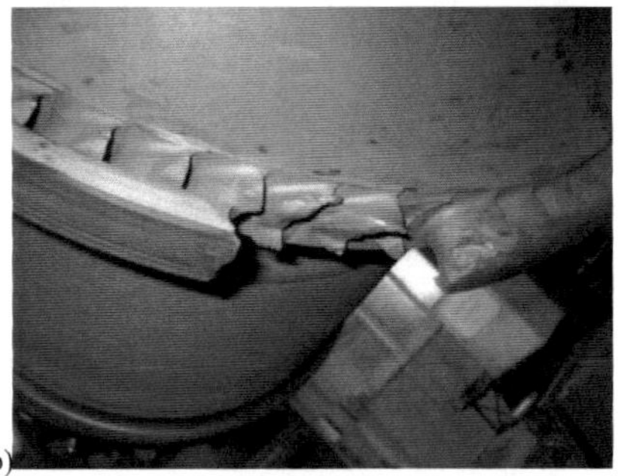

Fig. 1. Damages of control stage chamber after the first failure: a) the casing protection ring, b) the rotating wheel of control stage – two blades with shroud are completely destroy.

* Corresponding author: pawel.ziolkowski@imp.gda.pl

While research it was found that breakthroughs fragments of detached blades on control stage are fatigue. It could suggest that the primary cause of the failure was a rupture of the blades on the degree of control and its propagation leading to a break of the rotating element. The time, when entire process took place in, is difficult to determine. The broken off part of the blade came between the rotating wheel of control stage and the control ring (the casing protection ring), causing a damage of the catch pawl of this ring, which consequently has been destroyed (Fig. 1).

On the other hand, analysing the second failure it is clear that the microstructure of the shaft was not fatigue or weak because of operating. In places of microsections, the material had a homogeneous fine-grained structure and hardness. Based on the issued metrics, there was not observed any curvature of the shaft.

a)

b)

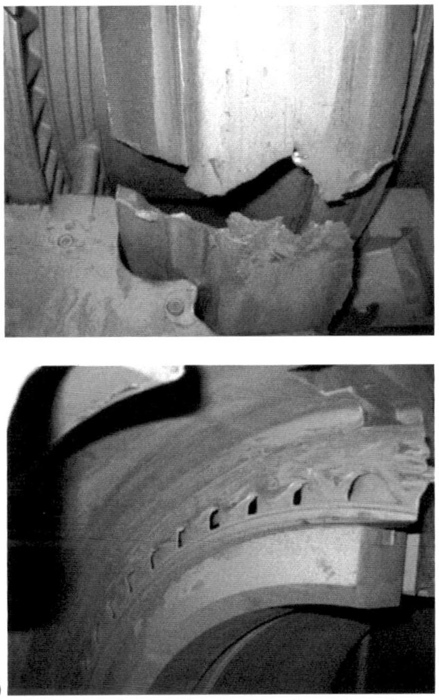

Fig. 2. Damages of control stage chamber after the second failure: a) the casing protection ring – huge crash and displacement, b) the rotating wheel of control stage.

Based on the survey of the manufacturer (producer), it has been stated that the shaft has to be repaired and then is suitable for further use. Because of the scope of damage all of the rotor blades have to replace by new ones (Fig.2). During the researches, it found out an impossibility of get off the blades from 4th to 8th rotor stages. Because, in order to allow to further operating of the shaft, we have to check the state of notches blade, the only way to dismantling those blades was their bore. During the failure 1-3 rotor degrees were completely destroyed, so there was not any way to their reconstruction. The damage of 9-19 degrees were also significant.

Also on the basis of the manufacturer research, it has been stated that repair of the damaged catch pawl in the casing for the directing segment does not guarantee the safe operation of the turbine. It was recommended to construct a new casings.

On the other hand, on the basis of Institute Fluid-Flow Machinery Polish Academy of Sciences (IFFM PAS) investigation, traditional experiment tools and classical computational tools indicated that none of the known mechanisms of failure appeared. Additionally, it was shown that classical mechanisms of degradation such as: cyclic plasticity, low-cyclic thermal fatigue, low-cyclic creep fatigue, high-cyclic fatigue induced by flow pulsations, excessive resonant forced vibration do not appear in the chamber of control stage.

Therefore, it was necessary to put forward a hypothesis as to the possible mechanisms of destruction and look for secondary and indirect causes. For example, one of the hypothetical mechanisms of degree destruction can be an introduction a small piece of metal to the sealing area. This could become due to human causes during the shut-down. It could also be metal shavings or corroded part of the boiler detached by an accidental blow of the hammer.

At the time of the failure, the wheel structures, for inputs of turbines for units 3 and 2, were different. Because of destruction of the whole ring of the input from unit 2, it was virtually impossible to determine the causes of failure. But due to differences in rotating wheel design is excluded this element as a focus failure, so the causes and course failure were different for the two blocks. Despite the fact that the existence of two such large failure at one time on machines with a relatively short period of operation, in combination with long-term operation of twin machines, makes us seek common denominator for these events. This may be the design of wheel ring or operating regime. It was decided to introduce structural changes and introduce additional measurement raising the safety of operating.

1.2 Constructive improvements

The main constructive change was to complete the measuring system with a new element to measure the temperature around the control stage chamber (Fig. 3).

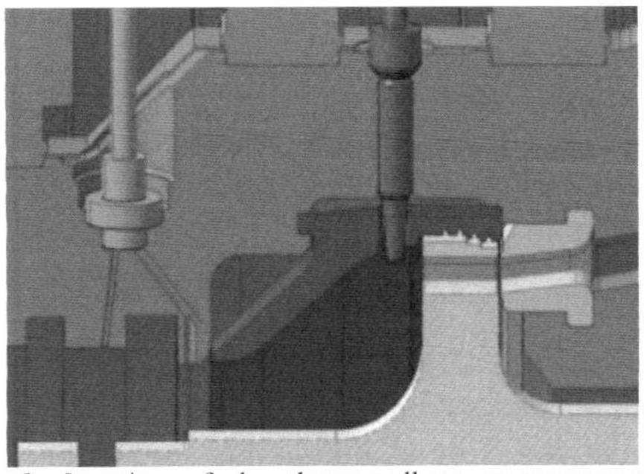

Fig. 3. Location of the thermowell to measurement of temperature in the control stage chamber.

In order to renew the control stage chamber and increase the supply of its security, the material of control ring (the casing protection ring) was modified.

The power plant complemented the measuring system and made amendments to the diagnostic system by using the "dual diagnosis": first the classical and the other based on "reference state calculation" turbine, designed for engineering supervision and performed in "forecasting repairs" to monitor the expenditure and degradation operating.

In order to improve the safety of operating and equip the turbines in numerical model of control stage working parameters, the IFFM PAS conducted following numerical simulations:

1. Geometry and numerical models to analysis of the vitality of control stage.
2. Three-dimensional (3D) flow analysis of control stage with taking into account full geometry of the nozzle boxes and inlets in order to determine parameters behind the nozzle boxes.
3. Three-dimensional (3D) flow analysis of warming up the disc of control stage with a shaft part basing on boundary conditions enumerated in the analysis above.
4. The analysis of thermal stress distribution accuring in the disc of control stage, the casing protection ring and in the part of rotor shaft.
5. The algorithm for the vitality loss of control stage lifetime in particular stages of unsteady turbine work.

Taking into account the imposed tasks and aims, the FSI scheme of calculations was adopted, which are presented in next part of this paper.

1.3 The governing equations

An important tool in the analysis of the entire considered issue (deformation, stress and temperature field) is the combination of CFD + CSD, often called FSI - Fluid Solid Interaction or Fluid Structure Interaction. On the basis of FSI (Fluid Solid Interaction) system of equations both for liquid and solid the three dimensional geometry have been calculated. The part below is taken from the methodology of Energy Conversion Department IFFM PAS [5,6,11] for describing fluid behaviour using the CFD equations. The starting point for CFD computation is formulation of universal set of mass, momentum and energy balance equations for the fluid, supplemented with equations for turbulence evolution k-ε in the form of:

$$\frac{\partial}{\partial t}\begin{Bmatrix}\rho\\ \rho\mathbf{v}\\ \rho e\\ \rho k\\ \rho\varepsilon\end{Bmatrix}+\mathrm{div}\begin{Bmatrix}\rho\,\mathbf{v}\\ (\rho\,\mathbf{v}\otimes\mathbf{v})+p\mathbf{I}\\ (\rho e+p)\mathbf{v}\\ \rho\mathbf{v}k\\ \rho\mathbf{v}\varepsilon\end{Bmatrix}=$$

$$=\mathrm{div}\begin{Bmatrix}0\\ \mathbf{t}^{res}\\ \mathbf{t}^{res}\,\mathbf{v}+\mathbf{q}^{c}\\ \mathbf{J}_{k}\\ \mathbf{J}_{\varepsilon}\end{Bmatrix}+\begin{Bmatrix}0\\ \rho\mathbf{b}\\ \rho S_{e}\\ \rho S_{k}\\ \rho S_{\varepsilon}\end{Bmatrix} \qquad (1)$$

where: : $\rho=\rho(\mathbf{x},t)$ - density depends, in general, on time t and location $\mathbf{x}$, $\mathbf{v}=v_i\mathbf{e}_i$ - velocity, $\mathbf{e}_i$ - direction vector, p - thermodynamic pressure, $\mathbf{I}=\delta_{ij}\mathbf{e}_i\otimes\mathbf{e}_j$ - unit tensor, δ_{ij} - Kronecker's delta, $\mathbf{t}^{res}=\mathbf{t}^{lam}+\mathbf{t}^{tur}$ - flux components of viscous stress – (laminar and turbulent, respectively), $\mathbf{b}$ - mass force of gravity, $e=u+\frac{1}{2}\mathbf{v}^2$ - sum of internal and kinetic energy , $\mathbf{q}^c$ - total heat flux, $\mathbf{J}_k$, $\mathbf{J}_\varepsilon$ - diffusive flux k and diffusive flux ε and sources S_k, S_ε . Then, using the CSD equations, we obtain following system:

$$\frac{\partial}{\partial t}\begin{Bmatrix}1\\ \rho\mathbf{v}\\ \rho e\\ \rho\mathbf{e}^{pl}\\ \rho\boldsymbol{\alpha}\\ \rho r\end{Bmatrix}+\mathrm{div}\begin{Bmatrix}0\\ \rho\,\mathbf{v}\otimes\mathbf{v}\\ \rho e\mathbf{v}\\ \rho\mathbf{e}^{pl}\otimes\mathbf{v}\\ \rho\boldsymbol{\alpha}\otimes\mathbf{v}\\ \rho r\mathbf{v}\end{Bmatrix}=$$

$$=\mathrm{div}\begin{Bmatrix}0\\ \boldsymbol{\sigma}\\ \boldsymbol{\sigma}\,\mathbf{v}+\mathbf{q}^{c}\\ 0\\ 0\\ \mathbf{J}_{r}\end{Bmatrix}+\begin{Bmatrix}0\\ \rho\mathbf{b}\\ \rho S_{e}\\ \rho S_{pl}\\ \rho S_{\alpha}\\ \rho S_{r}\end{Bmatrix} \qquad (2)$$

where: $\boldsymbol{\sigma}$ - tensor of solid stresses; $\mathbf{e}^{pl}$ - tensor of plastic strain; $\boldsymbol{\alpha}$ - kinematic hardening; r - isotropic hardening; $\mathbf{J}_r$- diffusive flux of r; $S_{pl}, S_{\boldsymbol{\alpha}}, S_r$ sources of plasticity , kinematic and isotropic hardening typical for the Chaboche model. The system of equations and the way of CSD and CFD solutions are analogical to each other. Both methods use the basic balances: mass, momentum and energy. However, the difference often occurs in the way of discretization by applying the finite element method (FEM) in CSD and the finite volume method (FVM) in CFD. It is a kind of numerical challenge to combine both methods to transfer boundary conditions.

2 Results

As mentioned above, analysis was conducted in order to reconstruct the working conditions of stage control node by the preparation of flow and constructional geometry, discretization the space, and then measure the boundary conditions during the operating of the block. The geometry used in analysis is presented on Fig.4, where: a) the area of control stage; b) the rotor with the control stage c) cutting from 1st blade and discretization of both liquid and solid and d) internal casing of HP, used to CSD strength analysis. The calculation area includes in the part of turbine inlet to the HP by nozzle boxes, the blades of control stage (64 blades in accordance with the profile used in this construction), the chamber control stage to the outlet of the first stage rotor palisade the unregulated part of the HP. So it shows four inlet pipelines, nozzle sectors, internal casing, control stage rotating wheel and the control stage chamber (with the casing protection ring), and first full-arc admission stage.

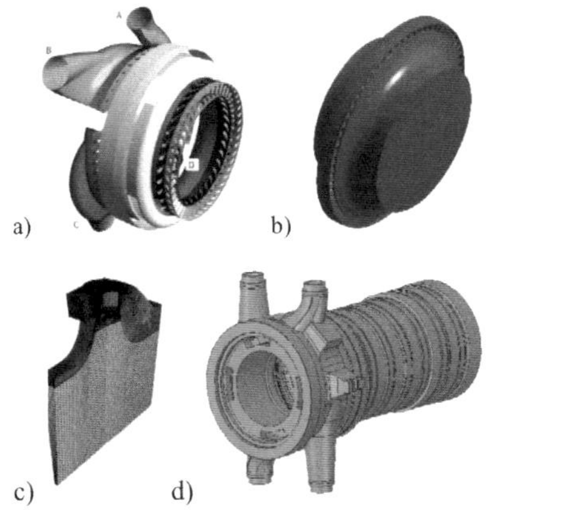

Fig. 4. The elements of geometry: a) fluid domena of the area of control stage; b) the rotating wheel of control stage; c) cutting from 1st blade and discretization of both liquid and solid; and d) internal casing.

Examples of the results of flow analysis are presented in Fig.5, which shows the cold start-up after 68 minutes with taking into account: a) the state of pathlines coming out of the nozzle box and passing through the chamber of control stage up to the first unregulated stage; b) the view of the local changes in the field of static pressure in both the axial cross-sectional and circumferential view; and c) swirling at the outlet of the nozzle up to the control wheel (velocity field). Thanks the obtained results it is possible to determine the force acting on the individual components of the turbine. The results from Fig. 5 correspond to the turbine operating with partial load. This operation mode differs from the nominal load by opening of two nozzle boxes only. The remaining two are fully closed by the stop valves and steam is flowing only through the boxes with a smaller number of nozzles. Consequently, higher non-uniformity of pressure Fig.5b) and velocity distribution Fig.5c) is observed for the steam leaving the two loaded boxes. It can be noticed that the flow forms the main core of higher velocity beginning at the exit from the boxes with a smaller number of nozzles. Next in the control stage chamber, fields of pressure and temperature are more uniform. As was mention before Fig.5a) presents the temperature fields at selected time step however additionally shows the state of pathlines coming out of the nozzle box and passing through the chamber of control stage.

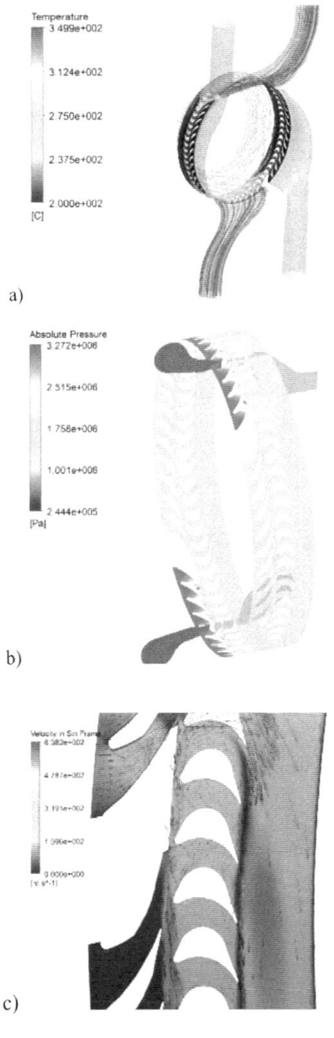

Fig. 5. CFD results after 68 minutes of cold start-up: a) pathlines with temperature distribution in the and nozzle boxes; b) pressure field in the circumferential view and axial cross-sectional and c) velocity field.

Another important information obtained from the CFD analysis is temperature distributions over time (Fig. 6) which could determine the speed of heating of the solid from flowing steam. Temperature fields during cold start-up in the following seconds, where a) the 1600s .; b) 2400s .; c) 5200s; d) 14400s has been

 DOI: 10.1051/e3sconf/20171304003

presented in Fig.6. Local temperature variations do not exhibit significant fluctuations. At this type of heating, after examining distribution of the flow parameters and its character, it can be stated that the flow at the control stage chamber is stable and does not exhibit symptoms of uncorrected start-up.

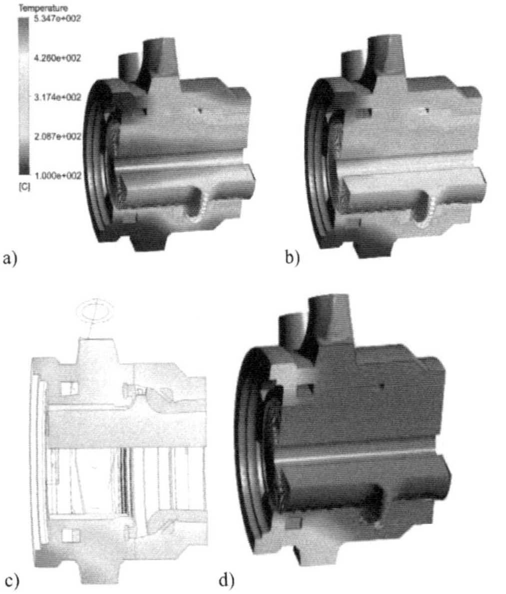

Fig. 6. Temperature fields during cold start-up in the following seconds, where a) the 1600s .; b) 2400s .; c) 5200s; d) 14400s.

The example of the effect on stress state is shown in Fig.7. Analysis of CSD aimed to reconstruct the stress state, and then in perform a study on the opportunity to increase the rate of change in power (start-up, shutdown, part load, increase and decrease of part load power) with respect to the referential data by using the modified regulation and discuss and compare these solutions, which are currently used in power plant. The steam forces calculated in CFD simulations were used as loads in mechanical integrity analysis. The calculated distributions of stress field resulting from the centrifugal forces at a rotating speed of 3000 rev/min are presented in Fig. 9. Based on this it can be stated that the static stress criteria are fulfilled.

It should be added that the kind used regulation was verified and confirmed its safety. The important issue while improving the safety of the turbine regulation is also an algorithm for accurate and reliable calculation of the temperature in control stage chamber based on the turbine measurement in one point. Starting from well known the measuring, the model is calibrated, and then the CFD tools are used to estimate the value of the remaining points. In turn, the algorithm counts the lifetime loss of control stage is based on FSI model, which allow to compare the critical point (the blade of the rotating wheel) with measurements made in points quite distant from the critical point. In other words, those tools allow to correlate measurements of pressure and temperature, as measured on the casing with the state of mechanical and thermal stresses in the rotating blade and any point on the internal casing and the control stage.

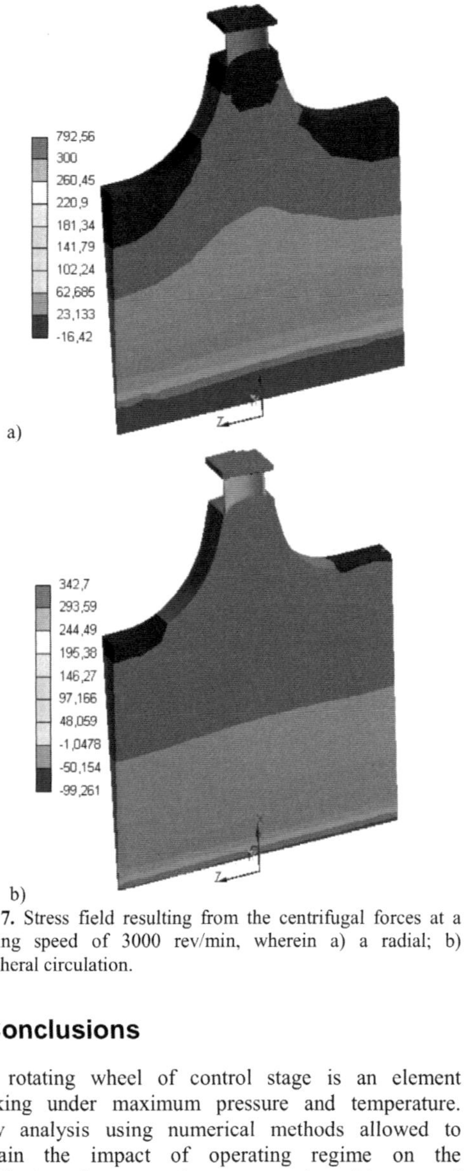

Fig. 7. Stress field resulting from the centrifugal forces at a rotating speed of 3000 rev/min, wherein a) a radial; b) peripheral circulation.

3 Conclusions

The rotating wheel of control stage is an element working under maximum pressure and temperature. Flow analysis using numerical methods allowed to explain the impact of operating regime on the distribution of power and temperature in turbine control stage. The analyses and tests have been showed that present method of control contributes a substantial increase in safety and lifetime endurance, while maintaining the flow parameters such as power, efficiency and mass flow rate.

The temperature of control chamber came as a main parameter related with safety, hence it is reasonable to on-line monitoring, on the basis of measurements at calibration point and other available points. To calculate temperature we dedicated a special algorithm.

Additionally, with the passes of temperature, pressure, stress and deformation, we created the algorithm counting viability of the control stage in sensitive points for safety.

References

1. Bzymek G., Badur J., *III Conference ZRE Katowice*, Krynica-Zdrój 13-15 April 2016, 73-102 (2016).
2. Banaszkiewicz M., Bzymek G., Radulski W., Energetyka **8** 519-524 (2015).
3. Badur J., Kornet S., Sławiński D., Ziółkowski P., Banaszkiewicz M., Rehmus-Forc A., *Energetyka* **10** 647-650 (2016).
4. Taler J., Węglowski B., Cebula A., *Energetyka* **9** 646-650 (2006).
5. Taler J., *Int. J. Heat Mass Transfer*, **35** (6) 1625-1634 (1992).
6. Badur J., Ziółkowski P., Sławiński D., Kornet S., *Energy* **92** 142-152 (2015).
7. Zima W., Nowak-Ocłoń M., Ocłoń P., *Energy* **92** 117-127 (2015).
8. Grądziel S., Majewski K., *Procedia Engineering* **157** 44 – 49 (2016).
9. Taler J., Węglowski B., Taler D., Sobota T., Dzierwa P., Trojan M., Madejski P., Pilarczyk M., *Energy* **92** 153-159 (2015).
10. Marek A., Okrajni J., *JMEPEG* **23** 31–38 (2014).
11. Banaszkiewicz M., *Int. J. of Fatigue* **73** 39–47 (2015).
12. Kosman W., *J. of Power Technologies* **95** 47–53 (2015).
13. Banaszkiewicz M., Rehmus-Forc A., *Eng. Failure Analysis* **51** 55–68 (2015).
14. Dobrzański J., Purzyńska H., Matusik M., *Energetyka* 737 (11) 761-764 (2015).
15. Banaszkiewicz M., *Appl. Thermal Eng.* **94** 763-776 (2016).
16. Banaś K., Badur J., Proc. *11th Int. Cong. On Thermal Stresses*, University of Salerno, Italy, 5-9 June 2016, 19-22 (2016).
17. Kowalewski Z., *IPPT PAN Publishers*, Warsaw, 2005.
18. Odahara S., Murakami Y., Inoue M., Sueoka M., *JSME Int J SerA* **48** (2) 109-117 (2005).

Increasing the Performance and Reliability of Power Boiler by Monitoring Thermal and Strength Parameters

Tomasz Sobota[1,*]

[1]Cracow University of Technology, Institute of Thermal Power Engineering, Al. Jana Pawła II 37, 31-864 Cracow, Poland

Abstract. The paper presents a method for determination of thermo-flow parameters for steam boilers. This method allows to perform the calculations of the boiler furnace chamber and heat flow rates absorbed by superheater stages. These parameters are important for monitoring the performance of the power unit. Knowledge of these parameters allows determining the degree of the furnace chamber slagging. The calculation can be performed in online mode and use to monitoring of steam boiler. The presented method allows to the operation of steam boiler with high efficiency.

1 Introduction

The fouling processes caused by slagging and ash deposits are strongly affected by the kind of coal [1-6]. They become more intense if coal is co-fired with biomass. In the case of the furnace chamber, the fouling processes are different on its individual walls. Depending on the furnace chamber aerodynamics, local slag overhangs may be formed in the burner area or at other locations on the chamber walls. In places where temperature is high, such as the platen superheater area, the superheater surface slagging may also be caused by melted drops of ash forming deposits on the surface of tubes [7]. If biomass is co-fired, due to the low melting point of ash, intense processes of ash deposit formation and slagging may occur on the superheater stages located in zones with a lower flue gas temperature.
Ash fouling processes also affect the lifetime of the boiler superheaters [8-10]. It should be noted that about 40% of the boiler failures are caused by damage to boiler superheaters.
For this reason, there is a need to develop an effective and efficient method of determination of thermo-flow parameters for steam boilers. The presented in this paper method allows to the operation of steam boiler with high efficiency.

2 Calculations of the boiler furnace chamber

The degree of fouling of the boiler furnace chamber walls may be assessed by an online mode determination of the flue gas temperature at the furnace chamber outlet T_{fe}, the heat flux absorbed by the furnace chamber walls $\dot{Q}_r$ and the heat effectiveness ψ of the furnace chamber walls. The outlet of the combustion chamber is located directly beneath of the platen superheater at a height of 29 m (Fig. 1a). Adopted position of the outlet section of the furnace is consistent with the industrial definition, which says that the outlet from the furnace is where the spacing of the platens is less than 2.0 m. The spacing of the platens is 0.7 m in the boiler analyzed in the paper.
The heat effectiveness ψ of the furnace chamber is defined as an average ratio of the heat flux absorbed by the water-wall to the heat flux falling on the water-wall of the combustion chamber.
Based on experimental data obtained from the testing of large power boilers, a formula containing parameter *M*, which characterizes the location where the maximum flame temperature occurs in the boiler was used [11]:

$$\frac{T_{fe}}{T_{ad}} = \frac{Bo^{0,6}}{M\,\varepsilon_{pal}^{0,6} + Bo^{0,6}} \tag{1}$$

Parameter M in Eq. (27) is a function of the kind of fuel (oil, gas or coal) and of the relative height at which burners are placed [11-13]. After temperature T_{fe} is calculated from (1), the average heat flux $\dot{Q}_r$ absorbed by the furnace chamber walls may be determined.
The degree of slagging of the furnace chamber walls may be assessed determining the waterwall heat efficiency coefficient The ψ coefficient gets smaller in proportion to slagging of the furnace chamber walls. If the furnace chamber walls are covered with a layer of slag, the temperature of the slag surface is high and the density of the heat flux $\dot{q}$ absorbed by the wall (i.e. the wall thermal load) is much smaller than in the case of a clean wall.
Slagging of the furnace chamber walls results in an increase in the flue gas outlet temperature T_{fe} and a decrease in the heat flux absorbed by the furnace chamber walls $\dot{Q}_r$.

* Corresponding author: tsobota@mech.pk.edu.pl

3 Determination of the average degree of the furnace chamber fouling

The waterwall heat efficiency coefficient ψ will be determined in an online mode from the following nonlinear equation:

$$\dot{m}_{par}^{zm} = \dot{m}_{par}^{obl}(\psi), \tag{2}$$

where: $\dot{m}_{par}^{zm}$ and $\dot{m}_{par}^{obl}$ denote properly measured and calculated mass flows of steam. The steam mass flow $\dot{m}_{par}^{obl}$ is a function of the waterwall heat efficiency ψ and it is calculated from the boiler evaporator energy balance. The symbol $\dot{m}_{par}^{zm}$ denotes the steam mass flow at the boiler drum outlet determined using the steam mass flow $\dot{m}_p$ measured at the boiler outlet and the water mass flows $\dot{m}_{w1}$ and $\dot{m}_{w2}$ fed into steam attemperator No. 1 and No. 2, respectively.

$$\dot{m}_{par}^{zm} = \dot{m}_p - \dot{m}_{w1} - \dot{m}_{w2}, \tag{3}$$

The condition of equality of the calculated and measured values of the steam mass flow allows an assessment of the degree of the furnace chamber fouling ζ.

The waterwall heat efficiency coefficient ψ is equal to the product of the waterwall shape factor x, which for tight walls in the OP-380 boiler is equal to unity, multiplied by the fouling coefficient ζ of the chamber furnace heating surfaces:

$$\psi = x\zeta, \tag{4}$$

where: $x = 1$ for membrane water-walls used in the boiler studied.

Determining ζ so that the calculated and measured values of the boiler efficiency are the same, it is possible to assess the average degree of the fouling of the walls of the furnace chamber.

Owing to this way of determining the value of ζ, the temperature of flue gases at the outlet from the furnace chamber T_{fe} may be calculated much more accurately.

If the fouling of the chamber walls is higher, the waterwall efficiency coefficient decreases: $\psi = \zeta$. The heat flux absorbed by the boiler evaporator $\dot{Q}_{par}$ gets smaller and the flue gas temperature at the furnace chamber outlet gets higher. As a result, there is a reduction in the steam mass flow generated in the evaporator and directed to the superheaters, which, at a higher flue gas temperature T_{fe}, causes an increment in the reheated steam temperature (at the same mass flow of fuel: $\dot{Q}_{pal}$). In order to maintain the reheated steam temperature at the set level, bigger mass flows of water fed into the reheated steam attemperators are necessary.

Also the flue gas temperature after individual superheater stages rises due to the fouling of the furnace chamber walls.

The following quantities:

- the waterwall heat efficiency ψ,
- the mass flow rates of water $\dot{m}_{w1}$ and $\dot{m}_{w2}$ injected into attemperators,
- the flue gas temperature after individual superheater stages

are strongly dependent on the degree of the fouling of the furnace chamber walls. If ψ is reduced, and $\dot{m}_{w1}$ and $\dot{m}_{w2}$, as well as T_{sp}, after individual stages reach permissible set values, the steam slag and ash blowers in the furnace chamber should be activated.

Apart from the above-mentioned indices characterizing the fouling of the boiler heating surfaces in terms of heat exchange, the increase in the flue gas subatmospheric pressure after the superheater stages located further down the convective duct (at a higher distance from the furnace chamber) is another significant symptom of the fouling of the boiler – mainly of the surfaces of superheaters. Increased aerodynamic resistance caused by ash fouling of superheaters leads to a higher power consumption of the fans. The measurement of the flue gas pressure and of the engine power capacity may also be used to assess the degree of ash fouling of superheaters.

4 Mathematical model of the boiler evaporator

a)

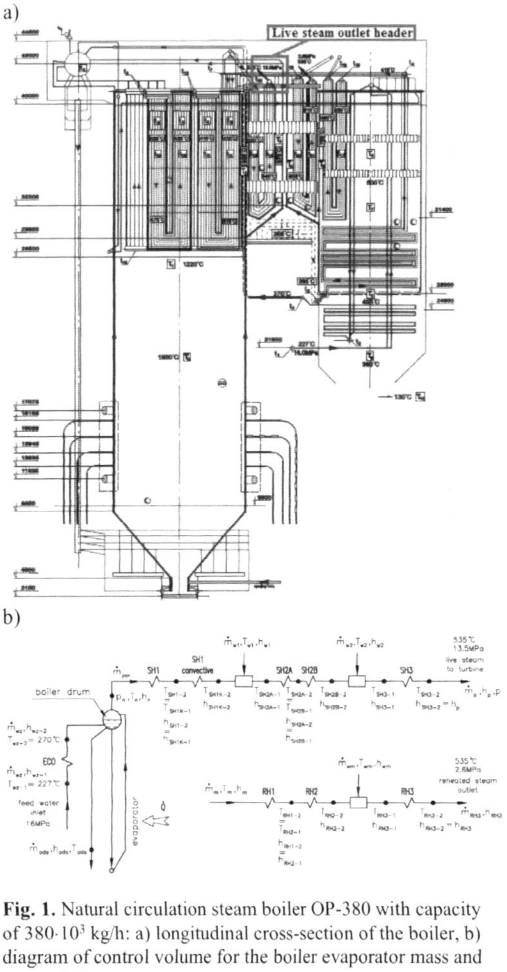

Fig. 1. Natural circulation steam boiler OP-380 with capacity of $380 \cdot 10^3$ kg/h: a) longitudinal cross-section of the boiler, b) diagram of control volume for the boiler evaporator mass and energy balance

The live steam mass flow is calculated from the boiler evaporator mass and energy balance (Fig. 1b):

$$\dot{m}_{wz} = \dot{m}_p - \dot{m}_{w1} - \dot{m}_{w2} + \dot{m}_{ods}, \qquad (5)$$

$$\dot{m}_{wz} h_{wz-2} + \dot{Q}_{par} = \\ = \dot{m}_{ods} \cdot h'(p_n) + (\dot{m}_p - \dot{m}_{w1} - \dot{m}_{w2}) \cdot h''(p_n), \qquad (6)$$

Substitution of (5) in (6) results in:

$$\dot{m}_p = \frac{\dot{Q}_{par}}{h''(p_n) - h_{wz-2}} - \dot{m}_{ods} \frac{h'(p_n) - h_{wz-2}}{h''(p_n) - h_{wz-2}} + \\ + \dot{m}_{w1} + \dot{m}_{w2} \qquad (7)$$

where $\dot{Q}_{par}$ is the heat flux transferred from flue gases to the evaporator via radiation and convection.

The heat flux $\dot{Q}_{par}$ may be calculated from the following expression:

$$\dot{Q}_{par} = \dot{Q} - \dot{m}_g \cdot c_{p,g}\Big|_0^{T_{fe}-273.15} (T_{fe} - 273.15), \qquad (8)$$

where $\dot{Q}$ is the heat flux brought to the furnace chamber with fuel and air, and which is defined as:

$$\dot{Q} = \dot{m}_g \cdot c_{p,g}\Big|_0^{T_{ad}-273.15} \cdot (T_{ad} - 273.15), \qquad (9)$$

The symbol T_{fe} denotes the flue gas temperature at the furnace chamber outlet.

It should be added that both efficiency and thermal calculations require the development of many functions and programs used for the calculation of thermal properties of water, saturated and superheated steam, and flue gases.

5 Determination of fuel mass flow rate

The calculation of the fuel mass flow rate is based on the fuel chemical composition or on the coal calorific value. The second approach is more appropriate for practical applications since the chemical composition of the coal is not determined continuously. Boiler efficiency is determined online by the indirect method [6, 14]. According to the ASME Performance Test Codes: PTC 4 - Fired Steam Generators [15] the stack gas temperature is corrected for air heater leakage based on O_2 difference between flue gas into air heater and flue gas O_2 leaving air heater. After correcting the temperature of the flue gas exiting the boiler the thermal efficiency of the boiler is calculated.

Based on the boiler efficiency η determined in an online mode, the coal mass flow is calculated from the definition of the boiler heat efficiency

$$\eta = \frac{\dot{Q}_n}{\dot{Q}_h} = \frac{(\dot{m}_p - \dot{m}_{w1} - \dot{m}_{w2})(h_p - h_{wz-2})}{\dot{m}_{pal} W_d} + \\ + \frac{[(\dot{m}_{w1} + \dot{m}_{w2}) h_p - \dot{m}_{w1} h_{w1} - \dot{m}_{w2} h_{w2}]}{\dot{m}_{pal} W_d} + \\ + \frac{\dot{m}_{ods}(h_n - h_{wz-2})}{\dot{m}_{pal} W_d} + \frac{\dot{m}_m (h_{RH1-2} - h_{RH1-1})}{\dot{m}_{pal} W_d} - \frac{\dot{m}_{wm} h_{wm}}{\dot{m}_{pal} W_d} \\ + \frac{\dot{m}_m (h_{RH2-2} - h_{RH2-1})}{\dot{m}_{pal} W_d} + \frac{(\dot{m}_m + \dot{m}_{wm})(h_{RH3-2} - h_{RH3-1})}{\dot{m}_{pal} W_d}, \qquad (10)$$

Heat balance across the superheater attemporators is used to determine spray water mass flows. The fuel mass flow rate at various boiler loads is determined from Eq. (10) taking into account that the boiler efficiency η is known,

$$\dot{m}_{pal} = \frac{(\dot{m}_p - \dot{m}_{w1} - \dot{m}_{w2})(h_p - h_{wz-2})}{\eta W_d} + \\ + \frac{[(\dot{m}_{w1} + \dot{m}_{w2}) h_p - \dot{m}_{w1} h_{w1} - \dot{m}_{w2} h_{w2}]}{\eta W_d} + \\ + \frac{\dot{m}_{ods}(h_n - h_{wz-2})}{\eta W_d} + \frac{\dot{m}_m (h_{RH1-2} - h_{RH1-1})}{\eta W_d} + \\ + \frac{\dot{m}_m (h_{RH2-2} - h_{RH2-1}) - \dot{m}_{wm} h_{wm}}{\eta W_d} + \\ + \frac{(\dot{m}_m + \dot{m}_{wm})(h_{RH3-2} - h_{RH3-1})}{\eta W_d}. \qquad (11)$$

Equations (10) and (11) are valid only for steady state conditions. The symbols: h_{wz-2}, h_n, h_{w1}, h_{w2}, h_{ods}, h_p, h_{RH3} in equations (10) and (11) denote, respectively, the enthalpy of: feed water after the economizer, steam saturated at the boiler drum pressure, injection water in superheated steam attemperators, brine, live steam at the boiler outlet, and reheated steam at the boiler outlet (Fig. 2).

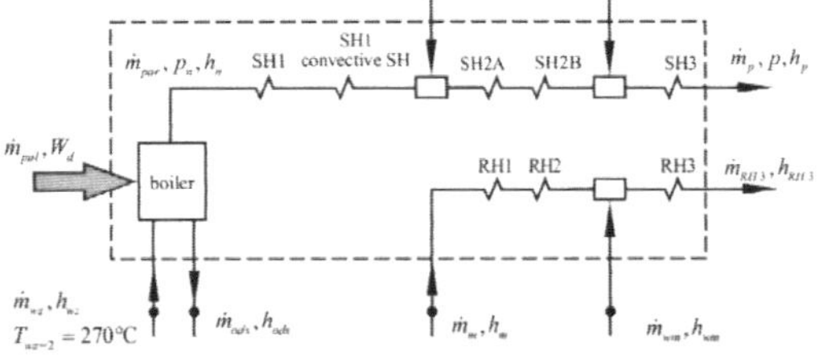

Fig. 2. Diagram of control volume for the boiler mass and energy balance

The calculations of the volume and mass flow rates of the flue gases are based on the determined fuel mass flow rate. Based on the determined excess air number λ using the expression: $\lambda = 21/(21-O_2)$, defined as the ratio of the air to the air flow rate required theoretically, the volume and mass flows of moist flue gases are found. The symbol O_2 denotes the volume fraction of oxygen in the flue gas.

6 Calculation of heat flow rates absorbed by superheater stages and monitoring the degree of fouling of superheaters and reheaters

The degree of fouling of the *i*-th superheater stage with ash ξ_i is calculated continuously from the following expression:

$$\xi_i = \frac{\dot{Q}_{i,z}}{\dot{Q}_{i,c}} \qquad (12)$$

where: $\dot{Q}_{i,z}$, in W, denotes the heat flow rate absorbed by the superheater stage affected by fouling, and $\dot{Q}_{i,c}$, in W, is the heat flow rate absorbed by a clean superheater stage. The heat flow rate $\dot{Q}_{i,c}$ at a given value of the excess air number λ is a function of the boiler load. After the boiler is thoroughly cleaned, the measurements of $\dot{Q}_{i,c}$ are made depending on the steam mass flow rate $\dot{m}_i$ flowing through a given superheater stage. The advantage of formula (12) is no need for measuring the temperature of the flue gas after the each superheater stage, which is especially difficult in the regions of high flue gas temperatures.

If the excess air ratio in the boiler or the slagging of the waterwalls of the combustion chamber vary greatly over time, the fouling factor calculated from the formula (12) can lead to less accurate results. An alternative way to assess the degree of ash fouling of a superheater stage is the ratio of the current overall heat transfer coefficient $U_{i,z}$ versus overall heat transfer coefficient for the clean surface $U_{i,c}$.

$$\xi_{U,i} = \frac{U_{i,z}}{U_{i,c}} \tag{13}$$

The overall heat transfer coefficient U_i related to the surface area A_i is calculated in the on-line mode using the following relationship:

$$U_i = \frac{\dot{Q}_i}{A_i \cdot \Delta T_{m,i}} \tag{14}$$

where the symbol $\Delta T_{m,i}$ denotes the mean logarithmic temperature difference (LMTD) between the flue gas and steam. The flue gas temperature at the outlet of the combustion chamber and before each superheater stage was calculated based on the flue gas temperature measured after the last superheater stage in the flue gas duct. The flue gas temperature is then low and easier to measure.

Assuming that the measured value of the flue gas temperature after the entire superheater in known, the flue gas temperature before each stage of the superheater including flue gas temperature at the outlet of the combustion chamber can be determined sequentially.

Knowing the heat flow rate $\dot{Q}_{i,c}$ absorbed by the clean i-th stage of the superheat at a given boiler load and heat flow rate $\dot{Q}_{i,z}$ absorbed by the fouled i-th stage at the same load of the boiler the fouling degree of the i-th superheater stage can be evaluated in the on-line mode.

Pre-setting the limiting value of the fouling degree ξ_i for i-th superheater stage, the frequency of sootblowers activating can be controlled. If the fouling degree ξ_{is} lowers below the prescribed limit then the sootblower is activated. Steam and water enthalpies were determined using International Steam Tables [16]. The heat flux measurements using different methods are presented in [17-20].

7 On-line monitoring of the steam boiler thermo-flow parameters

The ability to monitor the thermo-hydraulic parameters of steam power boiler is shown on the example of OP-380 unit used in CHP Krakow S.A. RAFAKO design OP-380 steam boiler is drum type, radiant, pulverized coal fired boiler, with natural circulation. Its characteristic is shown in Table 1.

Table 1. Characteristic of the OP-380 power boiler

Boiler type	With natural circulation
Firing system	Tangential
Output capacity	105,6 kg/s
Live steam temperature	540°C
Live steam pressure	13,9 MPa
Reheated steam temperature	540°C
Reheated steam pressure	2,65 MPa
Feed water temperature	227°C
Boiler efficiency	91%
Fuel type	Hard coal
Fuel low heating value	21,0 MJ/kg

The Figs. 3a and 3b shows the time changes of the fuel and live steam mass flow rates and also changes in the excess air ratio and the boiler efficiency. Fig. 3a shows the stable operation of the boiler close to nominal capacity. Throughout the time period the value of the excess air ratio does not change, and the boiler efficiency is high - in the range of 91.6% -92.0%.

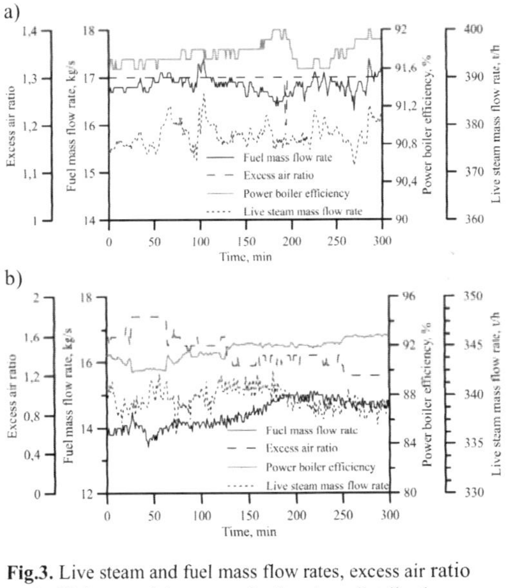

Fig.3. Live steam and fuel mass flow rates, excess air ratio and efficiency of the power boiler time distributions

A different situation is presented in Fig. 3b. This figure shows large time variation of the excess air ratio. In the first part of the analyzed interval excess air ratio reaches a value of approx. 1.6 which is accompanied by a decrease of the boiler efficiency up to 90%. It is

DOI: 10.1051/e3sconf/20171304004

associated with high temperature in the combustion chamber and hence a large heat losses in the flue gases. In the second part of the diagram 3b excess air ratio decreases to 1.2, which is a typical value for this type of boiler. This situation is reflected into an increase in boiler efficiency above 92%.

Heat flow rates received by the power boiler evaporator (Fig. 4), live steam superheater (Fig. 5) and steam reheater are also monitored.

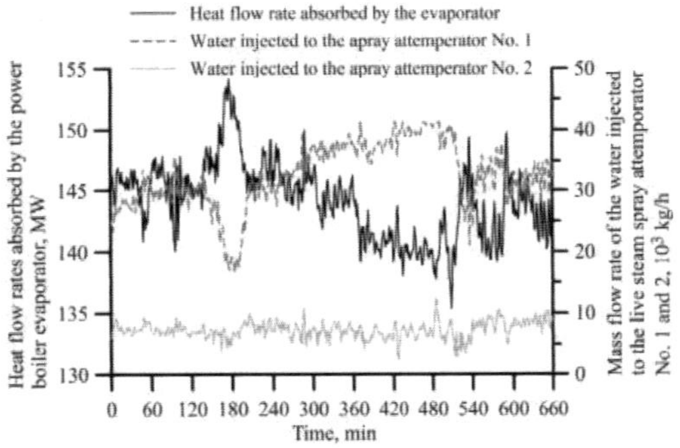

Fig. 4. Time distribution of the heat flow rate absorbed by the boilers evaporator

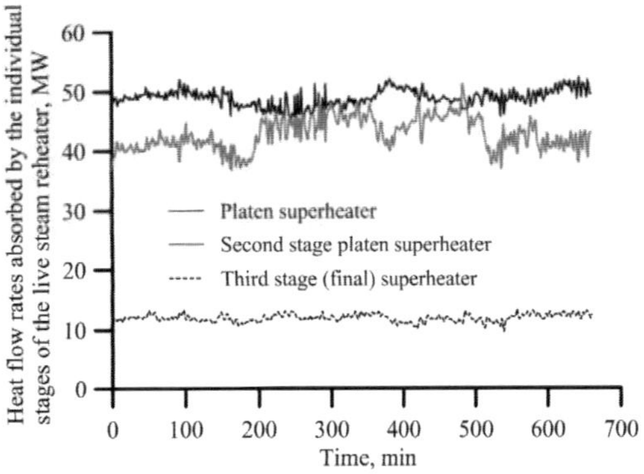

Fig. 5. Time distribution of the heat flow rates absorbed by the live steam superheater stages

On-line monitoring of the heat flux received by the steam boiler heating surfaces can be used to assess the fouling and slagging degree of the combustion chamber and live steam superheater. The increase in fouling of the combustion chamber is reflected by a decrease in live steam flow. This information can be a signal for the operator to start the boiler water or steam soot blowers.

8 Stress monitoring

High thermal stresses can occur in partially filled horizontal vessels. Thermal stresses in the cylindrical elements are caused by the occurrence of temperature differences both in the radial and circumferential direction. In practice, the thermal stresses due to temperature drop over the tube wall are monitored. The stresses caused by the occurrence of a temperature difference on the thick-walled component circumference are rarely monitored.

The calculations performed using the finite element method (FEM) show, however, that the temperature difference occurring at the periphery of the horizontal cylindrical element (header) can lead to much higher stresses than the stresses caused by the radial temperature difference.

During operation under transient conditions, for example, during power plant start-up and shut-down, there are significant temperature differences over the circumference of the horizontal pressure vessels [21-22]. This phenomenon is caused by the different heat transfer coefficients in the water and steam spaces. This takes place in large steam generator drums, superheater headers and steam pipelines. Water steam condensate collects at the bottom of the horizontal pressure elements, such as superheater headers and pipelines in case of malfunctioning drainage.

High thermal stresses caused by nonuniform temperature distribution on vessel circumference also occur in emergency situations such as fire of partially filled fuel tanks. Thermal stresses can limit the heating and cooling rates of temperature changes in power boiler thick-walled pressure components. The largest absolute value of thermal stresses appears at the inner surface. Direct measurements of these stresses are very difficult to take, since the inner surface is in contact with water or steam under high pressure. For that reason, thermal stresses are calculated in an indirect way based on measured temperatures at selected points, located on an outer thermally insulated surface of a pressure element.

Fig. 6. View of the experimental installation for testing the computer system for on-line monitoring thick-wall pressure components

In order to validate the developed method for stress monitoring, modern experimental installation was built (Fig. 6). In the design, particular emphasis was placed on ensuring that the conditions are similar to those that may occur in pressure components installed in power plants. In addition to measuring of the metal temperature, mass flow rate, and pressure of the steam were measured as a function of time. To measure temperature distribution on the outer surface of the steam header pre-calibrated thermocouples were used.

The experimental installation consists of the following elements: steam generation unit with steam output capacity 700 kg/h, installation for boiler feed water treatment, cascade-type deaerator, steam pressure reduction station (10 bar /4 bar /2 bar), boiler control system, steam header made of martensitic high alloy P91 steel.

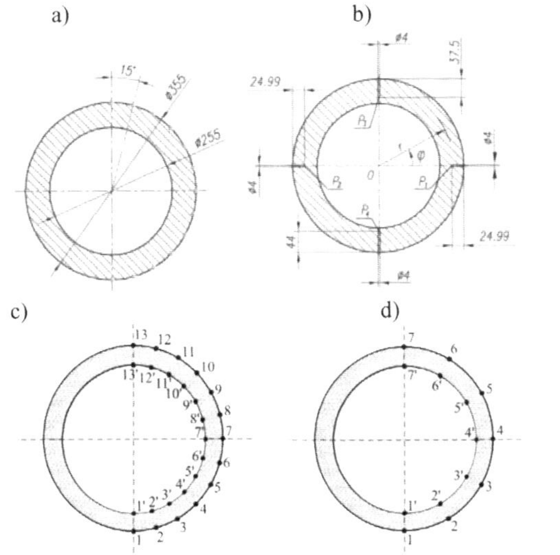

Fig. 7. Location of temperature sensors for measuring the header temperature: (a) thermocouples for temperature measuring at the outer surface of the steam header, (b) location of interior thermocouples used for experimental validation of the inverse procedure, (c) numbering of thirteen measuring points (1-13) and their counterparts at the inner surface, (d) numbering of seven measuring points (1-7) and their counterparts at the inner surface

The measurements of temperature carried out on the outer surface of the steam header will be used for the determination of the temperature and stress distributions at the wall cross-section including the inner and outer surfaces of the steam header using the inverse heat conduction methods. Points on the outer surface of a header, at which thermocouples are installed are shown in Figs. 7c and 7d. Thirteen temperature sensors are uniformly spaced every 15 degrees at half the circumference (Fig. 7c). If the temperature of the tube outer surface is measured at seven points a spacing of temperature sensors is 30 degrees.

The steam header was made out of the ferritic-martensitic high alloy P91 steel (0.08% C, 0.2% Si, 0.3 Mn, ≤0.02% P, ≤0.015% S, ≤0.04% Al, 8% Cr, 0.85÷1.05% Mo, 0.06÷0.1% Mo, ≤0.4% Ni). The following physical properties of the P91 steel have been adopted for calculations: specific heat, c = 486 J/(kg·K), thermal conductivity, k = 29 W/(m·K), density, ρ = 7750 kg/m^3, modulus of elasticity, E = 2.28·10^{11} Pa, Poisson's ratio, ν = 0.29, coefficient of linear thermal expansion β = 0.098·10^{-4} 1/K. The outer diameter d_{out}, wall thickness δ and the length L are: 355 mm, 50 mm and 3765 mm, respectively. Thirteen thermocouples NiCr-NiAl (K-type) were mounted every 15° on the half of the outer circumference at the distance of 2150 mm from the inlet of steam (Fig. 7a).

The actual temperatures measured at the outer surface of the steam header were used in the analysis.

To validate the inverse technique developed in the paper, four thermocouples were installed at the interior locations: $\delta_1 = \delta_2 = d_{out} - 0.50\delta$, $\delta_3 = d_{out} - 0.75\delta$ and $\delta_4 = d_{in} + 6$ mm (Fig. 7b), where the symbol δ denotes the tube thickness. These measurements were carried out with sheathed thermocouples NiCr-NiAl (K-type) with outer diameter of 3 mm. The temperature and thermal stress on the internal surface of the header are based on the metal temperature measured at the 13 or 7 points on the outer surface to evaluate the effect of the number of temperature measuring points on the accuracy of the solution of the inverse problem.

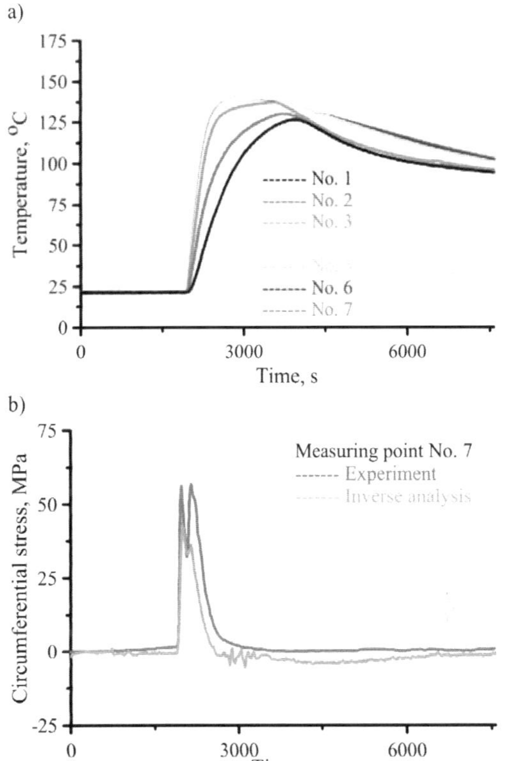

Fig. 8. Measured temperature at 7 points situated on the outer surface of the header (a) and comparison of the circumferential thermal stress obtained using inverse analysis with the experimentally determined stress using strain gauges (b)

Figure 8 illustrates a satisfactory agreement between measured and calculated stresses at the measuring point No. 7 (Fig. 7d).

Extensive comparisons of temperature and thermal stress distributions, obtained by measuring the temperature at 7 and 13 points showed that satisfactory accuracy is also

DOI: 10.1051/e3sconf/20171304004

achieved for seven measuring points. Therefore, the critical elements of the boiler pressure metal temperature is measured on the outer surface at seven points. This is also advantageous because of the lower cost of installing temperature sensors.

The temperatures measured at seven points (22-28) on the outer surface of the monitored cylindrical component (Fig. 9) were used for the determination of the temperature and stress distributions at the wall cross-section including the inner surface. Fig. 9 shows the finite volume mesh used to solve the inverse heat conduction problem in on-line mode. Based on the estimated temperature distribution at the cross-section, thermal stresses were computed using the FEM (Finite Element Method).

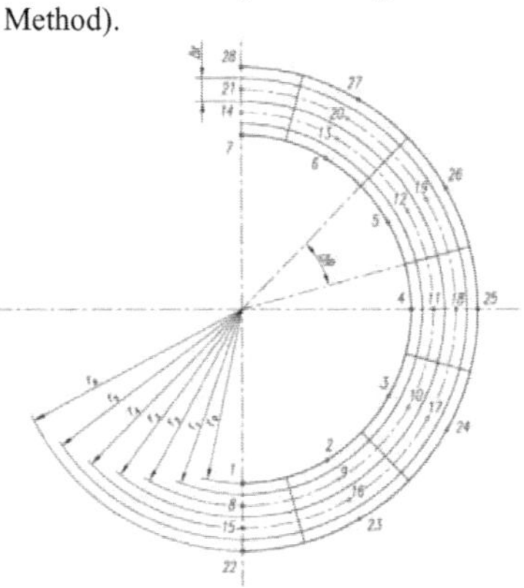

Fig. 9. Division of the steam header cross-section into finite volumes

a)

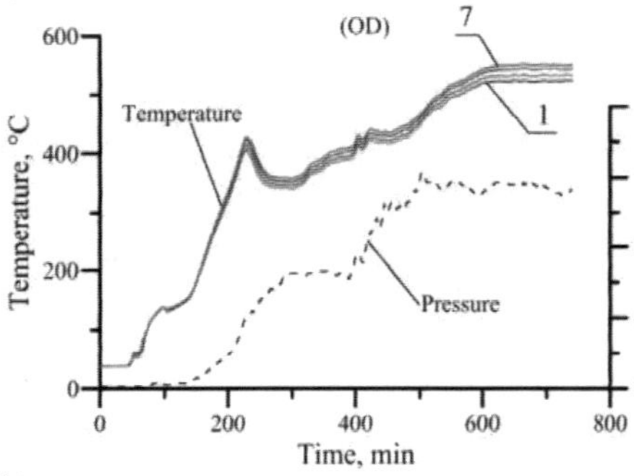

b)

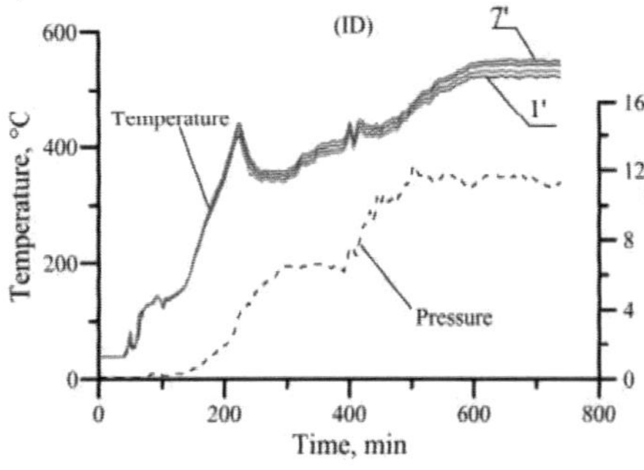

Fig. 10. Time distribution of the temperature measured on the outer surface OD (a) and calculated on the inner surface ID (b)

Selected results for a boiler start-up from a cold state are depicted in Figs. 10-13. Thermal and pressure caused stresses at monitored outlet header of the final superheater stage are shown in Figs. 12-13. The steam header has outer diameter of 324 mm and wall thickness of 70 mm, and is manufactured of 10 CrMo 9-10 steel. The chemical composition of 10 CrMo 9-10 steel shows Table 2.

Table 2. The chemical composition of 10 CrMo 9-10 steel

C	Si	Mn	P	S	Cr	Mo	Cu
0.08-0.14	≤ 0.50	0.40-0.80	≤ 0.02	≤ 0.01	2.00-2.50	0.90-1.10	≤ 0.30

The measured temperature histories are shown in Fig. 10a. The temperatures obtained from the solution of the inverse heat conduction problem are depicted in Fig. 10b.

Temperature variations on the header circumference for selected time points illustrates Fig. 11.

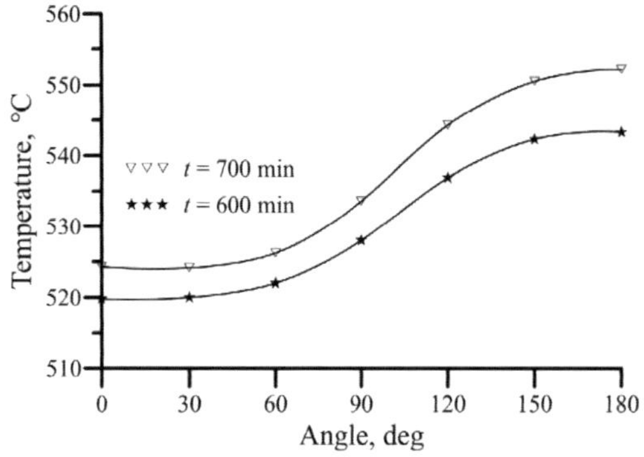

Fig. 11. Wall temperature on the outer surface of steam header as a function of an angle at different times

The temperature difference between the top and bottom of the header is within 25 K (Fig. 11).

After determining the temperature distribution the stresses due to temperature and inner pressure were calculated using the finite element method (FEM) of an unpierced cylinder with *SCC* = 1.0. Circumferential and longitudinal stresses at the inner header surfaces as functions of time are shown in Fig. 12a and 12b, respectively.

a)

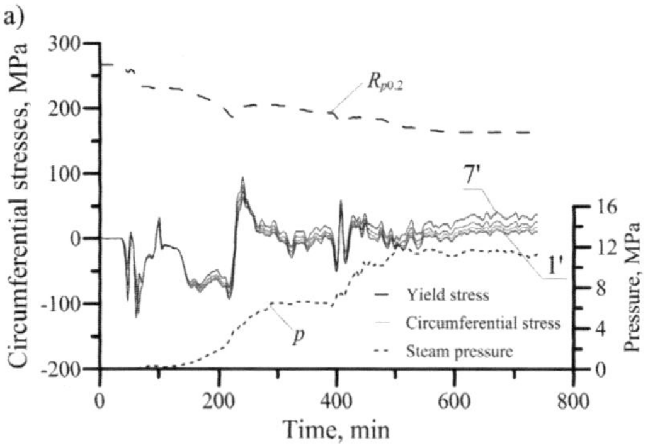

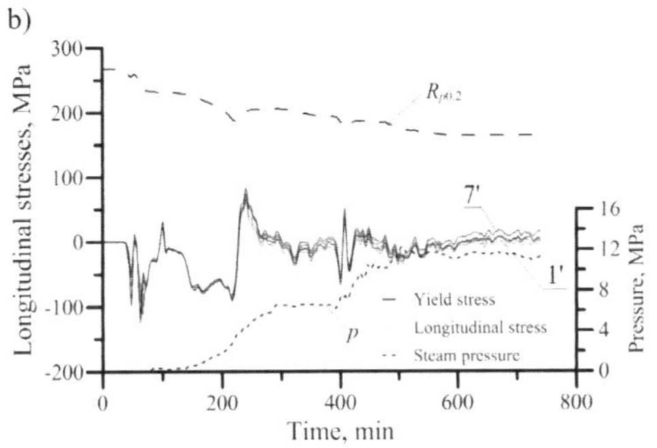

Fig. 12. Calculated circumferential (a) and longitudinal (b) stresses due to pressure and thermal loads at the inner surface of the outlet header of the final superheater; location of the points 1'-7' is shown in Fig. 7d with *SCC* = 1.0.

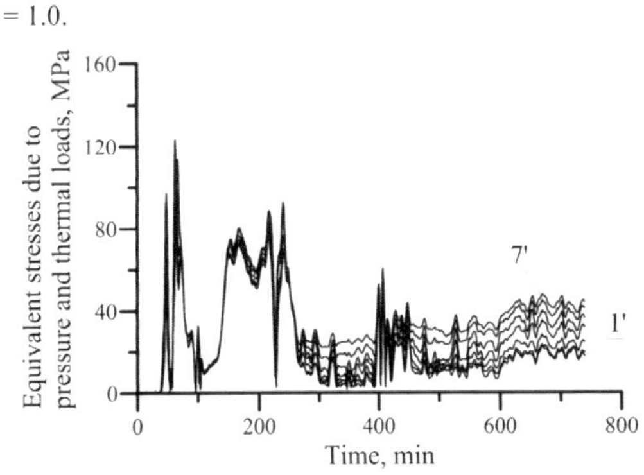

Fig. 13. Calculated equivalent stresses due to pressure and thermal loads at the inner surface of the outlet header of the final superheater; location of the points 1'-7' is shown in Fig. 7d.

Time changes of equivalent stresses at seven locations on the inner header surface are presented in Fig. 13. The analysis of the results shown in Fig. 12-13 can see that the value of the stresses affect not only the pressure but also heat loads.

The analysis of the results depicted in Figs 12-13 shows that the circumferential and longitudinal stresses are less than the yield strength. It should be noted, however, that the stresses on the edges of the holes are high (*SCC* = 2.5 or as in [17]) and may lead to plastic deformations, which in turn cause the thermal fatigue. The permissible tensile and compressive at the edge of the hole, calculated in accordance with EN 12952-3 [23] are respectively 323.97 MPa and 241.83 MPa. Also the maximum stress on the edge of the opening is smaller than the specified allowable stresses. There are two operating procedures to reduce effectively the longitudinal and circumferential temperature differences after the boiler has been built: properly draining the horizontal header and slowing down the start-up rate. Also, at the design stage the flexibility of the boiler can be improved by providing spring supports for the headers to allow their significant deflection. To avoid large temperature differences on the circumference of headers multiple 35 mm drain connections at the header invert should be provided to ensure good drainage.

Conclusion

A computer system for thermo-hydraulic monitoring of the steam boilers delivers in on-line mode the following operation parameters: heat losses and boiler efficiency, coal mass flow rates, fouling and slagging degrees of the combustion chamber and all superheater stages. The local and average slagging degrees of combustion chamber walls are determined experimentally. The monitoring systems performs the on-line heat transfer calculations of the boiler combustion chamber and heat flow rates absorbed by superheater stages.

An important task of the computer system developed is also determining thermal stresses in critical pressure parts of the boiler and to estimate their remnant life. The proposed computer system ensures economical and safe operation of the boiler in steady states and during load changes. Due to the development of wind farms blocks in thermal power plants should be capable to rapid starts. By installing developed system start-up time of the boiler can be greatly reduced without the risk of overheating superheater tubes and shortening service life of the boiler structural elements.

Since the degree of slagging of the combustion chamber and degree of ash fouling of superheater stages are calculated in on-line mode, the computer system can be used to control the activating frequency of slag and ash blowers.

The ability of the developed system to monitor thermal and flow parameters, thermal stresses and lifetime of critical pressure components has been shown on example of a steam boiler with natural circulation.

References

1. H. Bilirgen, Slagging in PC boilers and developing mitigation strategies. Fuel, **115**, pp. 618-624 (2014)
2. N.S. Harding, D.C. O'Connor, Ash deposition impact in the power industry, Fuel Processing Techn. **88**, pp. 1082-1093 (2007)
3. L.M. Romeo, R. Gareta, Hybrid System for Fouling Control in Biomass Boiler, Eng. Apps. of Artificial Intelligence, **19**, pp. 915-925 (2006)
4. L. M. Romeo, R. Gareta, Fouling Control in Biomass Boiler. Biomass and Bioenergy **33**, pp. 854-861 (2009)
5. A.U. Syed, N.J. Simms, J.E. Oakey, Fireside corrosion of superheaters: Effect of air and oxy-firing of coal and biomass. Fuel, **101**, pp. 62-73 (2012)
6. J. Taler, M. Trojan, D. Taler, Monitoring of Ash Fouling and Internal Scale Deposits in Pulverized Coal Fired Boilers, Nova Science Publishers, New York (2011)
7. M. Poniewski, A. Salij, J. Stępień, Impact of biomass co-firing on selected parameters of a 225MW power unit, J. of Power Techn., **95**, pp. 84-90 (2015)

8. H. Othman, J. Purbolaksono, B. Ahmad, Failure investigation on deformed superheater tubes. Eng. Failure Analysis, **16**, pp. 329-339 (2009)
9. A.K. Ray, Y.N. Tiwari, R.K. Sinha, P.K. Roy, S.K. Sinha, R. Singh, S. Chaudhuri, Remnant life assessment of service-exposed pendent superheater tubes. Eng. Failure Analysis, **9**, pp. 83-92 (2002)
10. P. Madejski, D. Taler,. Analysis of temperature and stress distribution of superheater tubes after attemperation or sootblower activation. Energy Conversion and Management, **97**, pp. 131-137 (2013)
11. N.V. Kuznetsov, V.V. Mitor, I.E. Dubovskij, and E.S. Karasina, *Thermal Calculations of Steam Boilers. Standard Method* (Energy, Moscow, 1973)
12. A.G. Blokh, *Heat Transfer in Steam Boiler Furnaces* (Hemisphere, Washington, 1988)
13. S. Kakaç, *Boilers, Evaporators, and Condensers* (Wiley, New York, 1991).
14. J. Taler (Ed.), *Thermal and Flow Processes in Large Steam Boilers. Modeling and Monitoring* (WNT, Warsaw, 2011)
15. *ASME PTC 4. Fired Steam Generators. Performance Test Codes* (The ASME, New York, 2013)
16. W. Wagner, H. J. Kretzschmar, *International Steam Tables Properties of Water and Steam Based on the Industrial Formulation IAPWS-IF97* (Springer-Verlag, Berlin, 2008)
17. J. Taler, D. Taler, P. Ludowski, Measurements of local heat flux to membrane water walls of combustion chambers, Fuel, **115**, pp. 70–83 (2014)
18. J. Taler, D. Taler,. Measurement of Heat Flux and Heat Transfer Coefficient. Chapter 1, in: *Heat Flux: Processes, Measurement Techniques and Applications*, G. Cirimele and M. D'Elia (Eds.) (Nova Science Publishers New York, pp. 1-104, 2012)
19. P. Duda, J. Taler, A new method for identification of thermal boundary conditions in water wall tubes. Int. J. Of Heat an Mass Transfer, Vol. **52**, Issues 5-6, pp. 1517-1524, (2009)
20. J. Taler, D. Taler, Measurements of Local Heat Flux and Water-Side Heat Transfer Coefficient in Water Wall Tubes, Chapter 1, in *Heat Transfer*, Editor : D. M. Salim Newaz Kazi (InTech, Rijeka – Shanghai, pp. 3-34, 2012)
21. J.M. Fetköter, D.J. Lauer, M. Wedemeyer, G. Engelhard, Replacement of Shutdown Cooling System and Repair of Nozzle Welds of Reactor Pressure Vessel in the Forsmark 1 + 2 Nuclear Power Plant, VGB PowerTech, Vol. **81**, No. **6**, pp. 62-64 (2001)
22. P. Rop, Drum Plus: a drum type HRSG with Benson benefits. Modern Power Systems, Vol. **30**, No. 10, pp. 35-40 (2010)
23. *EN 12952-3 Water-tube boilers and auxiliary installations – Part 3: design and calculation for pressure parts.* European Committee for Standardization, Brussels

Analysis of role of erosion process of coal particles in generation of loss of incomplete combustion in fluidized bed boilers

Piotr Pelka[1,*]

[1] Czestochowa University of Technology, Institute of Thermal Machinery, Armii Krajowej 21, Silesia 42-201 Czestochowa, Poland

Abstract. In the herein paper the results of research carried out in the Institute of Thermal Machinery of Technical University of Czestochowa have been presented. They concern the process of combustion of coal particles in a fluidized bed combustor. The tests were carried out on a test stand which allows modelling of conditions around the fuel particle in a real object. On the basis of the registered mass loss of the particles of Polish hard and brown coal chosen, the impact of the mass flow rate of inert material with different particle size distribution and different collision velocity with burning coal particle on the rate and level of its burning out was evaluated. The obtained results enables the estimation of the influence of both parameters on the intensity of erosion and determination of its role in the generation of fine particles elutriated out of the combustion chamber. The results also facilitated the verification of the methodology [1] of calculation described in literature relating to the total mass loss of coal particles as a result of combustion and erosion. According to this methodology, the slip velocity between the char particle and the fine bed solid is the main parameter of the erosion process alongside the attrition rate constant k_a.

1. Introduction

The loss of incomplete combustion depends on many factors. Alongside the coal type and the associated physical and chemical properties, both the construction type and operating factors of the fluidization process also have an impact on its level. This loss is divided into losses related with the presence of combustible gasses hydrocarbons and carbon monoxide in gaseous combustion products caused by unburned solid carbon in the ash. The unburned carbon loss U_{cl} can be calculated as follows [2]:

$$U_{cl} = \frac{X_c \cdot W_a \cdot 32790 \cdot 100}{HHV} \qquad (1)$$

where X_c is the fractional carbon in the solid waste, W_a is the ash per unit mass of the fuel feed and HHV is the heating value of coal. This loss has essential contribution to the heat balance especially in CFB boilers. Table 1 compares heat losses in conventional pulverized coal-fired boilers (PC) and CFB boilers [1]. It can be easily seen that the unburned carbon loss has a significant role in the heat balance, especially in the case of CFB boilers which can reach 2%. It should be emphasized that in this comparison, data from the CFB boilers of an uncooled hot cyclone was used. New generation objects with cooled separators have lower flue gas loss, therefore the percentage share of unburned carbon will be even higher in the total heat balance. Many factors influence the unburned carbon loss and consequently the amount of carbon in the flue ash in CFB boilers, of which the following are first and foremost distinguished: coal type, particle size distribution of coal, temperature in the bulk of particle, oxygen concentration, furnace design, type of fluidization.

Table 1. Comparison of heat losses in typical PC and CFB boilers [2].

Items	PC (%)	CFB (%)
Moisture in limestone	-	0.06-0.10
Calcination	-	1.69
Unburned carbon	**0.25**	**0.50-2.00**
Heat in dry flue gas	5.28	5.57-5.60
Moisture in fuel	1.03	1.03
Moisture from H_2 burning	4.16	4.19
Radiation and convection	0.03	0.30-0.80
Moisture in air	0.13	0.14
Sensible heat in boiler ash	0.03	0.09-0.76
Bottom ash	0.05	-
Total loss	10.96	12.9-16.31

The basic condition of the complete burning out of solid fuel is that the residence time of char particle τ_p must be only a little longer than the burnout time τ_s. Both of these time periods depend on the magnitude of the afore-mentioned parameters which hinders the precise determination of their values, particularly the fact that

* Corresponding author: pelka@imc.pcz.czest.pl

some of the afore-mentioned parameters undergo change together with the furnace height, i.e.: temperature, oxygen concentration, hydrodynamic conditions. The next parameter which complicates the analysis of the process is the comminution of coal particles which begins from the fuel feeding moment and lasts continuously to the moment of the elutriation of the particles from the furnace. These processes are fragmentation and erosion. On the one hand, the comminution improves the combustion conditions as a result of the increase of surface reaction of char particles as well as the continuous removal of the ash layer formed on the surface blocking the diffusion process. On the other hand, in the opinion of many authors [3,4], the erosion process as a result of collision between solid material and coal particles generates fine particles of less than 100 μm. The residence time of coal particles above this diameter is enough to complete the burning out of fuel particles but smaller particles have significantly reduced residence time in the combustion chamber [5]. Taking into account that the process runs in the whole volume of a furnace, in subject-related literature many opinions that the erosion process, also called attrition, is mainly responsible for generating the loss of incomplete combustion and consequently the increase of the unburned carbon in fly ash can be found [2,3,4,5].

2 Effect of erosion in char combustion model

The most popular way of determining the burning rate of coal particles in the phase of char combustion is the Levenspiel model [6], namely the shrinking unreacted core model. In CFB this model is particularly useful for the sake of high gas-solid relative velocity and abrading the ash layer on the char. The burning rate for the shrinking unreacted core model is given as follows [7]:

$$q = \frac{c_g}{1/h_m + 1/k_c} \tag{2}$$

where:

$$h_m = \frac{12\phi Sh D_g}{d_c R T_m} \tag{3}$$

and:

$$k_c = k_o \exp\left(-E/RT_s\right). \tag{4}$$

Additionally, with the aim of indicating the equation (2) the erosion mass loss of char particle burning in CFB boiler [8,9], the proposed new formula (5) is added.

$$\frac{dW_c}{dt} = qA + \frac{k_a U_g W_c}{d_{av}} \tag{5}$$

According to this formula the intensity of mechanical mass loss depends on the external surface area of char particle exposed to erosion W_c/d_{av}, superficial gas velocity U_g and attrition rate constant k_a. The range of the rate constant in combustion assisted attrition in a fast bed is within the range of $0.5\text{x}10^{-7}$ to $5\text{x}10^{-7}$ kg/s [3]. Its value may only be estimated during the experiment. In the case of conditions in CFB, the hydrodynamic force is proportional to the slip velocity between the char and the abrading bed material [9] $U_g=(U_c - G_s/\rho_b)$. This means that the slip velocity is the decisive parameter of the erosion process for coal particles of the same type and diameter.

3 Experimental Verification

In order to verify the erosion formula in the equation (5), the tests of analysis of mass loss of different coal (table 2) for different mass rate flow of inert material and different collision velocity was carried out. The research was conducted on a test stand facilitating the modelling of real combustion conditions in CFB particularly in the middle and upper zone of the commercial CFB plants [10].

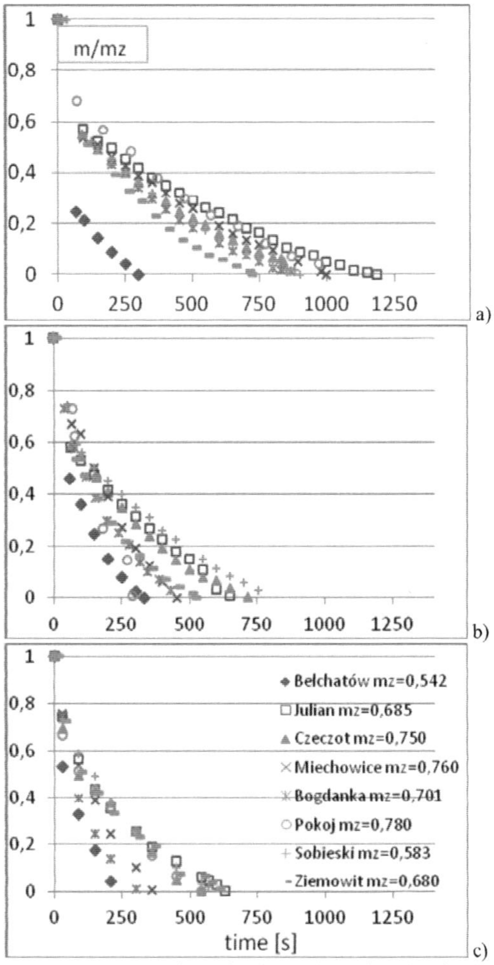

Fig.1. Relative mass loss of coal particles (m/mz) burning in air atmosphere a) and in flow of inert material for Gs=2.5 b) and Gs=5.0 c).

The velocity of solid material in the core of the diluted zone is close to u_p ~4m/s. In the test, a manually prepared coal particle of 10mm of chosen type was used. Details of the experiment and design of the test stand were described in [11]. The combustor chamber with measurements 80x100x120mm was powered by 4kW electric heaters and a temperature inside the combustor was 850C. In the area of the burning coal, gravitationally accelerated quartz sand was entered. The particle size diameter of the solid material was typical for the commercial fluidized bed (Fig.2 "medium"). Figure1 presents the registered relative mass loss of coal of different types obtained during combustion, firstly in air atmosphere without interaction of inert material and then in conditions modelling combustion process in a fast FB. The mass rate flow of solid material was in turn: Gs=0; 2.5; 5.0 kg/m^{2}s. The results of proximate analysis were presented in Table 2. The range of mass rate flow was taken on the basis of the measurement of concentration in commercial boilers [10]. The registered mass loss of coal particles presented in Fig.1 unequivocally indicates that the flow of inert material strongly accelerates the mass loss in relation to the registered in air atmosphere. The average rate of mass loss for the individual char of tested hard coals (after combustion of volatile matter) was within a range of Gs=0 from 3.8x10^{-7} (Julian mine) to 5.8x10^{-7}kg/s (Ziemowit mine). In the flow of inert material this range was clearly higher for Gs=2.5 from 5.4x10^{-7} (Sobieski mine) to 12.5x10^{-7}kg/s (Miechowice mine) and for Gs=5.0 from 6.8x10^{-7}(Ziemowit mine) to 15.2x10^{-7}kg/s (Miechowice mine). Thus, putting coal particles in the area of flow of inert material increased the rate of mass loss for Gs=2.5 average 74% and for Gs=5.0 107% in the relation to values registered for Gs=0. However, the following questions are still valid: Which of the following factors: erosion, parameters of coal or hydrodynamic conditions around the particle has the biggest effect in the accelerated mass loss observed? What is the contribution of the first term of the equation (2), chemical reaction and the second term erosion to the total mass loss of fuel coal? How does the particle size distribution and collision velocity affect the erosion intensity? In order to estimate the attrition rate constant k_a in equation (5), the determination of slip velocity U_g between the burning coal particle and the solid material is necessary. During the experiment, a single coal particle was still placed inside the combustion chamber on the handle of a tensometric branch scale. The collision velocity may change as a result of change in the mean particle diameter used in the inert material. In order to estimate the slip velocity, it is sufficient to solve the easy set of equations describing the free drop of spherical particle in the viscid medium. The model results of calculation for particles of different diameters are presented in Table.3. In the experiment the inert material with different particle size distribution is presented in Fig.2 and the characteristic parameters shown in Table 4 were used.

This time the tests were carried out for two different coal types: from Sobieski mine (type 31.2) and from Miechowice mine (type 32.1). These coals characterised the biggest mass loss difference measured during the combustion in flow of inert material. The experiment was conducted for different mass flow rates respectively for Sobieski mine *Gs*~4.0; 7.0 and 15kg/m^{2}s and for Miechowice mine *Gs*~1.5; 4.0 and 6.5kg/m^{2}s. In Fig.3 a-f, the obtained results were presented.

Table 2. Results of proximate analyses of tested coals.

Mine	Proximate analyses				
	V	W	A	FC	Q
	%	%	%	%	kJ/kg
Belchatow (lignite)	42.5	14.5	18.5	24.5	18460
Julian (hard)	37.5	7.6	5.9	49.0	29026
Ziemowit (hard)	26.9	7.6	20.7	44.8	21179
Sobieski (hard)	27.9	12.4	16.7	43.0	21558
Czeczot (hard)	29.3	9.5	14.6	46.6	20300
Miechowice (hard)	30.2	6.7	12.0	51.1	23000
Bogdanka (hard)	27.8	2.9	25.4	43.9	22542
Pokoj (hard)	31.5	1.9	2.7	63.9	31500

Table 3. Results of numerical simulations of collision velocity.

Particle diameter [μm]	Velocity [m/s]	***Re***
500	3.18	6.7
400	2.5	5.3
300	1.77	3.7
200	1.08	2.3
100	0.31	0.7

On the basis of registered mass loss of coal particles from Sobieski mine, it may be noted that in the case of all mass rate flow inert material the time differences for fine medium and coarse particle size distribution are similar and do not exceed 100s. In the phase of combustion of volatile matter the mass loss is almost the same. Only differences in the phase of char combustion are observed. The smallest effect on the erosion mass loss was noted for combustion in the flow of coarse inert material, for which the equivalent Sauter diameter was *Da*=485 μm. While the mass loss for medium and fine inert material was almost the same in the whole tested range of *Gs*, only in the case of the biggest mass flow rate (15kg/m^{2}s) in the last stage of combustion was there a slightly bigger effect for fine material registered. A similar course of the phenomenon was observed for Miechowice coal. The influence of the particle size diameter in the analyzed range is close and the different result observed is most likely from the impurity of the coal prepared. In the case of the low mass rate flow *Gs* =1.5kg/m^{2}s, the overall time of mass loss registered for coarse inert material was very close to that obtained during the test on medium material.

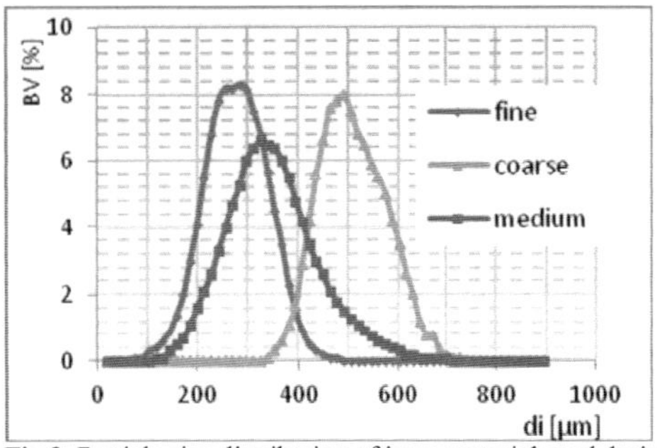

Fig.2. Particle size distribution of inert material used during the experiment.

Table 4. Parameters of used in experiment inert material.

Parametersof inert material [µm]	fine	medium	coarse
Dn – arithmetical diameter	127	169	49,1
Ds – surface diameter	165	219	126
Dv – volume diameter	189	248	197
Da – Sauter diameter	247	319	485
Dgeo – geometrical diameter	74,8	88,2	22,1
Dmed	117	186	16,3
Dmod	9,9	9,9	9,9

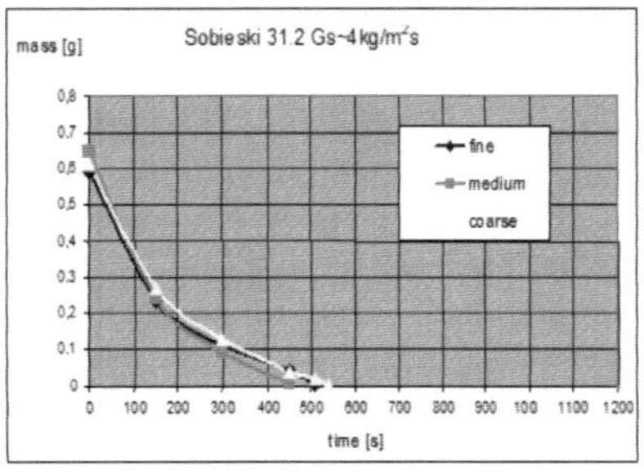

a)

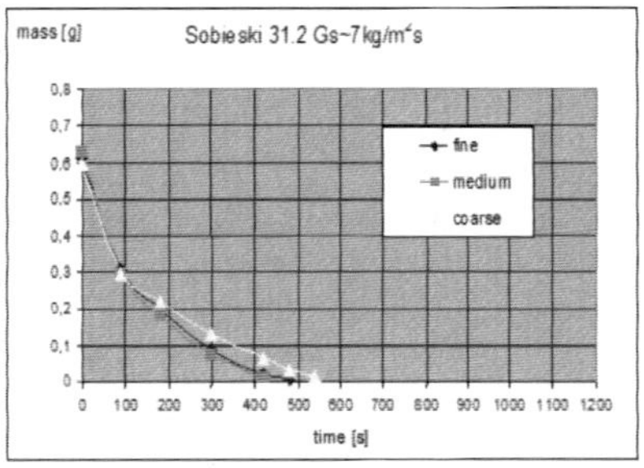

b)

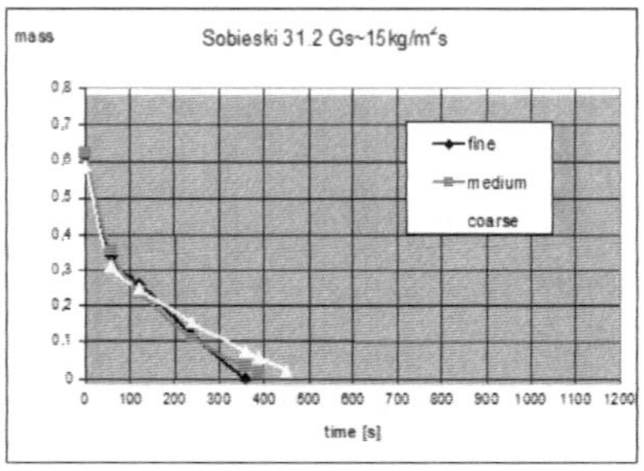

c)

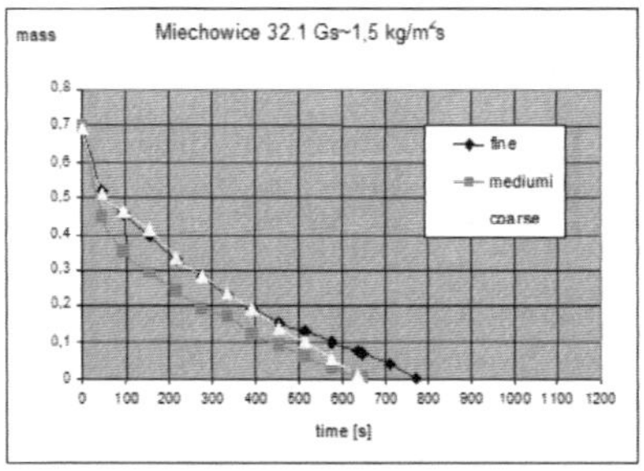

d)

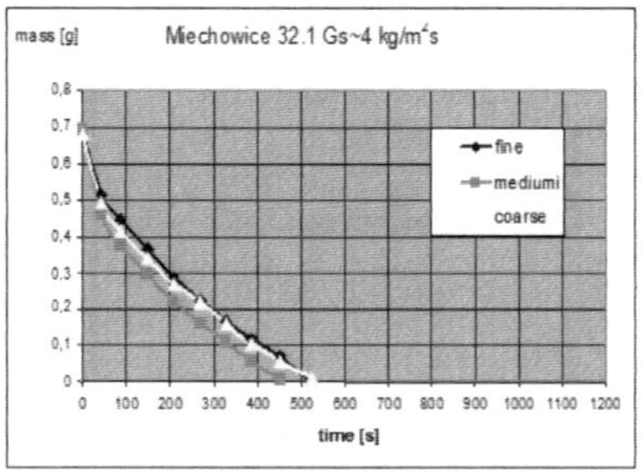

e)

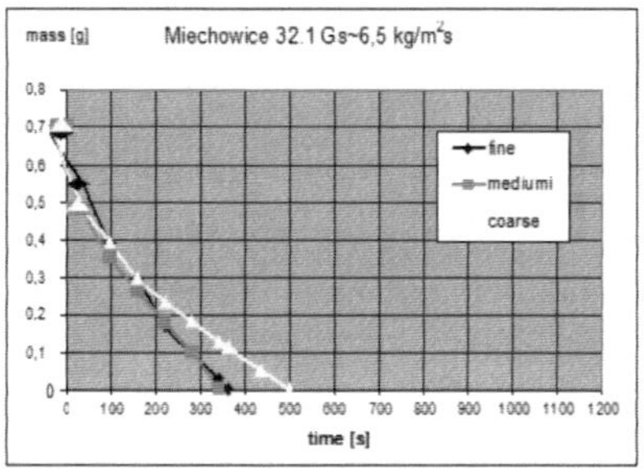

f)

Fig.3. Mass loss of coal particles: a, b, c from Sobieski and d, e, f from Miechowice mine during the combustion in different flow of solid and different particle size distribution.

With the increasing mass rate flow, the difference of overall time of mass loss for fine and medium material increases and for *Gs* =6.5kg/m^{2}s is about 150s in favour of medium material.

While exactly the opposite trend may be seen for fine inert material, which for the lowest mass flow rate distinguishes the smallest influence the erosion mass loss (time difference Δt_c =120s), and with an increase to *Gs* = 6.5 was almost the same as the biggest mass loss observed for the medium solid material. Thus, on the basis of the tests carried out, it may be stated that the particle size distribution in the chosen range is not important for the generation of erosion mass loss. The overall time difference of mass loss for all cases did not exceed 15% despite the Sauter diameter being twice bigger between the fine and coarse material. The one trend noted (in five out of six considered cases) is the smallest erosion mass loss caused by the fine solid material with the highest slip velocity between the char particles.

4 Conclusions

The research conducted confirmed that the combustion of coal fuel in the flow of inert material characterises much greater mass loss. The loss is caused as a result of oxidation reaction and the erosion process running simultaneously. The erosion process causes abrasion (attrition) of the ash layer forming during combustion and the external weakened char surface. The rate of mass loss for the parameters taken in the tests presented that match the conditions in the middle and upper zone of CFB boilers may be even more than 100% in relation to combustion without the influence of inert material. The results obtained in the second part of the research on materials with three different particle size distributions with the Sauter diameter were respectively: 247; 319; 485 μm which indicate that the size of solid particles and slip velocity between inert material and burning coal particle has no important meaning. It suggests that the proposed equation of attrition rate char in subject-related literature either incorrectly describes the real mechanism of mass loss or the contribution of erosion process in mass loss is negligible. The difference of the mass loss rate for both coal types and three particle size distributions applied do not exceed 15% and are of a random nature. It may be stated that the increasing mass loss rate observed in all the analyzed particle size distributions is due to the same factor. This is most likely the intensification of the diffusion of reactants and products in the reaction area on the burning coal particle surface. Hitting the coal surface and flowing in the area of burning coal inert material notably improves mass exchange. The increase of their size and slip velocity have a small influence on the process tested. The primary role of the erosion process is the attrition of ash forming on the surface reaction.

Nomenclature

A - external surface area of char, m^2
c_g - oxygen concentration in partial pressure, kPa
d_{av} - average carbon particle diameter, m
d_c - average diameter of char particle, m
D_g - molecular diffusivity of oxygen, m^2/s
E - activation energy, kJ/kmol
G_s - circulation rate in fast bed, kg/m^2s
h_m - mass transfer coefficient, $kg/m^2kPa\ s$
k_a - attrition rate constant
k_c - kinetic rate constant, kg/m^2 s Pa
k_o - frequency factor, kg/m^2 s Pa
m - mass of coal particle, kg
m_z - initial mass of coal particle, kg
q - specific burning rate, kg/m^2s
R - gas constant, 8.314 kPa m^3/kmol K
Sh - Sherwood number
T_s - temperature of the surface of the char, K
T_m - mean temperature of the diffusion layer around the char particle, K
U_c - char velocity, m/s
U_g - superficial gas velocity, m/s
W_c - overall mass loss, kg
ϕ - mechanism factor
ρ_b - bulk density of the fast bed, kg/m^3

References

1. P.K. Halder, Ph.D. Thesis, Technical University of Nova Scotia, (1989)
2. P. Basu, *Combustion and gasification in fluidized bed*. Taylor & Francis Group, (2006)
3. P. Basu, Chem. Eng. Sci. **54**, 5547 (1999)
4. R. Chirone, L. Massimilla, P. Salatino, *Prog. Energy Combust. Sci.* **17**, 297 (1991)
5. Z. Bis, *Kotły fluidalne, teoria i praktyka.* Czestochowa University of Technology, (2010)
6. O. Levenspiel, *Chemical reaction engineering (2nd Ed.)* New Delhi: Wiley Eastern Ltd. 368 (1984)
7. Basu P., Fraser S.A.: *Circulating fluidized bed boilers – design and operations*, 95 (1991)
8. I.B. Ross, Ph.D. Dissertation, University of Cambridge, U.K., (1979)
9. P.K. Halder, P. Basu, Chem. Eng. Sci., **47**, 527 (1992)
10. R. Sekret, *Warunki cieplno-przepływowe i emisje zanieczyszczeń w kotłach z cyrkulacyjną warstwą fluidalną dużej mocy*, Silesian University of Technology, **1694** Gliwice (2005)
11. P. Pelka, Combust. Flame **156**, 1604 (2009)

Experimental verification of a CFD model intended for the determination of restitution coefficients used in erosion modelling

Bartłomiej Hernik[1,*], *Marek* Pronobis[1], *Robert* Wejkowski[1], and *Wacław* Wojnar[1]

[1]Silesian University of Technology, Division of Boilers and Steam Generators, Konarskiego 20, 44-100 Gliwice, Poland

Abstract. Erosion caused by solid particles transported with the steam or flue gas has a negative impact on the power unit reliability and availability. The erosion rate depends inter alia on the restitution of the particle velocity upon impact. The restitution coefficients determine the angle of the particle reflection off the tube surface and the particle post-impingement velocity, i.e., they determine the direction of the particle path, which has a substantial impact on the erosion phenomenon inside the tube. An attempt is made herein to develop a method of determination of restitution coefficients by means of numerical modelling assisted by experimental testing on physical models that will be implemented further in the Ansys Fluent code. Such a numerical procedure will verify the model of erosion caused by particles of iron oxides. The erodent impingement angle α_1, the impingement velocity w_1, and the reflection velocity w_2 are measured using the Casio High-Speed Exilim EX-F1 camera, which enables filming at a high rate. The film is then processed graphically for "frame-by-frame" tracking. The following erodents were used in the testing: iron oxides, quartz sand with a different grain size (490, 1000, 1500, 2000 μm), and 1000 μm-diameter steel balls. The steel balls, due to their ideal shape, were treated as the comparative analysis reference standard. Erosion of three types of 5x10 cm plates was tested: a plasma-coated plate with an anti-erosion layer, an aluminium plate; and a steel sheet plate. Based on the restitution coefficient testing results, numerical simulations were performed of the particle reflection off the surface.

1 Introduction

Erosion losses in the boiler elements such as the heat exchanger tube banks located on the path of dust-laden flue gas have a considerable negative impact on the power unit reliability and availability. Erosion processes usually have a local character, that is, there are cases where on the same tube bank, some areas remain completely unaffected by them, whereas in others, the wall thickness gets close to the permissible limit value.

Erosion caused by solid particles transported with the steam may also occur inside tubes as well as in turbines. The stability of oxide layers on the inner surfaces of superheater tubes may be disturbed due to exfoliation or spallation of iron oxides. Because of changes in the tube temperature during boiler operation, stress arises in the oxide layers, causing their exfoliation. This phenomenon can be observed in superheaters and reheaters made both of low-chromium ferritic and high-alloyed austenitic steels [1-3]. Erosion depends on many factors, including the particle kinetic energy on impingement, the impingement angle, the mechanical properties of the erodent and of the material with which the erodent comes into contact, and the restitution of the particle post-impingement velocity. In case of two-phase gas-solid flows the rate of erosion and the restitution coefficients depend on the concentration of solid phase in the flow also.

2 Numerical studies on the determination of restitution coefficients

The process of accumulation of oxide layers on the inner surface of superheater tubes is disturbed in high steam temperatures because of exfoliation of oxides. Fig. 1 [**Błąd! Nie zdefiniowano zakładki.**] shows the result of such processes – the faults caused by the detachment of deposit fragments.

A correct simulation of the superheated steam two-phase flow with particles of iron oxides that takes account of the particle reflection off the tube wall cannot be performed without the knowledge of restitution coefficients. The coefficients determine the angle of the particle reflection off the tube surface and the particle post-impingement velocity, i.e., they determine the direction of the particle path, which has a substantial impact on the erosion phenomenon inside the tube.

The erosion rate and the velocity of particles (usually considered to be identical to the carrier gas velocity) are related by an exponential dependence [4-8]. Reference literature includes numerous works devoted to the problem of damage done to tubes by fly ash carried by flue gas. However, the erosion of the power unit elements caused by particles of iron oxides separated from the inner surface of the superheater/reheater tube walls due to exfoliation and carried by steam has not

* Corresponding author: bartlomiej.hernik@polsl.pl

been investigated in detail. An attempt is made herein to develop a method of determination of restitution coefficients by means of numerical modelling assisted by results of experimental testing on physical models that will be implemented further in the Ansys Fluent code [9]. Such a numerical procedure will verify the model of erosion caused by particles of iron oxides.

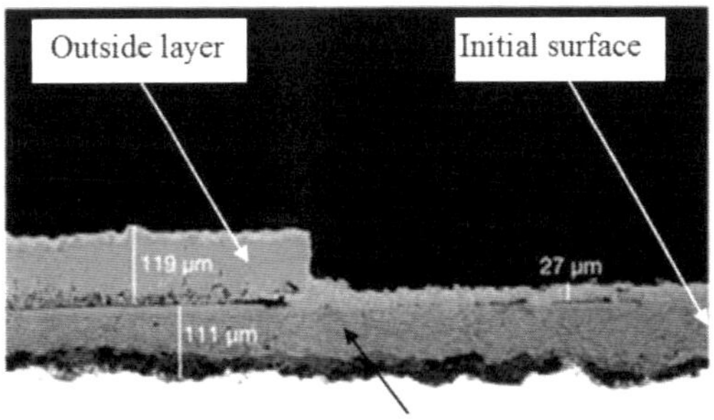

Fig. 1 The cross-section through oxide layers on a tube made of P92 steel (9% of Cr) after 6200 h in steam with a temperature of 650°C [**Błąd! Nie zdefiniowano zakładki.**].

2.1 Description of the reflection model used in the Ansys Fluent code

Numerical simulations make it possible to define the flow profile for the working medium and the erodent. They are also a source of information on the erodent particle inflow velocities in the area of erosion. The calculations were performed using the Ansys Fluent code and taking account of modifications based on physical experiments.

In order to make a precise prediction of trajectories of the erodent particles after their reflection off the surface, it is necessary to apply a reflection model, which involves the use of restitution coefficients e_t and e_n. The reduction in the reflection velocity compared to the impingement velocity is the effect of the erodent loss of kinetic energy. There is a direct relation between the difference in the system pre- and post-impingement kinetic energy and the erosion process intensity.

The coefficients are determined by measuring the erodent impingement angle α_1 to the eroded surface and by measuring the erodent impingement and reflection velocities: w_1 and w_2, respectively, (cf. Fig. 2).

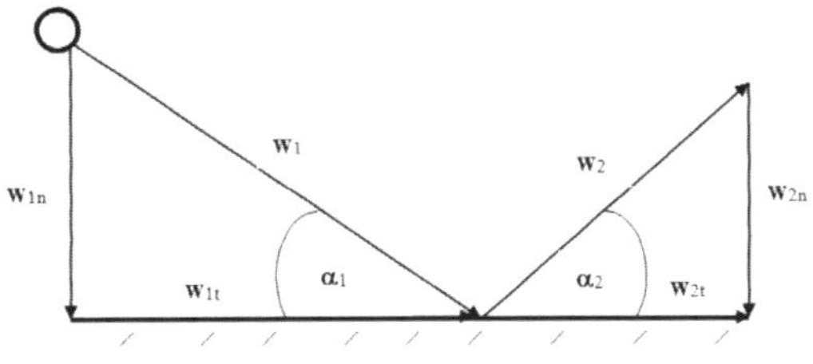

Fig. 2 Diagram of the particle reflection for boundary conditions of the restitution coefficient.

The normal restitution coefficient is defined as

$$e_n = \frac{w_{2n}}{w_{1n}} = \frac{w_2 \sin\alpha_2}{w_1 \sin\alpha_1} \quad (1)$$

whereas the tangential coefficient is

$$e_t = \frac{w_{2t}}{w_{1t}} = \frac{w_2 \cos\alpha_2}{w_1 \cos\alpha_1} \quad (2)$$

where:

w_1 – impingement velocity,
w_1 – reflection velocity,
α_1 – impingement angle,
α_2 – reflection angle,
w_n – normal-to-surface velocity component,
w_t – tangential-to-surface velocity component,
e_n – normal restitution coefficient, defined as the sum of momentum generated by the particle in the normal direction to the surface,
e_t – tangential restitution coefficient, defined as the sum of momentum generated by the particle in the tangential direction to the surface.

A normal or tangential restitution coefficient of 1 means that after the reflection the normal or tangential momentum of the particle is conserved (perfectly elastic collision), a coefficient equal to 0 means a perfect plastic collision (inelastic collision). In reality, for a real elastic-plastic collision, the coefficient value is included in the following interval: $0<e<1$. Studies indicate that the erodent particle impingement angle has a decisive effect on the restitution coefficient values. Restitution coefficients have an indirect impact on the erosion rate calculation results even though they do not appear directly in the erosion model used in this code. The Fluent code restitution coefficients determine the properties of the material with respect to its capacity to reflect the particle. Different correlations are put forward in the literature to describe restitution coefficients [**Błąd! Nie zdefiniowano zakładki.**, 10-13].

2.2 Description of the erodent and the testing stand

The erosion testing stand is presented in Fig. 3. The erodent-transporting medium is air pressurized in a compressor to about 5 bar.

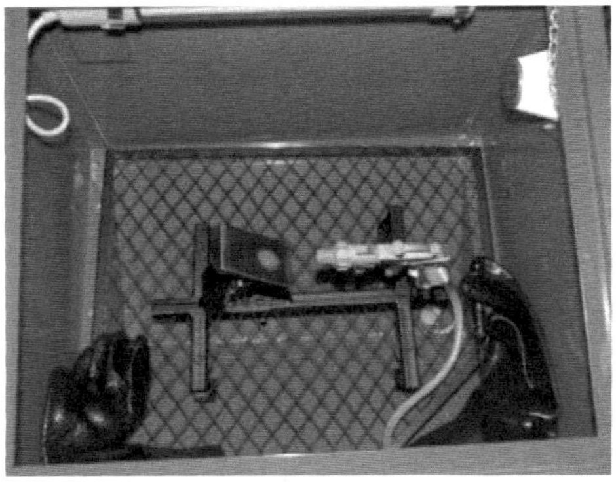

Fig. 3 Erosion testing stand.

The plate subjected to the erodent effect is set at an angle of 45°, which is typical in erosion testing of ductile

DOI: 10.1051/e3sconf/20171305001

metals (at this angle the material loss is the greatest). The erodent impingement angle α_1, the impingement velocity w_1, and the reflection velocity w_2 are measured using the High-Speed Exilim EX-F1 camera made by Casio, which enables filming at a high rate. The film is then processed graphically for "frame-by-frame" tracking. Three types of 5x10 cm plates were tested:

- a plasma-coated plate with an anti-erosion layer,
- an aluminium plate,
- a steel sheet plate.

The following erodents were used in the testing: iron oxides, quartz sand with a various grain size, and 1.0 mm-diameter steel balls. This enabled a comparison between a monodispersed erodent with a perfect geometry and two polydispersed ones with different density and hardness values. Quartz density is 2.65 g/cm³ with hardness 7 in Mohs' scale, whereas for haematite and magnetite these values range from 4.9 to 5.3 g/cm³ and from 5.5 to 6.5. Iron oxides are taken into consideration due to possible occurrence of exfoliation in the steam superheater tubes, which disturbs the stability of oxide layers on the tubes' inner surfaces.

Because it was impossible to obtain real iron oxides separated from steam superheaters, rust particles collected from steel elements were used for the erosion testing (Fig. 4).

Fig. 4 Example particles of iron oxides

Their composition was determined using a scanning electron microscope equipped with an energy dispersive X-ray spectrometer (the SEM EDS method).

Considering atomic masses of oxygen and iron and the oxide formula for three possible iron oxides, the following proportions of masses and atomic numbers listed in Table 1 are obtained (mass content – Wt% column, atomic content – At% column):

Table 1. Mass and atomic contents of oxygen and iron in oxides.

Oxide	Fe content		O content	
	Mass content Wt%	Atomic content At%	Mass content Wt%	Atomic content At%
Wustite FeO	77.78	50.00	22.22	50.00
Haematite Fe_2O_3	70.00	40.00	30.00	60.00
Magnetite Fe_3O_4	72.41	42.86	27.59	57.14

Table 2 presents the results of SEM EDS measurements for three samples of rust particles and the corrected results after elimination of small Al and Si admixtures and conversion to contents of iron and oxygen only.

Table 2. EDS results (atomic proportions) for 3 samples and their correction after conversion to contents of Fe and O only. M means Measurement and C means Correction.

Element	Sample MP1-1		Sample MP1-2		Sample MP1-3	
	M	C	M	C	M	C
O	57.73	58.86	53.5	54.53	60.67	61.47
Al	0.89	0.00	0.64	0.00	0.77	0.00
Si	1.03	0.00	1.24	0.00	0.53	0.00
Fe	40.35	41.14	44.62	45.47	38.04	38.54
Total	100	100	100	100	100	100

Comparing the proportions listed in Table 1 and those obtained by means of the SEM EDS method (cf. Table 2), it can be seen that the composition of the MP1-1 sample corresponds to a haematite/magnetite mixture with similar contents. In addition, slight amounts of aluminium and silicon were found in the sample. However, these contents are negligibly small in terms of the impact on erosive properties. A similar comparison made for the MP1-2 sample showed a slightly different composition corresponding to a magnetite/wustite mixture but with a higher content of magnetite (about 70 %). The result for the MP1-3 sample after elimination of Al and Si and conversion to contents of iron and oxygen only is the closest to haematite.

In summary, it may be stated that the oxide dust under analysis is a mixture of all three iron oxides and consists of larger contents of haematite and magnetite. From the perspective of erosion testing, such composition simulates well the material from austenitic superheaters exfoliated in the form of an epitactic layer. The layer is indeed mostly magnetite, but both the density (haematite: 4.9 – 5.3 g/cm³, magnetite: 5.2 g/cm³) and the hardness (5.5 – 6.5 in Mohs' scale) of the two oxides are similar.

2.3 Adopted calculation methodology

In graphical programs, the distance between individual points is given in pixels, which have to be converted to the real distance expressed in millimetres, covered by the erodent selected particles, to determine the real velocities of the particles in selected locations. The standard adopted for the conversion is a nozzle where the real distance between points *a* and *b* is 17 mm (Fig. 5).

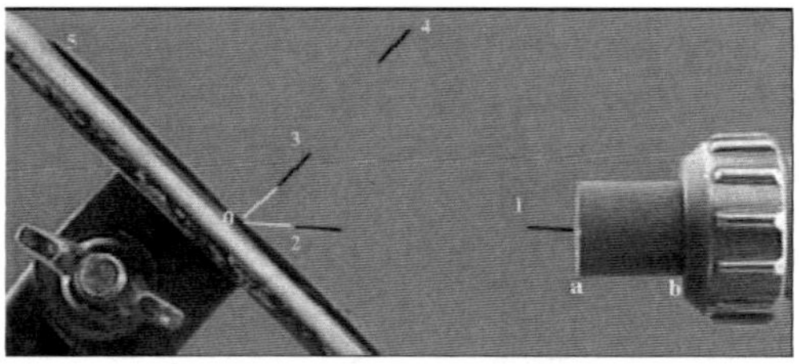

Fig. 5 Diagram of the test stand with marked characteristic points of the photographed particle trajectory.

This distance has an equivalent value expressed in pixels. The correction conversion from [pixels] to [mm] was calculated using the following relation:

$$v_{pix} = \frac{17}{a-b} = 0.2982 \qquad (3)$$

The equation describing the straight line passing through points **1** (x_1, y_1), **2** (x_2, y_2) takes the following form:

$$y = m_1 x + C_1$$
$$m_1 = \frac{y_2 - y_1}{x_2 - x_1} \quad C_1 = y_1 - m_1 x_1 \qquad (4)$$

The length of segment d_1 (the distance covered by the particle) between points ***1*** and ***2*** is:

$$d_1 = \sqrt{(x_2 - x_1)^2 + (y_2 - y_1)^2} \text{ [pixels]} \qquad (5)$$

$$d_1 = v_{pix}\, d_1 \quad \text{[mm]} \qquad (6)$$

If the time interval is known between subsequent frames of a film showing a particle ejected from the gun nozzle, it is possible to calculate the particle average velocity. The frame rate of the Casio High-Speed Exilim EX-F1 camera is 600 fps. This means that 1 frame moves in the time of 1/600 = 0.0016667 s. Knowing the distance and the time it takes for the particle to traverse that distance, the average velocity (w_1) at which the particle strikes against the plate mounted on the test stand may be calculated. Example subsequent phases of the erodent particles recorded with a specific time step are shown in Fig. 6.

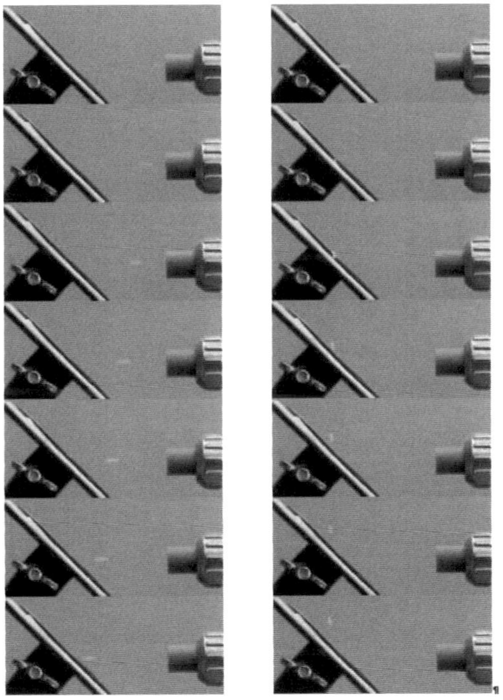

Fig. 6 Subsequent frames of a film showing the erodent motion before and after reflection off the plate.

$$w_1 = \frac{d_1}{1.6667\tau_1} \quad \text{[m/s]} \qquad (7)$$

where τ_1 is the number of time steps for the particle ejected from the nozzle.

A similar procedure is followed to determine the equation describing the straight line passing through points 3 (x_3, y_3), 4 (x_4, y_4) related to the reflected particle.

$$y = m_2 x + C_2$$
$$m_2 = \frac{y_4 - y_3}{x_4 - x_3} \quad C_2 = y_3 - m_2 x_3 \qquad (8)$$

$$w_2 = \frac{d_2}{1.6667\tau_2} \qquad (9)$$

Having two equations describing straight lines with slopes $m_1 \neq m_2$, the intersection point of the lines is found by solving a system of equations. The intersection point is the point of the particle reflection off the analysed plate. If the lines intersect, the following condition must be satisfied:

$$m_1 x + C_1 = m_2 x + C_2 \qquad (10)$$

which makes it possible to determine the reflection coordinate x_0:

$$x_0 = \frac{C_2 - C_1}{m_1 - m_2} \qquad (11)$$

In the next step, using the equation describing the line passing through points 1 and 2 or 3 and 4, the other coordinate y_0 is found. Coordinates x_0, y_0 relate to point 0 (reflection point – cf. Fig. 5).

In order to find the particle impingement and reflection angles, it is necessary to know the equation that describes the straight line lying on the plate. For this purpose, the coordinates are read of another point on the plate, e.g., 5 (x_5, y_5) – Fig. 5. The equation describing the straight line passing through points 0 and 5 has the following form:

$$y = m_3 x + C_3 \qquad (12)$$

As a result, the particle impingement and reflection angles, α_1 and α_2, may be calculated from the following relations:

$$tg\alpha_1 = \left| \frac{m_1 - m_3}{(1 + m_1 m_3)} \right| \qquad (13)$$

$$\alpha_1 = arc(tg\alpha_1)\frac{180}{\pi} \quad [°] \qquad (14)$$

$$tg\alpha_2 = \left| \frac{m_2 - m_3}{(1 + m_2 m_3)} \right| \qquad (15)$$

$$\alpha_2 = arc(tg\alpha_2)\frac{180}{\pi} \quad [°] \qquad (16)$$

Having the particle velocities and the impingement and reflection angles, restitution coefficients are calculated from the formulae (2 and 3).

2.4 Experimental determination of restitution coefficients of erosion for a selected erodent

The following erodents were used in the testing: iron oxides, quartz sand with a different grain size (490, 1000, 1500, 2000 µm), and 1000 µm-diameter steel balls. The steel balls, due to their ideal shape, were treated as the comparative analysis reference standard. In order to create an unequivocal state of reference, steel balls were used only for the steel plate impingement.

Due to the possibility of finding certain differences resulting from the difference in the nature of the materials used in the testing, a combination of different erodents and plates was taken into consideration for the other cases.

Sifting the erodent through many sieves with different mesh sizes, a wide range of the grain size of the analysed iron oxides was obtained. The grain size ranges adopted for the testing of iron oxides were as follows:

- 500 – 600 μm, average value: 550 μm,
- 600 – 800 μm, average value: 700 μm,
- 1000 – 1180 μm, average value: 1090 μm,
- 1180 – 1500 μm, average value: 1340 μm,
- 1500 – 2000 μm, average value: 1750 μm,
- 2000 – 3000 μm, average value: 2500 μm.

Unlike the sand or steel balls, for which a spherical shape of particles is assumed in the CFD calculations, the particles of iron oxides, especially in the case of larger grains, have a highly irregular shape. The erodent very often has the shape of irregular flat flakes. The shape of the tested erodent was analysed by means of microscopic observations. For grain sizes higher than 1200 μm, the erodent was treated as an irregular particle; for sizes lower than 1200 μm, it was treated as a regular particle. After an analysis of many large particles of iron oxides (grain size higher than 1200 μm), it was assumed that the grain width and thickness were 50% and 20% of its length, respectively.

Based on that, for irregular grains bigger than 1200 μm, the following was assumed:

- 1180–1500 μm, length/width/thickness 1.34 / 0.67 / 0.268 mm,
- 1500–2000 μm, length/width/thickness 1.75 / 0.875 / 0.35 mm,
- 2000–3000 μm, length/width/thickness 2.5 / 1.25 / 0.5 mm.

Restitution coefficients calculated according to formulae (1) and (2) for averaged variables affecting their values are listed in Table 5 – Table 17 together with a comparison to the CFD analysis results.

The results obtained during the test stand experiments show that the reflection angle α_2 for sand particles is included in the range (12 – 50). In the case of steel balls, the deviations are much smaller: α_2 = (33–42). For iron oxides, the reflection angle range is α_2 = (17–29). The comparison indicates that the smallest deviations in the reflection angle value occur for the steel balls, which results from their regular shape.

2.5 Numerical modelling of the erodent particle reflection off the surface

Based on the restitution coefficient testing results, numerical simulations were performed of the particle reflection off the surface. The gaseous phase (air) and the solid phase (erodent) flows are modelled together.

The geometrical model and the numerical mesh are made in the CFD Ansys Fluent commercial software [9] (Fig. 7). The mesh is structural in the entire area, and a hexagonal mesh composed of 519,000 nodes is used.

For the calculation of trajectories of moving particles, the flow of the gaseous phase is modelled in the Euler coordinate system, whereas the flow of the moving particle is modelled in the Lagrangian coordinate system. The flow area is estimated using the SIMPLE method and the k-ε real turbulence model. The interaction is used between the solid and the gaseous phase through the mass, energy, and momentum transfer. A spherical shape is assumed for the sand particles and the steel balls. For the iron oxides, a spherical shape of the erodent particles is assumed for three smaller diameters and a non-spherical shape is assumed for three bigger diameters.

The calculations were performed for averaged values of quantities obtained on the test stand for the erodent selected diameters and for three kinds of eroded materials. The CFD model boundary conditions are listed in Table 3 and Table 4.

Fig. 7 Model contour.

Table 3. CFD model boundary conditions for steel, aluminium, and steel with an anti-erosion coating; erodent: iron oxides.

Particle average diameter [mm]	0.55	0.70	1.09	1.34	1.75	2.5
Eroded material	**Steel**					
Impingement velocity w_1 [m/s]	6.20	5.76	5.18	4.34	4.41	3.77
Restitution coefficient e_n	0.26	0.32	0.38	0.36	0.33	0.23
Restitution coefficient e_t	0.59	0.59	0.62	0.58	0.63	0.66
Eroded material	**Aluminium**					
Impingement velocity w_1 [m/s]	7.33	6.28	4.04	4.95	3.86	4.00
Restitution coefficient e_n	0.30	0.34	0.39	0.32	0.27	0.26
Restitution coefficient e_t	0.67	0.69	0.71	0.66	0.62	0.59
Eroded material	**Plate with an anti-erosion coating**					
Impingement velocity w_1 [m/s]	5.83	5.89	4.65	5.76	4.26	3.75
Restitution coefficient e_n	0.34	0.30	0.38	0.24	0.33	0.33
Restitution coefficient e_t	0.59	0.54	0.63	0.60	0.53	0.54

Table 4. CFD model boundary conditions for steel, aluminium, and steel with an anti-erosion coating; erodent: sand, steel balls.

Erodent	**Sand**				**Steel balls**
Particle average diameter [mm]	0.49	0.10	1.50	2.00	1.00
Eroded material	**Steel**				
Impingement velocity w_1 **[m/s]**	6.37	5.85	8.76	9.53	8.72
Restitution coefficient e_n	0.46	0.46	0.63	0.41	0.55
Restitution coefficient e_t	0.73	0.72	0.76	0.69	0.71
Eroded material	**Aluminium**				
Impingement velocity w_1 **[m/s]**	6.15	5.73	6.98	6.17	-
Restitution coefficient e_n	0.40	0.33	0.37	0.41	-
Restitution coefficient e_t	0.81	0.77	0.73	0.71	-
Eroded material	**Plate with an anti-erosion coating**				
Impingement velocity w_1 **[m/s]**	7.54	6.53	7.70	5.85	-
Restitution coefficient e_n	0.38	0.54	0.50	0.70	-
Restitution coefficient e_t	0.76	0.73	0.58	0.59	-

The figures below present results of numerical calculations performed to determine the restitution coefficient (both tangential and normal) for different erodents. Fig. 8 and Fig. 9 presents an example trajectory of the erodent (iron oxide) particle, which is shown as the velocity before and after impingement on the eroded surface for steel and 6 diameters of the erodent. On the Fig. 8 in the top (T), the trajectory is shown for the 0.55 mm diameter. In the middle (M), the trajectory is shown for the diameter of 0.7 mm. In the bottom (B), the trajectory is shown for the diameter of 1.09 mm. On the Fig. 9 in the top (T), it is shown for the 1.34 mm diameter. In the middle (M)r, the trajectory is shown for the 1.75 mm diameter and in the bottom (B), it is shown for the diameter of 2.5 mm.

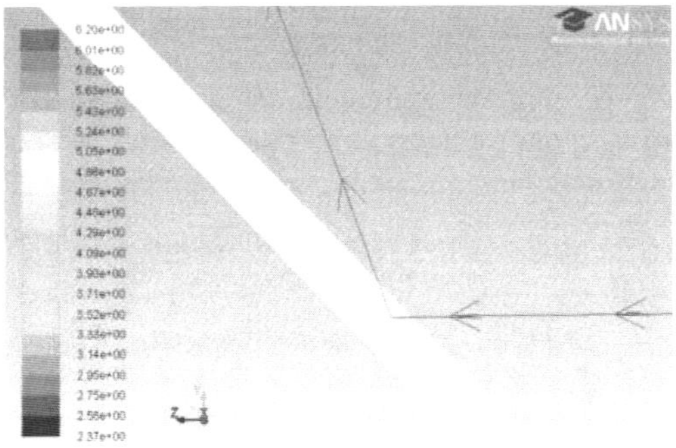

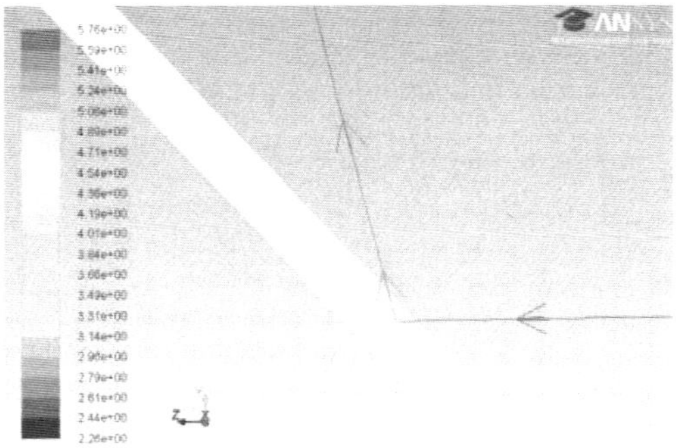

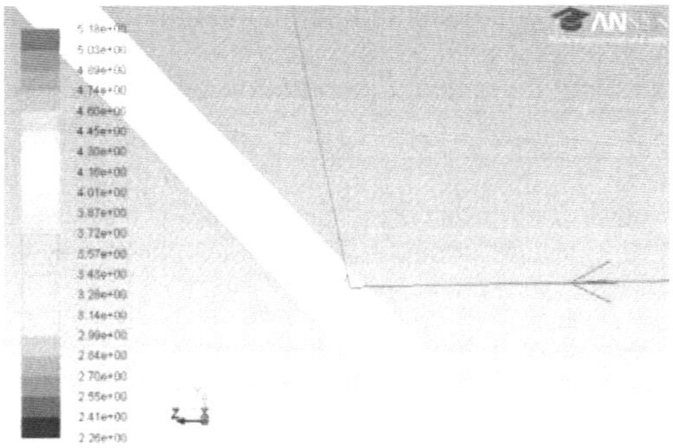

Fig. 8 Trajectory of the erodent (iron oxide) particle shown as the velocity before and after impingement on the eroded surface for steel and 6 diameters of the erodent: 0.55 (T), 0.7 (M), 1.09 (B).

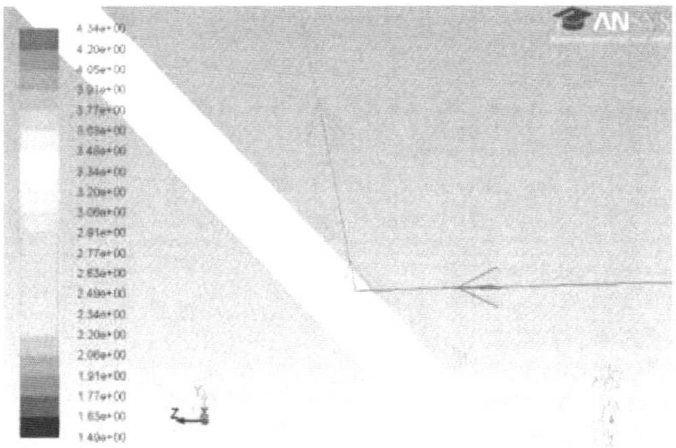

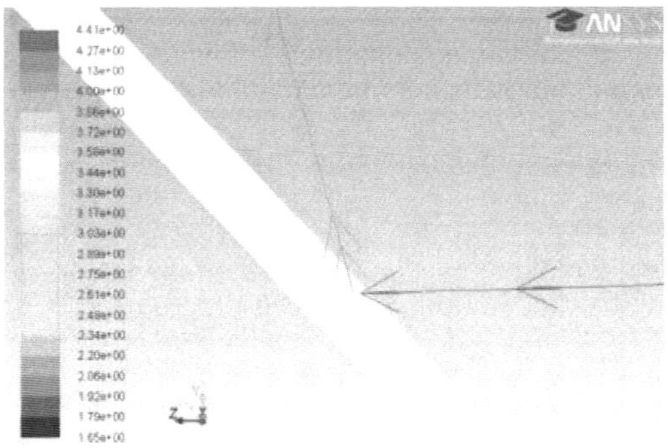

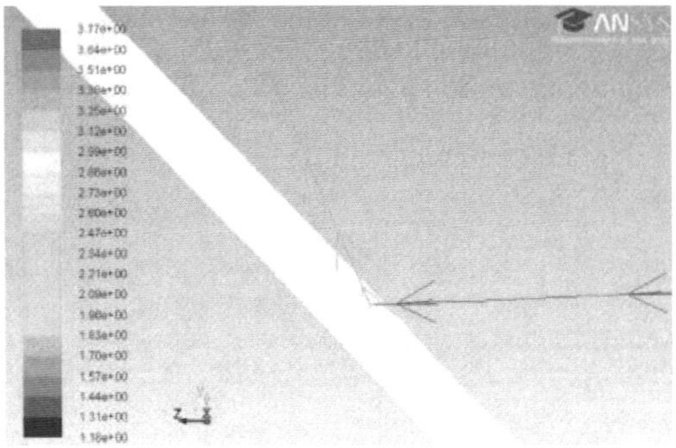

Fig. 9 Trajectory of the erodent (iron oxide) particle shown as the velocity before and after impingement on the eroded surface for steel and 6 diameters of the erodent: 1.34 (T), 1.75 (M), 2.5 (B).

Table 5 – Table 17 present the numerical modelling results together with the assumed values obtained from the test stand experiments for selected diameters of the erodent (iron oxide, sand, steel balls) and for three analysed samples (steel plate, aluminium plate, and plate with an anti-erosion coating).

Table 5. List of results obtained from laboratory testing for steel – erodent: iron oxide.

Erodent Iron oxide	EXPERIMENTS					
Particle diameter [mm]	0.55	0.70	1.09	1.34	1.75	2.5
Impingement velocity w_1 [m/s]	6.20	5.76	5.18	4.34	4.41	3.77
Reflection velocity w_2 [m/s]	2.95	2.81	2.73	2.15	2.31	1.95
Impingement angle $_1$	41.88	41.30	41.19	41.54	40.96	41.63
Reflection angle $_2$	21.73	25.60	27.98	28.79	24.55	17.31
Restitution coefficient e_n	0.26	0.32	0.38	0.36	0.33	0.23
Restitution coefficient e_t	0.59	0.59	0.62	0.58	0.63	0.66

Table 6. List of results obtained from CFD simulations for steel – erodent: iron oxide.

Erodent Iron oxide	CFD					
Particle diameter [mm]	0.55	0.70	1.09	1.34	1.75	2.5
Impingement velocity w_1 [m/s]	6.14	5.71	5.15	4.27	4.34	3.71
Reflection velocity w_2 [m/s]	2.76	2.65	2.62	2.03	2.14	1.76
Impingement angle α_1	45	45	45	45	45	45
Reflection angle α_2	23.9	29	31	32	28	20
Restitution coefficient e_n	0.26	0.32	0.37	0.36	0.33	0.23
Restitution coefficient e_t	0.58	0.57	0.62	0.57	0.62	0.63

Table 7. List of results obtained from laboratory testing for aluminium – erodent: iron oxide

Erodent Iron oxide	EXPERIMENTS					
Particle diameter [mm]	0.55	0.70	1.09	1.34	1.75	2.5
Impingement velocity w_1 [m/s]	7.33	6.28	4.04	4.95	3.86	4.00
Reflection velocity w_2 [m/s]	3.86	3.44	2.33	2.61	1.92	1.87
Impingement angle α_1	44.14	44.26	43.98	43.51	41.76	42.94
Reflection angle α_2	23.64	26.05	28.18	24.40	21.31	22.45
Restitution coefficient e_n	0.30	0.34	0.39	0.32	0.27	0.26
Restitution coefficient e_t	0.67	0.69	0.71	0.66	0.62	0.59

Table 8. List of results obtained from CFD simulations for aluminium – erodent: iron oxide

Erodent Iron oxide	CFD					
Particle diameter [mm]	0.55	0.70	1.09	1.34	1.75	2.5
Impingement velocity w_1 [m/s]	7.25	6.21	4.01	4.84	3.76	3.93
Reflection velocity w_2 [m/s]	3.73	3.34	2.29	2.5	1.79	1.79
Impingement angle α_1	45	45	45	45	45	45
Reflection angle α_2	24.9	26.92	29.0	26.36	23.31	22.77
Restitution coefficient e_n	0.31	0.34	0.39	0.32	0.27	0.25
Restitution coefficient e_t	0.66	0.68	0.71	0.65	0.62	0.59

Table 9. List of results obtained from laboratory testing for the plasma-coated plate with an anti-erosion layer – erodent: iron oxide

Erodent Iron oxide	EXPERIMENTS					
Particle diameter [mm]	0.55	0.70	1.09	1.34	1.75	2.5
Impingement velocity w_1 [m/s]	5.83	5.89	4.65	5.76	4.26	3.75
Reflection velocity w_2 [m/s]	2.92	2.69	2.50	2.76	1.94	1.72
Impingement angle α_1	39.83	39.75	39.97	41.61	41.42	41.02
Reflection angle α_2	25.95	25.12	26.78	19.57	28.65	27.94
Restitution coefficient e_n	0.34	0.30	0.38	0.24	0.33	0.33
Restitution coefficient e_t	0.59	0.54	0.63	0.60	0.53	0.54

Table 10. List of results obtained from CFD simulations for the plasma-coated plate with an anti-erosion layer – erodent: iron oxide

Erodent Iron oxide	**CFD**					
Particle diameter [mm]	0.55	0.70	1.09	1.34	1.75	2.5
Impingement velocity w_1 [m/s]	5.74	5.82	4.62	5.66	4.16	3.67
Reflection velocity w_2 [m/s]	2.76	2.54	2.4	2.58	1.84	1.64
Impingement angle α_1	45	45	45	45	45	45
Reflection angle α_2	29.65	29.15	31.14	22.19	31.35	31.17
Restitution coefficient e_n	0.34	0.30	0.38	0.24	0.33	0.33
Restitution coefficient e_t	0.59	0.54	0.63	0.60	0.53	0.54

Table 11. List of results obtained from laboratory testing for steel erosion – erodent: sand

Erodent Sand	**EXPERIMENTS**			
Particle diameter [mm]	0.49	1.00	1.50	2.00
Impingement velocity w_1 [m/s]	6.37	5.85	8.76	9.54
Reflection velocity w_2 [m/s]	3.93	3.53	6.20	5.41
Impingement angle α_1	43.16	44.58	40.79	44.34
Reflection angle α_2	30.87	32.37	35.87	29.94
Restitution coefficient e_n	0.46	0.46	0.63	0.41
Restitution coefficient e_t	0.73	0.72	0.76	0.67

Table 12. List of results obtained from CFD simulations for steel erosion – erodent: sand

Erodent Sand	**CFD**			
Particle diameter [mm]	0.49	1.00	1.50	2.00
Impingement velocity w_1 [m/s]	6.22	5.8	8.74	9.52
Reflection velocity w_2 [m/s]	3.8	3.5	6.1	5.4
Impingement angle α_1	45	45	45	45
Reflection angle α_2	32	33	39	30
Restitution coefficient e_n	0.46	0.46	0.62	0.40
Restitution coefficient e_t	0.73	0.72	0.77	0.69

Table 13. List of results obtained from laboratory testing for aluminium erosion – erodent: sand

Erodent Sand	**EXPERIMENTS**			
Particle diameter [mm]	0.49	1.00	1.50	2.00
Impingement velocity w_1 [m/s]	6.15	5.73	6.98	6.17
Reflection velocity w_2 [m/s]	3.92	3.38	4.04	3.64
Impingement angle α_1	44.81	45.32	44.56	43.84
Reflection angle α_2	25.99	23.37	26.74	28.93
Restitution coefficient e_n	0.39	0.33	0.37	0.41
Restitution coefficient e_t	0.80	0.78	0.72	0.71

Table 14. List of results obtained from CFD simulations for aluminium erosion – erodent: sand

Erodent Sand	**CFD**			
Particle diameter [mm]	0.49	1.00	1.50	2.00
Impingement velocity w_1 [m/s]	6.02	5.68	6.95	6.15
Reflection velocity w_2 [m/s]	3.85	3.37	4.02	3.56
Impingement angle α_1	45	45	45	45
Reflection angle α_2	26	23	27	30
Restitution coefficient e_n	0.40	0.33	0.37	0.41
Restitution coefficient e_t	0.81	0.77	0.73	0.71

Table 15. List of results obtained from laboratory testing for erosion of the plasma-coated plate with an anti-erosion layer – erodent: sand

Erodent Sand	**EXPERIMENTS**			
Particle diameter [mm]	0.49	1.00	1.50	2.00
Impingement velocity w_1 [m/s]	7.54	6.53	7.70	5.58
Reflection velocity w_2 [m/s]	4.60	4.26	4.22	3.76
Impingement angle α_1	43.40	42.08	42.63	43.42
Reflection angle α_2	25.33	33.47	38.49	48.54
Restitution coefficient e_n	0.38	0.52	0.46	0.70
Restitution coefficient e_t	0.76	0.72	0.57	0.59

Table 16. List of results obtained from CFD simulations for erosion of the plasma-coated plate with an anti-erosion layer – erodent: sand

Erodent Sand	**CFD**			
Particle diameter [mm]	0.49	1.00	1.50	2.00
Impingement velocity w_1 [m/s]	7.45	6.48	7.66	5.83
Reflection velocity w_2 [m/s]	4.47	4.16	4.15	3.77
Impingement angle α_1	45	45	45	45
Reflection angle α_2	27	36	41	49
Restitution coefficient e_n	0.39	0.53	0.50	0.69
Restitution coefficient e_t	0.76	0.73	0.58	0.60

Table 17. List of results obtained from laboratory testing and CFD simulations for steel erosion; erodent: steel balls

Erodent – Steel balls	**EXPERIMENTS**	**CFD**
Particle diameter [mm]	1.0	1.0
Impingement velocity w_1 [m/s]	8.72	8.72
Reflection velocity w_2 [m/s]	5.54	5.52
Impingement angle α_1	43.89	45
Reflection angle α_2	36.74	37.43
Restitution coefficient e_n	0.55	0.54
Restitution coefficient e_t	0.71	0.71

Attention should be drawn to slight discrepancies between the values of the impingement and reflection angles α_1 and α_2, respectively, obtained from experiments and from the CFD numerical model (Table 5 – Table 17). Assuming that restitution coefficients on the plate wall in the CFD analysis are identical to coefficients according to the experiments, a similar difference between the impingement/reflection angles is obtained from numerical and experimental investigations. If, however, the numerical model impingement angle is changed from $\alpha_1 = 45^{\circ}$ (the angle of the plate inclination on the test stand) to the angle used in the analysis, a different reflection angle is obtained but the CFD restitution coefficients remain unchanged.

The angle of 45° is assumed as a typical value for which the maximum erosion loss occurs for steel.

3 Summary and conclusions

- In case of two-phase gas-solid flows the rate of erosion and the restitution coefficients depend on the concentration of solid phase in the flow. During experimental determination of restitution coefficients it is impossible to keep the initial parameters (impingement velocity w_1 and angle α_1) at a steady level. Therefore, the physical experiments don't bring full information about dependence of e_t and e_n on parameters influencing the phenomenon of particle reflection off the eroded surface. In this situation it was necessary to create the numerical model enabling correction of process conditions in such a way that the influence of particular parameters is identifiable.
- Comparing Tables 5 – 17, it can be seen that good accordance is obtained between the results of numerical modelling and experimental testing, which proves the correctness of the numerical model of the restitution coefficient determination. This means that numerical modelling may be a useful tool for multi-variant analyses of the particle reflection off the surface.
- Based on the value of the restitution coefficient tangential component e_t, it is found that regardless of the erodent particle size or the erodent type, the biggest values of e_t occurred for the aluminium plate erosion. The lowest values of e_t occurred for the plasma-coated plate.
- Analysing the results listed in the tables presented above (Table 11 - 16), a distinct impact of the tested material plastic deformation can be observed. For example, if the impingement of sand particles with the diameter of 1.0 mm on various types of the eroded material is investigated, it may be stated that by steady impingement angle α_1 the lowest reflection angle α_2 occurs in the case of the aluminium plate (the the most ductile material), while the highest for the plasma-coated plate (the most brittle material). The mean value occurs by reflection of sand from the steel.
- Regardless of the size of particles and the type of erodent the highest values of e_t are for the aluminium plate
  - sand/aluminium $e_t = 0{,}71 \div 0{,}81$
  - sand/steel $e_t = 0{,}69 \div 0{,}77$
  - sand/plasma coating $e_t = 0{,}58 \div 0{,}76$
  - iron oxides/aluminium $e_t = 0{,}59 \div 0{,}71$
  - iron oxides/steel $e_t = 0{,}58 \div 0{,}66$
  - iron oxides/plasma coating $e_t = 0{,}53 \div 0{,}63$
- For the sand/steel plate system, the impingement velocity was w_1 = 5.85 m/s, and the reflection velocity was w_2 = 3.53 m/s (Table 11). If the sand impingement velocity was the same as in the case of erosion of the aluminium plate, i.e., $w_1 = 5.73$, the linearly converted sand reflection velocity w_2 would equal 3.46 m/s, whereas the obtained value for the aluminium plate is 3.38 m/s (Table 13), i.e., it is smaller - because aluminium is softer than steel.
  For the plate with plasma-coated anti-erosion layer, the impingement velocity was $w_1 = 6.53$ m/s, and the reflection velocity was w_2 = 4.26 m/s (Table 15). If the sand impingement velocity was the same as in the case of erosion of the steel plate, i.e., $w_1 = 5.85$, the linearly converted sand reflection velocity w_2 would equal 3.94 m/s, whereas the obtained value for the steel plate is 3.53 m/s (Table 11). The post-impingement velocity would be lower because steel is softer than the anti-erosion coating.

References

1. I. G. Wright, M. Schütze, S.R. Paterson, P.F. Tortorelli, R.B. Dooley, *Progress in prediction and control of scale exfoliation on superheater and reheater alloys* (EPRI Inter. Conf. on Boiler Tube and HRSG Tube Failures and Inspections, San Diego, 2-5 November 2004)
2. R. Knödler, S. Straub, VGB PowerTech **10,** (2008)
3. M. Pronobis, W. Wojnar, Eng. Fail. Anal. **32,** 54–62 (2013)
4. Н. Кузниецов, Теплоэнергетика **4** (1955) [Teploenergetika **4** (1955)]
5. J. Świrski, *Prace Instytutu Energetyki,* Book **1**, (Warszawa 1975)
6. M. Pronobis, *Modernizacja kotłów energetycznych [Modernisation of steam generators]* (WNT Warszawa 2002)
7. E. Raask, *Erosion wear in coal utilization* (Hemisphere Publishing Corporation, ISBN 3-540-18601-8, 1988)
8. V. Meuronen, *Lappeenranta University of Technology, Research Papers* **64**, (Lappeenranta 1997)
9. Ansys Fluent. Computational Fluid Dynamics, Ansys Inc., USA, www.ansys.com.
10. Forder, Wear, 183-193 (1998)
11. T. Grant, W. Tabakoff, J Aircraft **12,** 471–547 (1975)
12. K. Ahlert, *Effects of Particle Impingement Angle and Surface Wetting on Solid Particle Erosion on AISI 1019 Steel.* (M.S. Thesis, University of Tulsa. 1994)
13. J. K. Edwards, B. S. McLaury, S. A. Shirazi, *Evaluation of alternative pipe bend fittings in erosive service.* (In Proceedings of ASME FEDSM'00: ASME 2000 Fluids Eng. Division Summer Meeting, Washington D.C., June 2000)

Shape optimization of a thick-walled power boiler component

Piotr Duda[1,*], *Mirosław W.* Mrzygłód[2]

[1]Cracow University of Technology, Institute of Thermal Power Engineering, Al. Jana Pawła II 37, 31-864 Kraków, Poland
[2]Opole University of Technology, Department of Mechanics and Machine Design, Mikołajczyka 5, 45-271 Opole, Poland

Abstract. This paper presents a methodology and successful application of structural optimization of a T-pipe under transient thermal and mechanical loads. In order to find the optimal shape of a thick-walled power boiler component, a parametric FE model and the evolutionary algorithm (EA) are applied. The power boiler start-up and shutdown curves are based on the TRD 301 guidelines. Maximum total stresses are assumed as optimization constraints. The obtained geometry is by about 18.6% lighter than the original one due to thinning of the walls. Maximum tensile and compressive stresses in the modified geometry are smaller than in the original one during the whole cycle. Additionally, lower total stress values are recorded during heating and cooling processes. Therefore, these transient processes can be accelerated and the shutdown and start-up losses can be reduced.

1 Introduction

High thermal stresses are created in power boiler components during their operation. The cyclic nature of these stresses causes low-cyclic fatigue and may lead to fracture. Manufacturers of such devices advise keeping the heating and cooling rates within prescribed limits. Nevertheless, extended start-up and shutdown times cause considerable losses in the electricity generation process. Rapid transient operations of power boilers, on the other hand, involve high temperature differences and thermal stresses in their components [1]. Methods of finding optimum operating parameters, which can ensure safe heating and cooling processes, can be found in [2, 3, 4, 5]. Reducing the thickness of the boiler component walls while maintaining the stress level during the cycle within established limits can bring material savings. Furthermore, it can improve the thermomechanical characteristics of the device. A component with thinner walls can be heated and cooled more quickly.

Most state-of-the-art methods now available for structural optimization of the power unit equipment refer to the steady state. Wen et al. [6] present an example of structural optimization of cellular cores of metallic sandwiches subjected to laminar forced convection at fixed pumping power. Xie et al. [7] employ a globally convergent method of moving asymptotes (GCMMA) in lightweight optimization of a corrugated sandwich panel using heat transfer and thermomechanical analyses.

This paper presents a method of lightweight optimization of the power boiler components subjected to transient thermomechanical loads. The power boiler start-up and shutdown curves are based on the TRD 301 code [8]. Maximum total stresses are assumed as optimization constraints.

2 Method formulation

The geometry of a power boiler component determines the level of stresses arising during its operation. A reduction in the component thickness may result in an increase in stresses caused by internal pressure but – on the other hand – it should reduce thermal stresses. This leads to the conclusion that the component wall thickness can be reduced without increasing total stresses. Using thinner walls in a component may produce material savings and make it possible to carry out transient processes at a faster rate.

In order to find the optimal shape of a thick-walled power boiler component, a parametric FE model and the evolutionary algorithm (EA) are applied (see Fig. 1). This method is based on mechanisms of biological evolution. The EA performs a multi-directional search of the solution space, protecting the population of potential solutions and the exchange of information between them. It is possible to adjust the EA operation depending on the characteristics of the optimization object. This can be done by changing the algorithm control parameters, such as the population size *PJ*, the parameter of probability of crossing operations p_c, the parameter of probability of mutation p_m, the number of generations N [9, 10].

The EA is designed to solve single and multicriteria optimization problems for nonlinear programming problems. The single-criterion optimization is formulated to find the vector of decision variables (e.g. dimensions of a thick-walled boiler component)

$$\mathbf{y} = \{ y_1 \quad y_2 \quad \dots \quad y_N \} \tag{1}$$

* Corresponding author: pduda@mech.pk.edu.pl

Continuous Nonlinear Programming with the mapping from binary string to real number for the variable **y** is used. For all decision variables, the upper and lower constraints values are applied:

$$\underline{y_i} \le y_i \le \overline{y_i} \quad \text{for } i = 1,...,N \tag{2}$$

The optimal solution should satisfy the following inequality constraint

$$\max\left(\frac{\sigma_{3\min}(\mathbf{y})}{\breve{\sigma}_a}, \frac{\sigma_{1\max}(\mathbf{y})}{\hat{\sigma}_a}\right) \le 1 \tag{3}$$

where $\sigma_{3\min}(\mathbf{y})$ denotes the lowest third principal stress in the entire component during the heating process, $\sigma_{1\max}(\mathbf{y})$ is the highest first principal stress during the cooling process, $\breve{\sigma}_a$ and $\hat{\sigma}_a$ denote allowable stresses for the heating and cooling operations. Both values $\sigma_{3\min}(\mathbf{y})$ and $\breve{\sigma}_a$ are compressive stresses and are negative. Tensile stresses $\sigma_{1\max}(\mathbf{y})$ and $\hat{\sigma}_a$ are positive.

For the objective function, the structure volume $V(\mathbf{y})$ is chosen. The vector of decision variables **y** should be found to minimize the objective function $V(\mathbf{y})$.

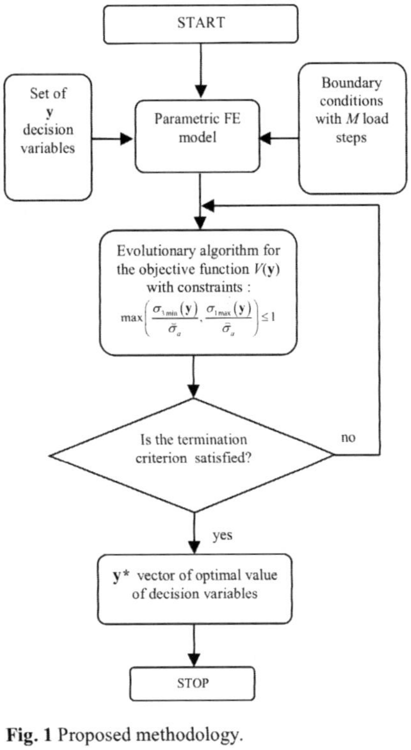

Fig. 1 Proposed methodology.

3 Numerical example

The proposed method will be used for a T-pipe shape optimization. The T-pipe is a part of a live and pre-heated steam pipeline in a 360 MW power unit. It was designed for pressure p_w=18 MPa and steam temperature T_w=540°C. It weighs 1378 kg. The T-pipe geometry is shown in Fig. 2. The T-pipe is made of alloy steel 14 MoV63. Temperature-dependent thermal and mechanical properties are considered [11].

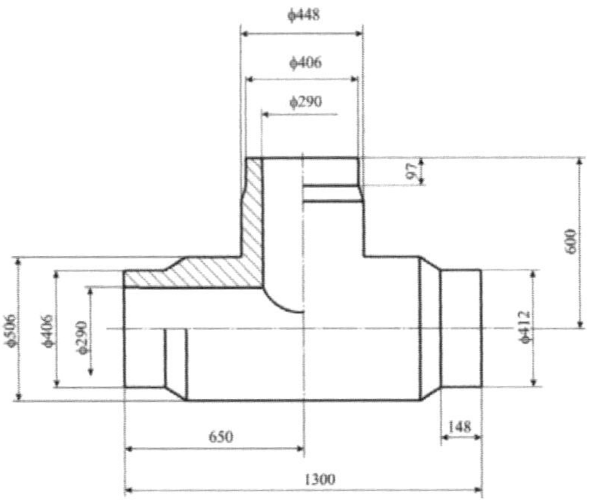

Fig. 2 T-pipe geometry.

TRD301 standards [8] can be used to propose the heating or cooling method for the selected T-pipe. They assume a quasi-steady state and one-dimensional temperature distribution through the wall of the structural element. The calculated allowable stresses for heating and cooling operations have values of $\breve{\sigma}_a = -214.4$ MPa, $\hat{\sigma}_a = 303$ MPa, respectively. The following allowable rates of temperature changes are obtained:

$v_{T1(TRD)} = 2.119$ K/min - the beginning of heating,

$v_{T2(TRD)} = 2.993$ K/min - the end of heating,

$v_{T1(TRD)} = -2.119$ K/min - the beginning of cooling,

$v_{T2(TRD)} = -2.993$ K/min - the end of cooling.

If the pressure at the beginning of the heating process equals 0 MPa and the pressure at the end of heating equals p_w, the steam temperature can be calculated from the following relation:

$$\frac{dT_f}{dt} = v_{T1(TRD)} + \frac{v_{T2(TRD)} - v_{T1(TRD)}}{p_w} p_f(T_f) \tag{4}$$

where p_f is independent of temperature or should be calculated as saturation pressure at T_f., The steam temperature for the cooling process can be obtained in a similar way. The calculated steam temperature and pressure during heating and cooling operations are presented in Fig. 3.

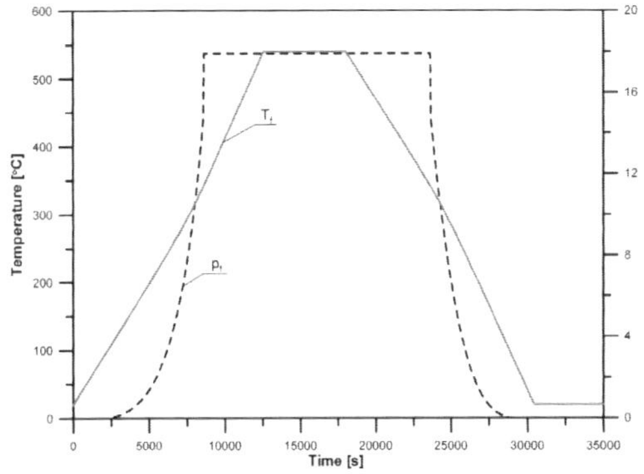

Fig. 3 Steam temperature and pressure based on TRD 301.

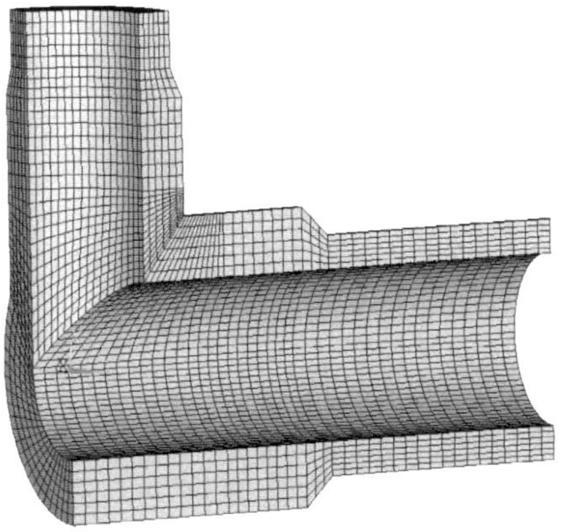

Fig. 4 T-pipe division into finite elements - original shape.

A quarter of the T-pipe, which is divided into 20-node finite brick elements according to symmetry conditions, is shown in Fig. 4. The temperature and stress distribution in this component is calculated by means of the finite element method using the ANSYS software [12]. The heat transfer coefficient on the inner surfaces of the T-pipe is assumed as 1000 W/m^{2}K [13]. The stress model is constrained to ensure symmetry conditions and it allows the possibility of unrestricted lengthening in the direction of the horizontal and vertical pipe. Additionally, the surfaces connecting the T-pipe to the pipeline must be plane.

The maximum value of equivalent stress (Huber-Mises-Henky) and the first principal stress caused by pressure and temperature occur after about 23100 s and equal $\sigma_{eq\,max}$=246.8 MPa and $\sigma_{1\max} = 229.2$ MPa, respectively. The maximum compressive stress can be estimated based on the minimum value of the third principal stress. The minimum value of the third principal stress occurs after about 10300 s and equals $\sigma_{3\min} = -139$ MPa. Throughout the entire cycle, the allowable stresses, as per the TRD 301 standard, $\check{\sigma}_a = -214.4$ MPa, $\hat{\sigma}_a = 303$ MPa are not exceeded. Principal stresses $\sigma_{1\max} = 229.2$ MPa and $\sigma_{3\min} = -139$ MPa will be assumed as the bounds of constraints during the T-pipe shape optimization.

The optimization problem is defined to minimize the volume (mass) of the structure assuming the following allowable stresses: $\check{\sigma}_a = -139$ MPa, $\hat{\sigma}_a = 229.2$ MPa. The optimization process involves a search for the solution in a 6-dimensional design variable space (4 radii and 2 lengths of the T-pipe reinforcements – see Fig. 5).

The following range of variation is proposed for the decision variables: y_1 = {215 mm, 265 mm}, y_2 = {215 mm, 240 mm}, y_3= {450 mm, 550 mm}, y_4 = {450 mm, 550 mm}}, y_5= {190 mm, 215 mm}, y_6 = {190 mm, 215 mm}.

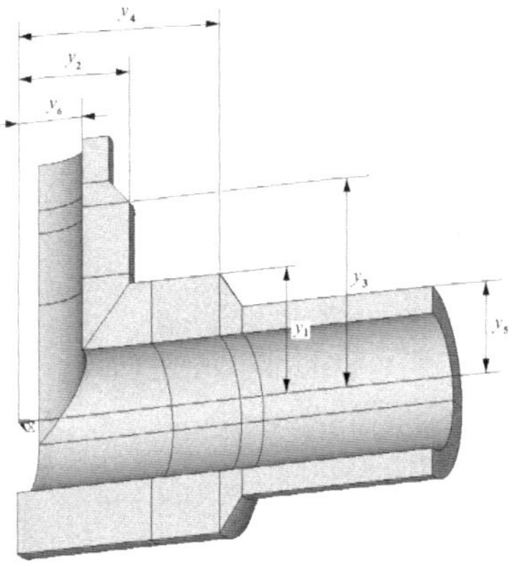

Fig. 5 6-dimensional design variable space of the EA shape optimization.

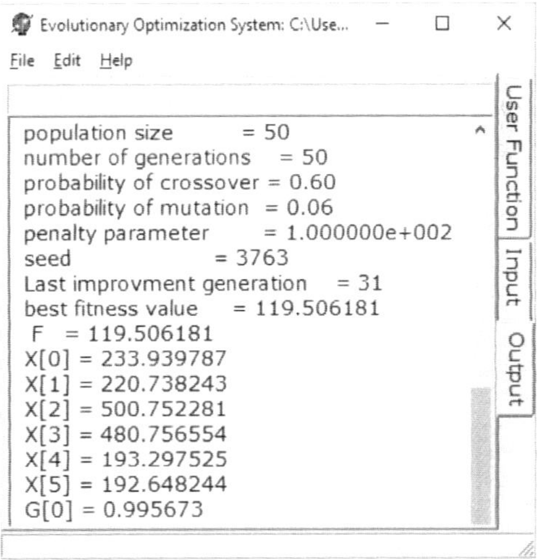

Fig. 6 Optimization result in the EOS.

The parametric FE model is subjected to the evolutionary algorithm (EA) using the Evolutionary

Optimization System (EOS) software [9]. Moreover, the following steering parameters of the EOS are selected: population size $PJ = 50$, number of generations $N = 50$, probability of crossover $p_c = 0.6$, probability of mutation $p_m = 0.06$. The EOS software is connected to the ANSYS FE program to perform batch processing calculations of the objective function value and stress constraints. The best solution $\mathbf{y}^*$ ($y_1^* = 233.94$ mm, $y_2^* = 220.74$ mm, $y_3^* = 500.75$ mm, $y_4^* = 480.76$ mm, $y_5^* = 193.3$ mm, $y_6^* = 192.65$ mm) is about 18.6% lighter than the original one. The calculated values are presented in Fig. 6, where F is the objective function $V(\mathbf{y})$ and X[0]...X[5] are all decision variables $y_1...y_6$. The variable G[0] is the constraint, which is defined by (3). The modified T-pipe can be seen in Fig. 7.

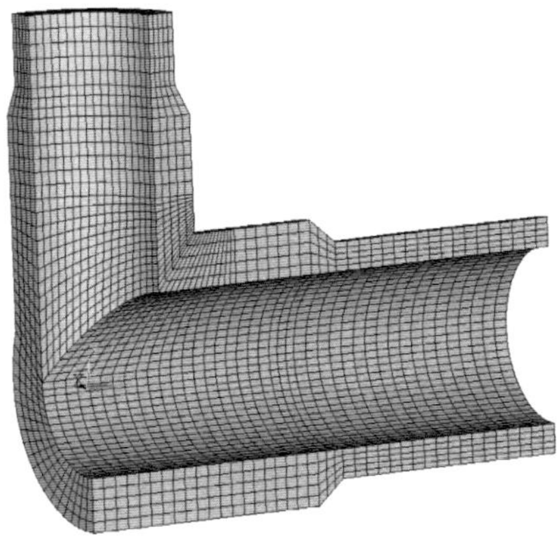

Fig. 7 Division of the modified T-pipe into finite elements - optimal shape.

Fig. 8 shows $\sigma_{1\max}(\mathbf{y})$ - maximum values of first principal stresses in the whole T-pipe over the load time history for the original and optimal shapes. Fig. 9 presents $\sigma_{3\min}(\mathbf{y})$ - minimum values of third principal stresses in the T-pipe for the two geometries.

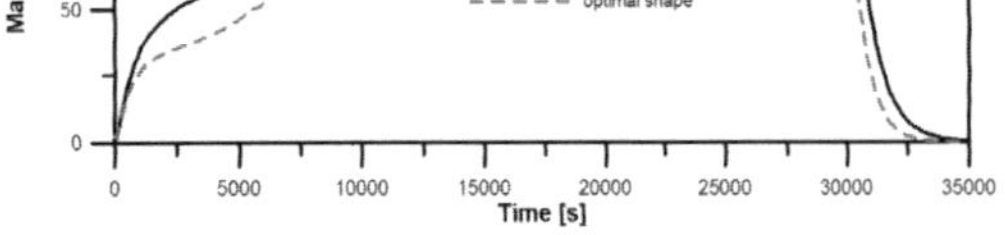

Fig. 8 Maximum first principal stress values in the entire structure depending on time.

The shape modification does not cause an increase in the maximum stress values in time and space. The highest recorded first principal stress for the optimal shape equals 228.2 MPa and is lower than the maximum value for the original shape, which equals 229.2 MPa. Additionally, the value of $S_{1\,max}$ for the modified geometry is lower than for the original one during almost the whole heating and cooling process (0 s – 12900 s and 22900 s – 35000 s). Maximum compressive stresses in the modified geometry are smaller than in the original T-pipe in the following time periods (0-12900 s and 23700-35000 s). Lower total stress values recorded during heating and cooling offer the possibility of speeding up these transient processes and reducing losses related to the boiler shutdown and start-up.

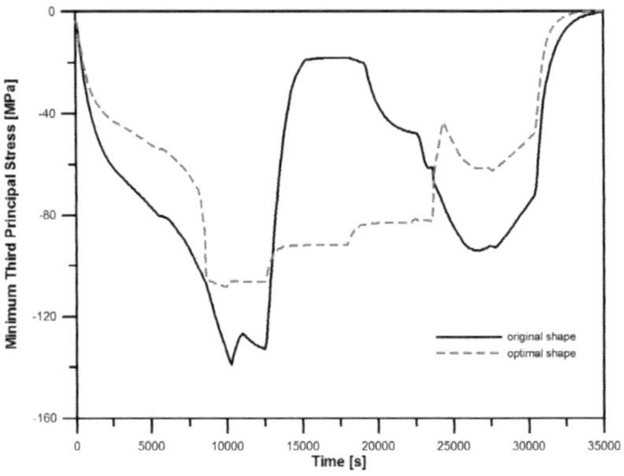

Fig. 9 Minimum third principal stress in the entire structure depending on time.

4 Conclusions

A methodology and successful application of structural optimization of a T-pipe under transient thermal and mechanical loads are presented. The obtained geometry is by about 18.6% lighter than the original one due to thinning of the walls. This result can be considered as significant, especially in view of the fact that this structure has been in use for many years and must have been optimized by designers before. Maximum tensile and compressive stresses in the modified geometry are smaller than in the original one during the whole cycle. Additionally, lower total stress values are recorded during heating and cooling processes. Therefore, these transient processes can be accelerated and the shutdown and start-up losses can be reduced.

The proposed methodology can be applied in any industry where heating and cooling processes take place.

References

1. T.S. Kim, D.K. Lee, S.T. Ro, Analysis of thermal stress evolution in the steam drum during start-up of a heat recovery steam generator, Applied Thermal Engineering, 20 (11) (2000) 977-992.

2. P. Duda, R. Dwornicka, Optimization of heating and cooling operations of steam gate valve, Structural and Multidisciplinary Optimization, 40 (2009) 529-535.
3. P. Duda, D. Rząsa, Numerical method for determining the allowable medium temperature during the heating operation of a thick-walled boiler element in a supercritical steam power plant, International Journal of Energy Research, 36 (2012) 703–709.
4. K. Krüger, M. Rode, R. Franke, Optimization of boiler start-up using a nonlinear boiler model and hard constraints, Energy, 29 (12–15) (2004) 2239–2251.
5. P. Dzierwa, J. Taler, A new method for determining allowable medium temperature during heating and cooling of thick-walled boiler components. Proceedings of the Congress on Thermal Stresses, Taipei, 2, (2007) 437-440.
6. T. Wen et al., The Structural Optimization of Two-Dimensional Cellular Metals Cooled by Forced Convection, International Journal of Heat and Mass Transfer, 50 (2007) 2590–2604.
7. G.N. Xie, Q. Wang, Thermomechanical optimization of lightweight thermal protection system under aerodynamic heating. Applied Thermal Engineering, 59 (2013) 425–434.
8. TRD 301, Technische Regeln für Dampfkessel, Carl Heymans Verlag, Köln und Beuth-Verlag: Berlin, (2001).
9. A. Osyczka et al., Evolutionary optimization system (EOS) for design automation. In: Burczynski T, Osyczka A (Editors) Evolutionary methods in mechanics. Kluwer, Dordrecht, (2004).
10. M. Mrzyglod, A. Osyczka, Parallel Computing for Design Optimization with Computationally Expensive Functions Using Evolutionary Algorithms, in B.H.V. Topping, (Editor), Proceedings of the Eighth International Conference on the Application of Artificial Intelligence to Civil, Structural and Environmental Engineering, Civil-Comp Press, Stirlingshire, UK, Paper 26, (2005), DOI:10.4203/ccp.82.26.
11. F. Richter, Physikalische Eigenschaften von Stählen und ihre Temperaturabhängigkeit. Mannesmann Forschungberichte 930, Düsseldorf, (1983).
12. ANSYS Theory Guide, Release 14, (2013).
13. P. Duda, Numerical and experimental verification of two methods for solving an inverse heat conduction problem, International Journal of Heat and Mass Transfer 84 (2015) 1101–1112.

Analysis of coal dust combustion and gasification in the cyclone furnace

Robert Zarzycki[1,*], and *Zbigniew* Bis[1]

[1]Czestochowa University of Technology, Department of Energy Engineering, ul. Brzeznicka 60a, 42-201, Poland

Abstract. This study presents the design and operation of the cyclone furnace fuelled with coal dust. The main function of the furnace is coal dust gasification. The combustible gases that mainly contain CO can be used to feed a pulverized coal-fired boiler. The results of numerical calculations presented in the study demonstrated that cyclone furnace can operate over a wide range of fuel flow rates: from the conditions of coal dust combustion, which ensure maintaining the cyclone furnace in a state of hot reserve to the conditions of coal dust gasification, which allow for production of CO. Gasification process helps control temperature under conditions of elevated oxygen concentration. The results allow for the use of the furnace to improve flexibility and coal dust operation through the reduction in its technological minimum.

1 Introduction

The examinations aimed to limit the environmental impact of the use of fossil fuels for electricity and heat generation have been conducted for many years. New technologies are being explored to help utilize chemical energy contained in the fuel more effectively. The analysed processes include oxy-fuel combustion [1,2,3,4], on-ground and underground gasification [5] or construction of high-performance power units [6,7].

One of the solutions which ensure further use of coal and steam power units with pulverized coal-fired boilers is to develop cyclone furnaces that are operated under oxy-combustion conditions with coal gasification [8]. Combustible gases generated during gasification (CO, H_2, CH_4) and supplied to the pulverized coal-fired boiler can be burnt in the boiler or, after preparation, used for other chemical processes as a raw material. Development of the renewable energy based on the use of wind causes rapid changes in production of electricity. Consequently, present and new energy power unit must be characterized by greater flexibility of changes in the load and possibly low minimum load.

This study presents the concept and opportunities for operation of the cyclone furnace for coal gasification to feed pulverized coal-fired boiler with the option of changing its load.

2 Cyclone furnace

Cyclone furnaces are power-generating devices with fuel combustion or gasification processes occurring in a strong eddy flow of gases. These furnaces have a relatively small combustion chambers compared to the equipment used in the energy sector. This allows for reaching high temperatures during the combustion process or during fuel gasification and for utilization of the furnaces for melting of the ash contained in the fuel.

Due to high temperatures that occur in cyclone furnaces, it is recommended to replace air with other oxidizing agents without nitrogen in order to limit NO_x emissions. The use of recirculation of oxygen-rich flue gas or, additionally, water steam, substantially limits the emissions of NO_x.

Combustion in the oxygen-rich atmosphere leads to a considerable increase in temperature, with its values exceeding thermal strength of the ceramic materials used for construction of the furnace. Therefore, with higher oxygen concentrations, the temperature that can be obtained has to be controlled through: flue gas recirculation, combustion in the substoichiometric conditions (gasification), injection of water or water steam in order to reduce temperature (evaporation of water, overheating the steam) or the use of water steam in the process of coal gasification. Control of temperature through gasification of fuel inside the cyclone furnace helps obtain combustible gases composed mainly of CO and H_2 with contents that depend on process conditions. Strong eddy makes it possible to elongate time of fuel stay in the combustion or gasification zone, thus limiting the waste of incomplete burning. With strong eddy motion, ash separation and vitrification in cyclone furnace is also feasible. After moving to the pulverized coal-fired boiler, hot combustible gases generated in the process of gasification (CO, H_2) can support its operation or stabilize coal dust combustion process at low load. It is possible to supply gases generated to cyclone furnace in order to maintain the hot reserve state.

* Corresponding author: zarzycki@is.pcz.czest.pl

3 Process of coal dust combustion and gasification in cyclone furnace

The process of coal dust combustion and gasification was performed in a cyclone furnace composed of two chambers: upper (PC2) and lower (PC1) [9,10,11]. The chamber PC2 is cylindrical. Its upper part features tangentially installed channels which supply fuel pneumatically in a stream of CO_2 (recirculated flue gas). The chamber PC2 is also cylindrical. It is comprised of several steps with gradually reduced diameters. There are nozzles on lateral surfaces of the chamber PC1 which allow for supply of the "driving" gas with assumed composition of O_2 and CO_2. The chambers PC1 and PC2 are connected with a channel with a plunger used to separate the flow of flue gas from the chamber PC1 to PC2 and the fuel flow (carbon residue) from the chamber PC2 to PC1 (Fig. 1).

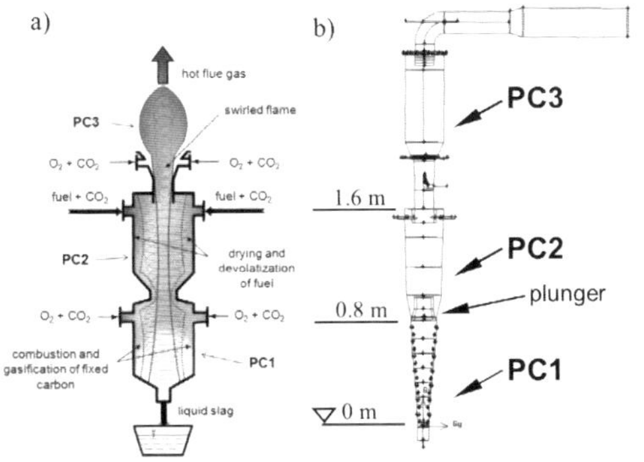

Fig. 1. Diagram (a) and view (b) of the model of a cyclone furnace with after-burner chamber.

The design of the cyclone furnace allows for:
- heating, drying and gasification of fuel in the chamber PC2;
- coal combustion or gasification process in the chamber PC1 which can be performed in the atmosphere of recirculated flue gas (CO_2 and O_2).

In order to ensure effective fuel gasification, it is necessary to burn certain part of carbon residue in order to generate the required amount of heat for maintaining and control of endothermic reactions.

The results of numerical calculations of the coal dust combustion and gasification process in the atmosphere of O_2 and CO_2 represent part of the extensive experimental research and numerical calculations performed within the Strategic Programme „ADVANCED TECHNOLOGIES FOR ENERGY GENERATION" [12,13,14,15,16,17,18].

The geometry of the cyclone furnace was designed using the Gambit software (Fig. 1b). Furthermore, the ANSYS FLUENT 14 software was also used for the calculations. Calculations of gas and fuel flow were performed using the Reynolds Stress turbulence model. Discrete Phase Model software was employed to model coal grain flow. Coal dust combustion and gasification was modelled using Species Transport model, which allows for modelling of chemical reactions both in the solid phase and gaseous phase [19,20,21,22,23]. Calculations were based on the radiation model termed Discrete Ordinate (DO). Calculations of the combustion process and gasification of coal dust with replacement diameter of 500 μm were carried out for the fuel with physicochemical parameters presented in Table 1. For simplification purposes, it was adopted that the fuel does not contain sulphur. Reaction rate constants were derived from the studies [20,21]. Table 1 presents the parameters of the fuel used in the study.

Table 1. Results of technical and elemental analysis of the fuel used for simulations (dry state).

Technical analysis	[-]	Elemental analysis	[-]
VM	0.45	C	0.85
FC	0.45	H	0.1
A	0.05	O	0.04
M	0.05	N	0.01

Calculations for combustion and gasification of coal dust were described with the following reactions:
- Reaction of volatile matter combustion

$$x_1 \cdot VM + x_2 \cdot O_2 = y_1 \cdot CO + y_2 \cdot H_2O + y_3 \cdot N_2 \quad (1)$$

where x_1, x_2, y_1, y_2, y_3 are stoichiometric coefficients
- Reaction of oxidation of carbon oxide

$$CO + 0.5O_2 = CO_2 \quad (2)$$

- Reaction of oxidation of fixed carbon (FC)

$$C_{(s)} + 0.5O_2 = CO \quad (3)$$

- Boudouard's reaction

$$C_{(s)} + CO_2 = 2CO \quad (4)$$

- Synthesis of water gas

$$C_{(s)} + H_2O = CO + H_2 \quad (5)$$

- Hydrogen oxidation reaction

$$H_2 + 0.5O_2 = H_2O \quad (6)$$

Conditions of coal dust gasification process modelling were as follows: flow of coal dust with mean grain size of 500 μm of 0.2 to 2 kg/s supplied tangentially with two inlets located at the opposite sides of the chamber PC2 (Fig. 1) in the CO_2 with temperature of 350 K and the velocity of 3 m/s. Following a helical line, the fuel moves downwards in the chamber PC2 as it is heated, dried and devolatilized through the effect of hot flue gas that flows in the furnace axis and contact with hot walls of the chamber PC2. Fuel gasification occurs through partial combustion of fixed carbon (FC) in the chamber PC1 [10,12,13,14,15,17,18]. The driving nozzles were used to supply (tangentially) the driving gas composed of the mixture of O_2/CO_2 with volumetric content of 60/40% with the velocity of 3 m/s and temperature of 600 K.

The numerical model takes into consideration the heat transfer through furnace walls to its surroundings. Actual thickness of individual components of the laboratory furnace was also assumed during calculation of heat loss, with coefficients of heat penetration and conduction for the materials. It was assumed that air temperature around the furnace is 300 K.

4 Analysis of coal dust combustion and gasification for variable fuel flow rate

The aim of the results presented in this study was to determine conditions of cyclone furnace operation at variable fuel flow rate in order to examine opportunities for control of the stream. For this purpose, a series of numerical simulations were performed for the flow rate supplied to the coal dust furnace ranging from 0.2 to 2 g/s.

Figures 2-6 present distribution of selected process values for three mass loads with the fuel flow rate to the furnace (0.2, 0.8 and 2 g/s). The driving gas which is tangentially supplied along the level of the chamber PC1, characterized by the O_2/CO_2 ratio of 60/40%, causes a strong eddy motion of the gas and fuel. Due to the effect of the centrifugal force, fuel is accumulated near the walls of the chambers PC1 and PC2, which is presented in Fig. 2.

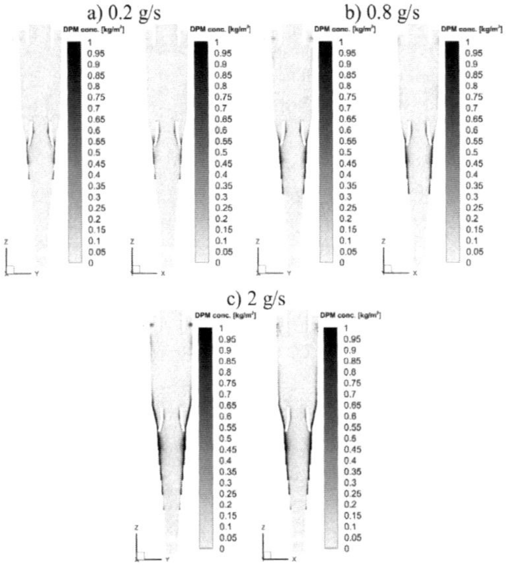

Fig. 2. Distributions of fuel concentration.

It can be observed that the increase in the mass flow rate of the fuel supplied to the furnace leads to the increase in the thickness of its layer on the walls and its transport towards the lower part of the furnace. The maximal fuel flow rate can be determined based on the maximal flow rate of the fuel supplied to the furnace and gasified without generation of the loss of incomplete combustion. The use of the stepped chamber PC1 allows for maintaining a substantial fuel mass in a state of eddy. The zone where fuel is present significantly depends on the amount of the oxidant. If the fuel flow rate increases, the demand for the oxidant is higher (see Fig. 3). The zone with elevated oxygen concentration reduces with the increase in fuel flow rate (Fig. 3). Increased oxygen content in the chamber PC2 can be also observed for the flow rate of 0.2 g/s, which reflects the presence of substoichiometric conditions in the chamber PC1 (see Fig. 3a). This leads to the increase in the concentration of CO_2 and low values of CO in the chamber PC2 (Figs. 5a and 6a). The highest temperature is observed in the upper part of the PC1 chamber (see Fig. 4a), reaching the chamber PC2.

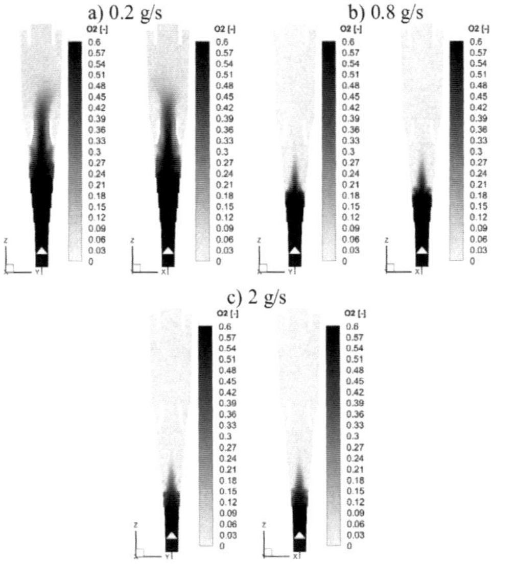

Fig. 3. Distributions of molar fractions of O_2.

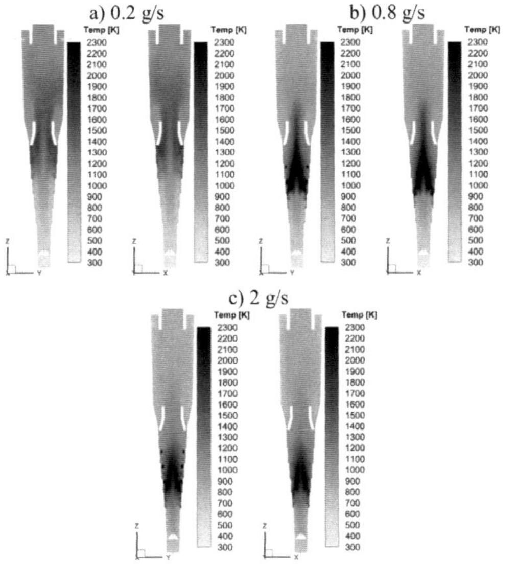

Fig. 4. Temperature distributions.

For the presented load of the furnace with the fuel flow rate of 0.8 and 2 g/s, the zone of high oxygen concentrations is significantly reduced and moving towards the lower part of the chamber PC1 (see Figs. 3b, 3c). In these cases, the process in the chamber PC1 occurs under substoichiometric conditions and is conductive to coal gasification process. The area of the highest temperatures is shifted to the central part of the chamber PC1 (Figs. 4b, 4c). High concentration of O_2 (Figs. 3b, 3c) and CO_2 (Figs. 6b, 6c) (resulting from the gas supplied) is observed below this area, whereas above

 DOI: 10.1051/e3sconf/20171305003

the area of high temperature, the content of O_2 declines virtually to zero (Figs. 3b, 3c), while the content of CO_2 decreases (Figs. 6b, 6c) as a result of the reaction (6) and the CO content rises (Figs. 5b, 5c).

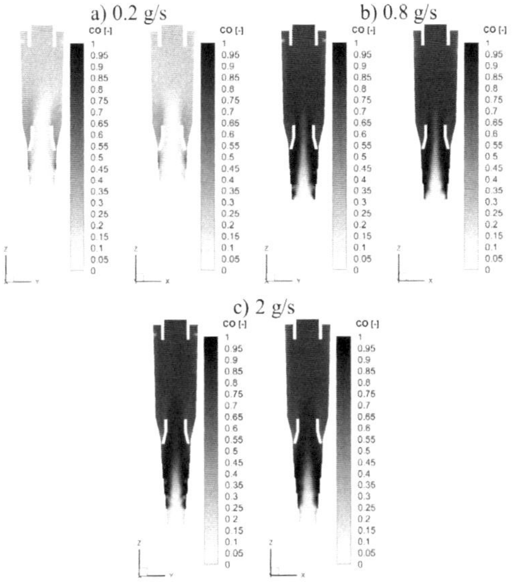

Fig. 5. Distributions of molar fractions of CO.

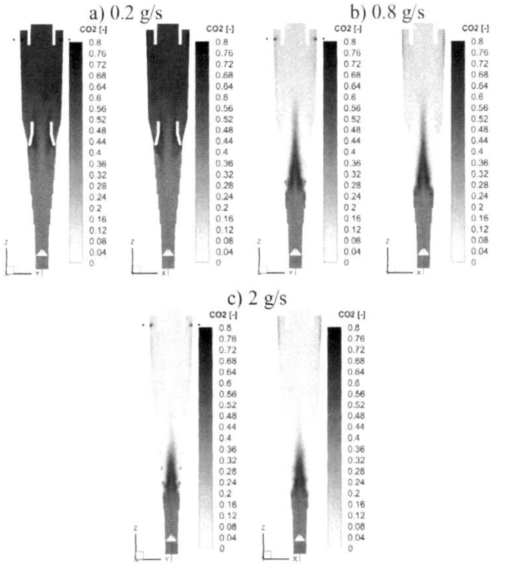

Fig. 6. Distributions of molar fractions of CO_2.

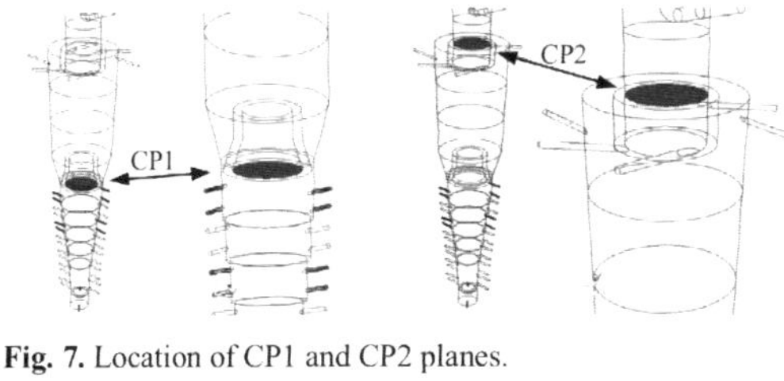

Fig. 7. Location of CP1 and CP2 planes.

These observations are supported by the profiles of mean values of selected parameters at the outlets from the chambers PC1 and PC2 (Fig. 7). The contents of O_2 presented in Fig. 8a show that the combustion process in the chamber PC1 occurs for the adopted aerodynamic conditions up to the fuel flow rate of ca. 0.4 g/s. This observation is also supported by the fact that the temperature maximum occurs in the planes CP1 and CP2 (Fig. 8b). Furthermore, a relatively low CO content (Fig. 9a) and high CO_2 content (Fig. 9b) are noticeable for fuel flow rates below 0.4 g/s. Very low H_2 contents are also observed in this area (Fig. 10a) in the planes CP1 and CP2, resulting from oxidation of hydrogen by the oxygen contained in the gases. A close to zero content of volatile matter in the plane CP1 (resulting from their oxidation) is also observed (see Fig. 10b).

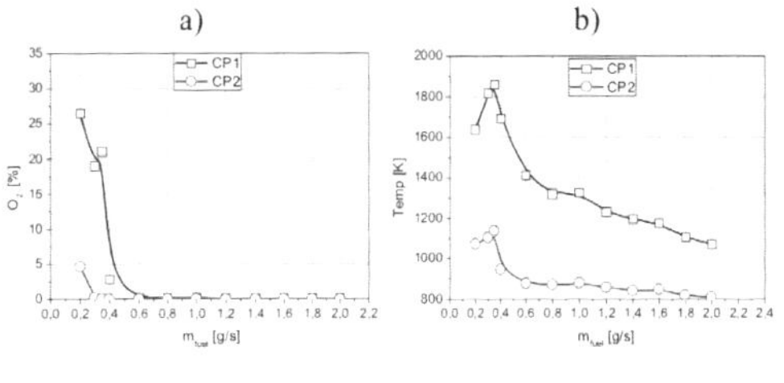

Fig. 8. Changes in mean values of (a) O_2 content and (b) temperature in the planes CP1 and CP2 versus fuel flow rate.

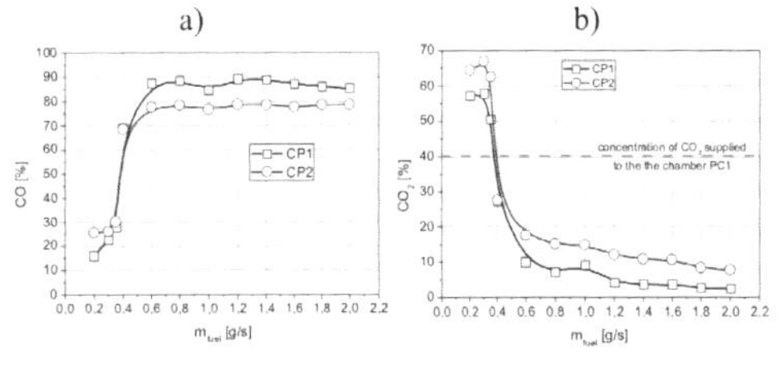

Fig. 9. Changes in mean values of (a) CO content and (b) CO_2 content in the planes CP1 and CP2 versus fuel flow rate.

For the fuel mass flow rate of over 0.4 g/s in the control plane CP1 and CP2, O_2 concentration is close to zero (Fig. 8a). A significant decline in temperature (Fig. 8b) in the CP1 plane is observed, resulting, on the one hand, from the movement of the gasification front and, on the other hand, from the intensive heat consumption during endothermic reactions. It is noticeable that, regardless of the flow rate for the gasified fuel stream, CO content (Fig. 9a) at the outlet from the chamber PC2 reaches 80%. Furthermore, a decline in CO_2 content (Fig. 9b) in the planes CP1 and CP2 is also observed for increasing fuel mass flow rates. This results from the increased demand for the flow rate of the gasifying medium for higher fuel flow rates. Figs. 10a and 10b illustrate the concentration of H_2 and volatile matter (VM), respectively. For the fuel flow rate of over 0.4 g/s, the analysed values are nearly linear, which means that their amount is proportional to the flow rate of the fuel supplied and, apart from the initial area (below 0.4 g/s), these substances do not take part in the processes of gasification and oxidation.

For the fuel flow rates from 0.2 to 1.6 g/s, no fuel outflow from the PC1 to the slag bath was found, which suggests complete fuel conversion. It was found that for the fuel flow rate of 1.8 and 2 g/s, non-reacted fuel leaves the chamber PC1 at the flow rate of 0.0275 g/s and 0.0498 g/s, which represents ca. 1.5% and 2.5% of the flow rate of the supplied fuel.

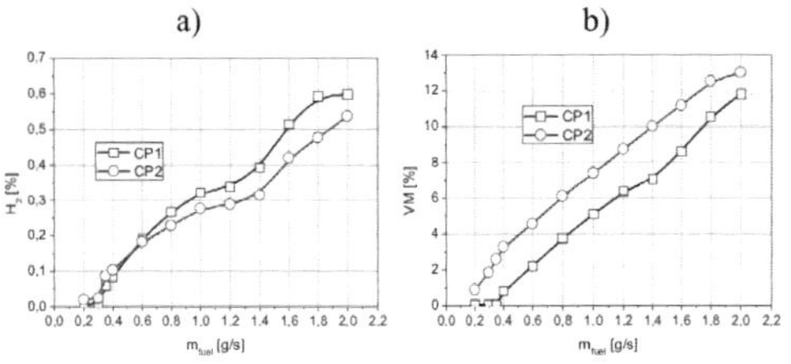

Fig. 10. Changes in mean values of (a) H_2 content and (b) VM in the planes CP1 and CP2 versus fuel flow rate.

5 Conclusion

The results of numerical calculations presented in the study demonstrated that, through changes in the mass flow rate of the coal dust supplied to the cyclone furnace, it is possible to obtain the convenient conditions for its combustion or gasification. Consequently, coal dust combustion can be performed at low mass flow rates, which causes that the cyclone furnace can be maintained in a state of hot reserve, thus generating hot flue gas and combustible gases with low CO contents (ca. 30%). Furthermore, at elevated fuel flow rates, coal dust gasification can be easily performed, with the CO concentrations at the outlet from the chamber PC2 reaching 80% and high content of volatile matter (ca. 13%). The combustible gases supplied to the chamber of the pulverized coal-fired boiler can be used to stabilize coal dust combustion process at low boiler load. They can replace the operation of the dust burners or ignition burners at low load while opening up the opportunities for maintaining the boiler in the hot reserve state and ensuring lower levels of the minimum load in the pulverized coal-fired boiler.

Acknowledgements

BS/PB-404-301/11

References

1. W. Nowak, T. Czakiert, *Spalanie tlenowe dla kotłów pyłowych i fluidalnych zintegrowanych z wychwytywaniem CO_2*, Wydawnictwo Politechniki Częstochowskiej, Częstochowa, (2012)
2. W. Nowak, W. Rybak, T. Czakiert, *Spalanie tlenowe dla kotłów pyłowych i fluidalnych zintegrowanych z wychwytywaniem CO_2. Kinetyka i mechanizm spalania tlenowego oraz wychwytu CO_2*, Wydawnictwo Politechniki Częstochowskiej, Częstochowa, (2013)
3. W. Nowak, M. Chorowski, T. Czakiert, *Spalanie tlenowe dla kotłów pyłowych i fluidalnych zintegrowanych z wychwytywaniem CO_2. Produkcja tlenu na potrzeby spalania tlenowego*, Wydawnictwo Politechniki Częstochowskiej, Częstochowa, (2014)
4. W. Nowak, M. Ściążko, T. Czakiert, *Spalanie tlenowe dla kotłów pyłowych i fluidalnych zintegrowanych z wychwytywaniem CO_2. Doświadczenia z instalacji pilotowych i perespektywy dla instalacji demonstracyjnych*, Wydawnictwo Politechniki Częstochowskiej, Częstochowa, (2015)
5. T. Borowiecki, J. Kijeński, J. Machnikowski, M. Ściążko, *Czysta energia, produkty chemiczne i paliwa z węgla - ocena potencjału rozwojowego*, Instytut Chemicznej Przeróbki Węgla, Zabrze, (2008)
6. T. Chmielniak, H. Łukowicz, *Modelowanie i optymalizacja węglowych bloków energetycznych z wychwytem CO_2*, Wydawnictwo Politechniki Śląskiej, Gliwice, (2015)
7. A. Rusin, *Wybrane aspekty eksploatacji i diagnostyki bloków energetycznych nowych generacji*, Wydawnictwo Politechniki Śląskiej, Gliwice, (2015)
8. R. Zarzycki, R. Kobyłecki, M. Kratofil, Z. Bis, *Kierunki rozwoju palenisk cyklonowych*, Polityka Energetyczna tom 17, z. 4. 2014. Wyd. Instytutu GSMiE PAN, Kraków, pp. 249 – 261
9. Z. Bis, R. Kobyłecki, R. Zarzycki, *Urządzenie do spalania i zgazowania paliw stałych*, Zgłoszenie wynalazku P.411668, 2015
10. R. Zarzycki, Z. Bis, R. Kobyłecki, *The concept of coal burning in a cyclone furnace*, IX ICCHME, 23-26 May 2016, Cracow, Poland, no. 492
11. R. Kobyłecki, R. Zarzycki, Z. Bis, *Przedpalenisko cyklonowe dla spalania i zgazowania pyłu węglowego*, Wydawnictwo ITC, Energetyka Gazowa tom 2, 2016, pp. 277-285
12. R. Zarzycki, Z. Bis, *Modelling of the process of coal dust combustion in a cyclone furnace*, IX ICCHME, 23-26 May 2016, Cracow, Poland, no. 493
13. R. Zarzycki, Z. Bis, *Modelling of coal dust gasification in a cyclone furnace under oxy-fuel combustion conditions*, IX ICCHME, 23-26 May 2016, Cracow, Poland, no. 494
14. R. Zarzycki, P. Warzecha, R. Kobyłecki, Z. Bis, *Modelowanie matematyczne i symulacje numeryczne tlenowego spalania węgla oraz zgazowania w przedpalenisku cyklonowym*, Spalanie tlenowe dla kotłów pyłowych i fluidalnych zintegrowanych z wychwytem CO_2, seria Monografie nr 301, Częstochowa 2015, pp. 310 - 334
15. R. Zarzycki, Z. Bis, *Modelowanie procesu spalania i zgazowania pyłu węglowego w palenisku cyklonowym*, Wydawnictwo ITC, Energetyka Gazowa tom 2, 2016, s. 361-376
16. R. Zarzycki, M. Kratofil, D. Pawłowski, M. Ścisłowska, R. Kobyłecki, Z. Bis, *Analiza wyników obliczeń numerycznych przepływu pyłu węglowego oraz gazu w palenisku cyklonowym*, Polityka Energetyczna, t. 16, zeszyt 3, 2013, s. 301-312

17. R. Zarzycki, Z. Bis, *Modelling of coal dust gasification using the water steam in a cyclone furnace*, Proceedings of the 4th International Conference on Contemporary Problems of Thermal Engineering, 14-16 September 2016, Katowice, Poland, ISBN 978-83-61506-36-2, p. 861-871
18. R. Zarzycki, M. Kratofil, D. Pawłowski, M. Ścisłowska, R. Kobyłecki, Z. Bis, *Analiza wyników obliczeń numerycznych przepływu pyłu węglowego oraz gazu w palenisku cyklonowym*, Polityka Energetyczna, t. 16, zeszyt 3, 2013, s. 301-312
19. D. Toporov, P. Bocian, P. Heil, A. Kellermann, H. Stadler, S. Tschunko, M. Förster, R. Kneer, *Detailed investigation of a pulverized fuel swirl flame in CO_2/O_2 atmosphere*, Combustion and Flame (2008) vol. 155, pp. 605-618
20. M. Vascellari, G. Cau, *Numerical simulation of pulverized coal oxycombustion with exhaust gas recirculation*, Proceeding of CCT2009 Fourth International Conference on Clean Coal Technologies. Dresden, Germany (2009)
21. L. Chen, M. Gazzino, A.F. Ghoniem, *Characteristics of pressurized oxy-coal combustion under increasing swirl number*, 35th International Technical Conference on Coal Utilization & Fuel Systems. Clearwater, Florida (2010)
22. P. Warzecha, A. Bogusławski, *Symulacje numeryczne spalania pyłu węglowego w atmosferze O_2-CO_2*, Archiwum Spalania, 2012, Vol. 12, nr 3 pp. 145-151
23. ANSYS Fluent Theory Guide, 2011

Swirl chamber for vitrification of fly ashes

Robert Zarzycki[1,*]

[1]Czestochowa University of Technology, Department of Energy Engineering, ul. Brzeznicka 60a, 42-201, Poland

Abstract. The study presents the concept of a swirl chamber used for vitrification of fly ashes. It assumes the use of coal dust in the process of fly ash melting. The coal dust supplied to the swirl chamber and gasified in the atmosphere of O_2, CO_2 and H_2O allows for obtaining combustible gases composed of CO and H_2, which are burnt with the pneumatically supplied fly ash. The above process allows for obtaining a product in the form of a molten slag which does not contain coal grains. The study presents numerical calculations for the process of combustion and gasification of coal dust and opportunities for ensuring adequate parameters in the fly ash melting zone. The combustible gases obtained during the process of gasification can be supplied to the chamber of a pulverized-bed boiler.

1 Introduction

The use of fossil fuels for production of electricity and heat leads to the emissions of CO_2, SO_x and NO_x and dust in the form of fly ashes. Development of the flue gas allows for a substantial reduction in the emissions of pollutants (including dusts) to the atmosphere. Fly ashes are used to fill the excavations in mines and are often discharged to landfill sites. They can also be used to manufacture building materials (cement, granular materials), as they have to meet specific requirements specified by the constructional standards PN-EN 450-1:2012, PN-EN 206:2014-04 concerning mainly coal content. In many cases, fly ashes generated in the combustion processes contain, apart from considerable amounts of carbon, also CaO and other components in the amounts that do not allow for their industrial use and which cannot be safely stored. Therefore, fly ash may represent a serious problem for electricity and heat power plants. One of the solutions proposed for this problem is vitrification and the use of the product for production of construction materials. This study presents the concept of a melting chamber for vitrification of fly ashes.

2 Fly ash vitrification

Vitrification is a technological process that consists in controlled heat supply [1] for vitrified substances in order to separate gases and combustible substances (e.g. carbon) and then to melt the remaining mineral substance. Using the controlled cooling of the liquid phase, ceramic materials with desirable mechanical properties and structure can be obtained. Process of fly ash vitrification can be compared to glass production in a glass furnace, with high temperatures leading to melting of the mineral substance. Apart from the supply of the sufficiently high temperature, the process of melting of the mineral substance in fly ashes requires supplying the required amount of heat for phase transitions. The heat source is electricity (resistance heating and plasma heating) or heat obtained through combustion of fuel (natural gas, burning oil) in rotary furnaces and cyclone furnaces [2,3,4]. The energy of around 2 MJ/kg [1,2] is needed for melting the mineral matter in fly ashes. In the case of plasma and arc methods, the energy demand for vitrification was evaluated as ranging from 3.6 to 18 MJ/kg [3]. The vitrification processes based on the use of electric power are mainly performed on a smaller scale due to high energy costs. However, the use of chemical energy contained in the fuel for vitrification and, simultaneously, for e.g. supply of hot combustible gases to pulverized-bed furnace improves energy efficiency and ensures a substantial reduction in unit costs of the process. Using fly ashes with higher carbon content for vitrification allows for utilization of the internal chemical energy while reducing the demand for fuel for the vitrification process.

The basic components of fly ashes include silica (SiO_2), aluminium compounds (e.g. Al_2O_3), iron compounds (e.g. Fe_2O_3), calcium compounds (CaO) and magnesium compounds (MgO) [1]. The mixtures of these oxides reach melting temperatures ranging from 1,200 to 1,800°C. Melting temperature can be reduced through addition of fluxes such as CaO or FeO [1]. In many cases, CaO is present in fly ashes in a natural form. The use of the fluxes helps control vitrification process in order to ensure complete ash melting. Therefore, the temperatures in the melting temperature ranging from 1,500-2,000°C will allow for melting of fly ashes and obtaining the slag in the molten form.

* Corresponding author: zarzycki@is.pcz.czest.pl

3 The concept of operation of the chamber for vitrification of fly ashes

As indicated above, the precondition for vitrification of fly ashes is to ensure adequate level of temperature and supply of heat necessary for heating and, consequently, melting of the mineral substance contained in the ash.

The required conditions of the vitrification process can be obtained through combustion of fuel in the form of the natural gas, fuel oil or coal dust. From the standpoint of the easiness of control of the combustion process, the best fuels are those in the liquid and gaseous forms. However, they are characterized by substantially higher unit energy costs compared to coal dust. On the other hand, the use of coal dust in the vitrification process may lead to the presence of non-combusted coal grains (molten slag), which substantially deteriorates properties of the product. The solution for this problem is offered by coal dust gasification in the strong eddy stream of oxidant and then controlled combustion of the part of combustible gases with pneumatically supplied fly ashes. This solution allows for the use of cheap fuel such as coal dust (from hard coal or brown coal) in the vitrification process and the fly ash melting process can be performed using combustible gases, which limits the carbon content in the product to the minimum level. The vitrification process has to be performed at temperatures which exceed melting points of the compounds contained in the ash. It is known that NO_x emissions in such temperatures are very high due to the presence of nitrogen contained in the air. However, vitrification in the oxy-fuel atmosphere can substantially reduce NO_x emissions. Using the oxy-fuel atmosphere may also improve the adiabatic combustion temperature, which is essential from the standpoint of the process. However, the opportunities for temperature control due to thermal strength of materials of the vitrification chamber should also be taken into consideration. It can be performed using three methods:

- Combustion process in substoichiometric conditions: coal dust gasification - CO generation;
- Combustion process in substoichiometric conditions with presence of water steam: coal dust gasification - H_2 generation;
- The use of flue gas recirculation.

These methods help control temperature and composition of the combustible gas obtained through gasification of the coal dust. Apart from its basic function, the proposed vitrification chamber can also be used as a system for coal dust gasification and supplying gaseous fuel to a pulverized-bed furnace.

Figure 1 presents the diagram of a chamber for fly ash vitrification. The chamber is composed of the stepped chamber SC1, with the ignition burner fuelled with methane and designed in the roof. The inlets of coal dusts are distributed centrically around the burner. The nozzles that supply gas are installed on side walls (tangentially), with the gas leading to formation of a strong eddy of coal dust and allowing for controlled combustion and gasification. The used of the stepped chamber SC1 allows for elongation of the time of coal dust presence while reducing the falling rate. The products which leave the chamber SC1 include gases containing CO and H_2 and insignificant amounts of CO_2 and H_2O.

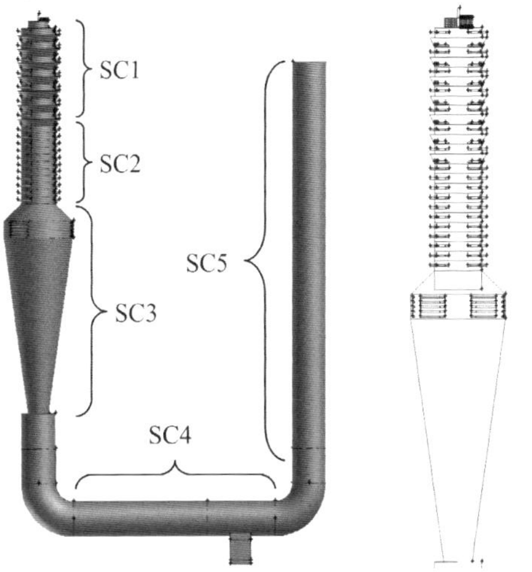

Fig. 1. The design of the chamber for vitrification of fly ashes.

It is also possible that tiny coal dust grains and ashes flow out from this chamber. The role of the cylindrical chamber SC2 with tangentially installed nozzles is to burn and gasify the tiniest fractions of coal dust and partial combustion of the gases generated in the SC1 chamber in order to ensure optimal conditions for the gasification process. No coal dust grains (only gaseous compounds and ash) are expected to be present at the outlet from the chamber SC2. The chamber SC3 below the SC2 is used as a chamber for melting of the fly ash, which, through nozzles installed in its upper part, is transported pneumatically to its interior. Gas composition (O_2, CO_2) can be used to control the degree of gas combustion for the gas generated in the chambers SC1 and SC2 and the temperature in the chamber SC3. The molten slag flows to the horizontal chamber SC4 in the form of a bath filled with molten slag with high-temperature gas flying over its surface. The second part of the chamber SC4 features outlet of molten slag. Hot gases containing combustible compounds are supplied through the channel SC5 to the chamber in the pulverized-bed furnace for afterburning.

4 Modelling of coal dust gasification process

The Reynolds Stress turbulence model was used due to strong gas eddy and fuel eddy in the swirl chamber. Modelling of flow of coal grains was based on the Discrete Phase Model, whereas coal dust combustion was performed using Species Transport model, which allows for modelling chemical reactions both in the solid phase and gaseous phase [5,6,7,8,9]. Calculations were based on the radiation model termed Discrete Ordinate (DO).

Calculations of the combustion process and gasification of coal dust with replacement diameter of 500 μm were carried out for the fuel with physicochemical parameters presented in Table 1. For simplification purposes, it was adopted that the fuel does not contain sulphur. Reaction rate constants were derived from the studies [5,6]. Table 1 presents the parameters of the fuel used in the study.

Table 1. Results of technical and elemental analysis of the fuel used for simulations (dry state).

Technical analysis	[-]	Elemental analysis	[-]
VM (volatile matter)	0.45	*C*	0.85
FC (fixed carbon)	0.45	*H*	0.1
A (ash)	0.05	*O*	0.04
M (moisture)	0.05	*N*	0.01

Calculations for combustion and gasification of coal dust were described with seven reactions:

- Reaction of volatile matter combustion

$$x_1 \cdot VM + x_2 \cdot O_2 = y_1 \cdot CO + y_2 \cdot H_2O + y_3 \cdot N_2 \quad (1)$$

where x_1, x_2, y_1, y_2, y_3 are stoichiometric coefficients

- Reaction of oxidation of carbon oxide

$$CO + 0.5O_2 = CO_2 \quad (2)$$

- Reaction of oxidation of fixed carbon (FC)

$$C_{(s)} + 0.5O_2 = CO \quad (3)$$

- Boudouard's reaction

$$C_{(s)} + CO_2 = 2CO \quad (4)$$

- Synthesis of water gas

$$C_{(s)} + H_2O = CO + H_2 \quad (5)$$

- Hydrogen oxidation reaction

$$H_2 + 0.5O_2 = H_2O \quad (6)$$

- Methane oxidation reaction

$$CH_4 + 1.5O_2 = CO + 2H_2O \quad (7)$$

5 Analysis of the results of calculations for the coal dust combustion and gasification processes

The results of calculations presented in this chapter were obtained for the flow rate that is supplied with the nozzles installed near the methane ignition burners for coal dust of 15 g/s. The gas for SC1, SC2 and SC3 chambers was supplied with the rate of 20 m/s with temperature of 600 K through the nozzles with diameter of 20 mm. The volumetric fractions of the gas supplied to the chamber SC1 were 40% for O_2 and 60% for H_2O. The gas supplied to the chamber SC2 contained 30% of O_2 and 70% of CO_2, whereas the gas supplied to the chamber SC3 was characterized by 40% of O_2 and 60% of CO_2. The nozzle installed in the roof of the chamber SC1 was used to supply the mixture of 80% of O_2 and 20% of CH_4 in order to initiate the coal gasification process and its stabilization.

Adoption of the above compositions of the gaseous mixtures resulted from the assumptions concerning coal dust gasification in the chamber SC1, where CO and H_2 are produced. Gasification or combustion of tiny fuel grains in the atmosphere of gases generated in the chamber SC1 in order to limit their flow to the chamber SC3 were assumed for the chamber SC2. The final combustion and gasification of the tiniest fuel grains and combustion of a part of the products of coal gasification in order to obtain the temperature required for fly ash heating and melting (depending on fly ash composition) occur in the chamber SC3, with the fly ash supplied with the gas stream flowing to this chamber.

The results of numerical computations for gas and fuel flow and the process of its gasification are presented in Figures 4, 5 and 6. Supplying the gas with tangentially installed nozzles allows for generation of the swirling motion of the gas (see Fig. 2a) and the coal dust inside the chamber (Fig. 2b). With the effect of the centrifugal force, the rotating coal dust is collected mainly near the walls of the chambers SC1 and SC2 (see Figs. 2b and 2c). The steps of the chamber SC1 (Figs. 2b and 2c) allow for maintaining the eddy of the greater fuel mass near the wall, which substantially extends the time of its remaining in the gasification zone. Tiny and the tinniest grains are transported with the flow to the chamber SC2, where strong eddy is also maintained (see Fig. 2b). With the gaseous mixture with oxygen supplied to the chambers SC1 and SC2 and with adequate temperatures, it is possible to perform combustion and gasification of the fuel, which can be observed in Fig. 2d that presents the parameter responsible for fuel conversion rate. It is noticeable that the highest values of this parameter occur near the walls of the chambers SC1 and SC2 in the location of gas supply. The values of this parameter in the chamber SC2 are reduced, which reflects fuel conversion from the solid state to gaseous products of combustion or gasification.

The fuel supplied in the upper part of the SC1 chamber (see Fig. 2c) is quickly heated by the central methane burner flame and the burnt and gasified fuel present near the walls. Fuel devolatilization process is presented in Figs. 3a and 3b which present the rate of devolatilization process and volumetric fraction of volatile matter, respectively. Through fragmentation of fuel to the form of dust and its eddy motion, devolatilization occurs very intensively (see Fig. 3a), which is reflected by high concentration of volatile matter (Fig. 3b). While increasing the distance from the location of fuel supply, the concentration of volatile matter declines through their partial oxidation and dilution in the gaseous component of the coal gasification process in the chamber SC1.

The process of fuel gasification requires adequately high temperature and the amount of heat for endothermic reaction. Consequently, the gas with oxygen is supplied at the level of the chambers SC1 and SC2. Partial combustion of coal dust and gasification products occurs in this case in order to maintain the assumed temperature (see Fig. 4a). The amount of oxygen supplied to SC1 and SC2 is substoichiometric (Fig. 3c) and allows only for maintaining the assumed temperature that guarantees performing the processes of coal dust gasification. The function of the SC3 chamber is to melt fly ash. For this purpose, gas is supplied to the chamber SC3 with the oxygen content (see Fig. 3c) such that the conditions that ensure melting of fly ash are obtained (see Fig. 4a). Then, a part of flue gas is burnt in the chambers SC1 and SC2, which is reflected by the increase in CO_2 and H_2O concentrations in the upper part of the chamber SC3. The temperature in the chamber SC3 can be adjusted by changing oxygen concentration in the supplied gas so that the performance of this chamber (connected with the amount of the vitrified fly ash) can be controlled.

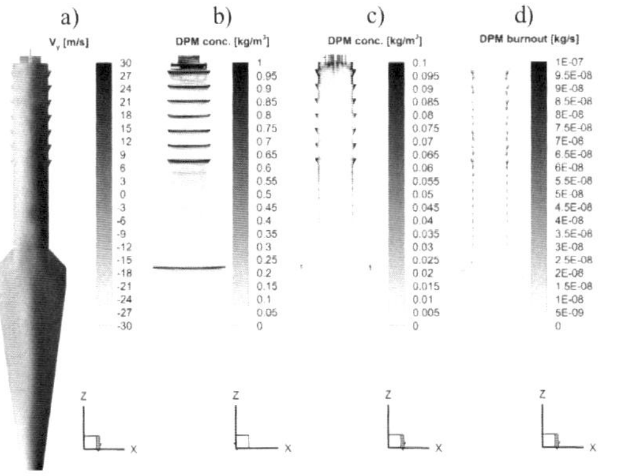

Fig. 2. Distribution of (a) peripheral velocity component, (b) fuel concentration on chamber walls and (c) in vertical cross-section of the chamber and (d) combustion rate.

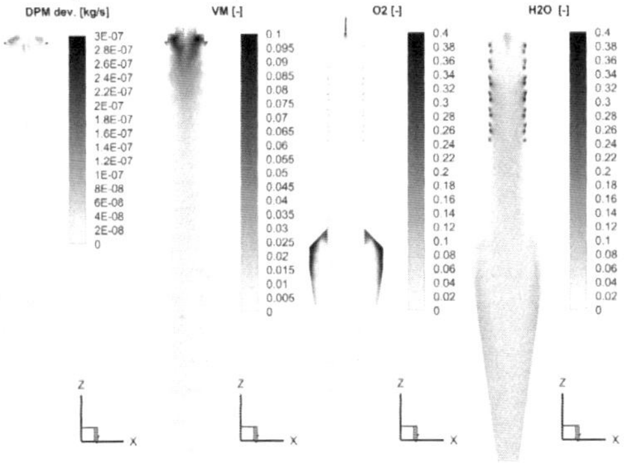

Fig. 3. Distribution of (a) devolatilization rate, (b) molar fractions for volatile matter, (c) molar fractions for O_2 and (d) molar fractions for H_2O.

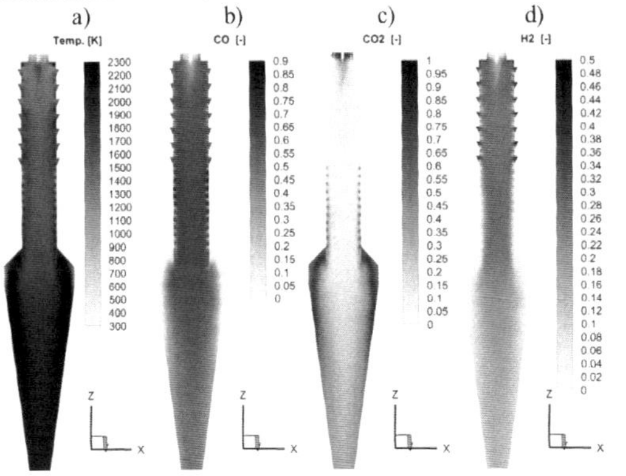

Fig. 4. Distribution of (a) temperature, (b) molar fractions for CO, (c) molar fractions for CO_2 and (d) molar fractions for H_2.

Supplying the gas with content of O_2 and H_2O to the chamber SC1 (see Fig. 3d) yields CO through fuel gasification (see Fig. 4b, Table 2) and H_2 (Fig. 4d, Table 2). Supplying the mixture of O_2 and CO_2 to the chamber SC2 allows for control of temperature through combustion of a certain part of the gases generated during gasification and the tiniest grains of the coal dust.

Table 2 presents means values of volumetric fractions of the main gaseous components and mean temperatures at the outlet from the chambers SC1, SC2 and SC3. The results demonstrated generation of CO and H_2 in the SC1 chamber as a result of coal dust gasification with the mixture of O_2 and H_2O. A decline in concentration of CO and H_2 is observed at the outlet from the chamber SC2, resulting from their partial combustion and leading to the temperature increase. Further decline in CO and H_2 concentrations is observed at the outlet from the chamber SC3, which is connected with combustion for the purposes of generation of heat necessary for melting of fly ash. Hot gases that leave the chamber SC3 and contain combustible components in the form of CO and H_2 can be supplied to the pulverized-bed furnace chamber in order to ensure afterburning.

Mean temperatures in the chambers SC1, SC2 and SC2 were 1,579 K, 1,645 K and 2,131 K, respectively. The mean temperature in the chamber SC3 ensures the required conditions of fly ash melting.

Table 2. Volumetric fractions of gaseous components in cross-sections located at the outlet from the chambers SC1, SC2, SC3.

At the outlet from the chamber	CO [%]	CO_2 [%]	H_2 [%]	H_2O [%]	O_2 [%]	VM [%]	Temp. [K]
SC1	63.32	2.31	24.67	8.83	0.42	0.45	1658.8
SC2	55.65	21.79	16.45	4.29	1.42	0.4	1789.9
SC3	43.44	36.94	12.93	6.39	0	0.3	2160.4

6 Conclusion

The concept of the system presented in this study allows for the use of the fuel in the form of the coal dust for the process of fly ash melting. With the division of the system into three chambers, the process of coal dust gasification can be successfully performed in the chamber SC1 while yielding high concentrations of CO and H_2. In order to ensure complete combustion of coal dust in the chamber SC2, the coal dust is burned or gasified with simultaneous combustion of a part of combustible gases while increasing the temperature. The role of the chamber SC3 is to increase the temperature to the expected level that depends on the composition of fly ash (its melting temperature) through combustion of a part of volatile gases (CO and H_2). The temperature in the chamber SC3 can be controlled through adjustment of oxygen concentration in the supplied gas. Similar method can be used to control fly ash melting efficiency. Hot gases that leave the chamber SC3 and contain CO and H_2 can be supplied to the pulverized-bed furnace in order to ensure afterburning.

Acknowledgements

BS/PB-404-301/11

References

1. W. Kordylewski, Ł. Robak, *Witryfikacja odpadów i popiołów*, Gospodarka Paliwami i Energią, nr 7, 2002, pp. 18-21
2. http://www.spalanie.pwr.wroc.pl/badania/publikacje /witryfikacja.PDF {access 2016.09.12}
3. A. Huczko, H. Lange, G. Chojecki, D. Dziadko, *Plazmowa witryfikacja odpadów o charakterze pylistym*, Przemysł Chemiczny, Wydawnictwo SIGMA-NOT, T. 79, nr 10, 2000, pp. 333 - 335
4. http://www.spalanie.pwr.wroc.pl/badania/publikacje /modyfikacja%20popiolu.pdf {access 2016.09.12}
5. D. Toporov, P. Bocian, P. Heil, A. Kellermann, H. Stadler, S. Tschunko, M. Förster, R. Kneer, *Detailed investigation of a pulverized fuel swirl flame in* CO_2/O_2 *atmosphere*, Combustion and Flame (2008) vol. 155, pp. 605-618
6. M. Vascellari, G. Cau, *Numerical simulation of pulverized coal oxycombustion with exhaust gas recirculation*, Proceeding of CCT2009 Fourth International Conference on Clean Coal Technologies. Dresden, Germany (2009)
7. L. Chen, M. Gazzino, A.F. Ghoniem, *Characteristics of pressurized oxy-coal combustion under increasing swirl number*, 35th International Technical Conference on Coal Utilization & Fuel Systems. Clearwater, Florida (2010)
8. P. Warzecha, A. Bogusławski, *Symulacje numeryczne spalania pyłu węglowego w atmosferze* O_2-CO_2, Archiwum Spalania, 2012, Vol. 12, nr 3 pp. 145-151
9. ANSYS Fluent Theory Guide, 2011

AUTHOR INDEX

EDP Sciences – Web of Conferences
17, Avenue du Hoggar
Parc d'Activité de Courtabouf
BP 112
F-91944 Les Ulis Cedex A
France

ISBN 978-1-5108-3653-2

CURRAN ASSOCIATES INC.
proceedings.com